FREE

Free Study Tips DVD

In addition to the tips and content in this guide, we have created a FREE DVD with helpful study tips to further assist your exam preparation. **This FREE Study Tips DVD provides you with top-notch tips to conquer your exam and reach your goals.**

Our simple request in exchange for the strategy-packed DVD is that you email us your feedback about our study guide. We would love to hear what you thought about the guide, and we welcome any and all feedback—positive, negative, or neutral. It is our #1 goal to provide you with top-quality products and customer service.

To receive your **FREE Study Tips DVD**, email freedvd@apexprep.com. Please put "FREE DVD" in the subject line and put the following in the email:

a. The name of the study guide you purchased.

b. Your rating of the study guide on a scale of 1-5, with 5 being the highest score.

c. Any thoughts or feedback about your study guide.

d. Your first and last name and your mailing address, so we know where to send your free DVD!

Thank you!

The HiSET Tutor Study Guide

HiSET 2020 Preparation Book with Practice Test Questions for the High School Equivalency Test [Updated for the New 2020 Exam Outline]

APEX Test Prep

Table of Contents

Test Taking Strategies...1

FREE DVD OFFER 4

Introduction to the HiSET...5

Language Arts: Reading...7

Comprehension 7

Inference and Interpretation 12

Analysis 20

Synthesis and Generalization 29

Practice Questions 34

Answer Explanations 52

Writing 60

Organization of Ideas... 60

Language Facility 63

Writing Conventions 69

Practice Questions 81

Answer Explanations 95

Essay Prompt 101

Mathematics 103

Numbers and Operations on Numbers 103

Measurement/Geometry 121

Data Analysis/Probability/Statistics 141

Algebraic Concepts 162

Practice Questions 194

Answer Explanations 208

Science 216

Life Science 216

Physical Science 243

Earth Science 259

Practice Questions 264

Answer Explanations 277

Social Studies 284

History 284

Civics/Government 301

Economics..307

Geography...311

Practice Questions...319

Answer Explanations..334

Test Taking Strategies

1. Reading the Whole Question

A popular assumption in Western culture is the idea that we don't have enough time for anything. We speed while driving to work, we want to read an assignment for class as quickly as possible, or we want the line in the supermarket to dwindle faster. However, speeding through such events robs us from being able to thoroughly appreciate and understand what's happening around us. While taking a timed test, the feeling one might have while reading a question is to find the correct answer as quickly as possible. Although pace is important, don't let it deter you from reading the whole question. Test writers know how to subtly change a test question toward the end in various ways, such as adding a negative or changing focus. If the question has a passage, carefully read the whole passage as well before moving on to the questions. This will help you process the information in the passage rather than worrying about the questions you've just read and where to find them. A thorough understanding of the passage or question is an important way for test takers to be able to succeed on an exam.

2. Examining Every Answer Choice

Let's say we're at the market buying apples. The first apple we see on top of the heap may *look* like the best apple, but if we turn it over we can see bruising on the skin. We must examine several apples before deciding which apple is the best. Finding the correct answer choice is like finding the best apple. Some exams ask for the *best* answer choice, which means that there are several choices that could be correct, but one choice is always better than the rest. Although it's tempting to choose an answer that seems correct at first without reading the others, it's important to read each answer choice thoroughly before making a final decision on the answer. The aim of a test writer might be to get as close as possible to the correct answer, so watch out for subtle words that may indicate an answer is incorrect. Once the correct answer choice is selected, read the question again and the answer in response to make sure all your bases are covered.

3. Eliminating Wrong Answer Choices

Sometimes we become paralyzed when we are confronted with too many choices. Which frozen yogurt flavor is the tastiest? Which pair of shoes look the best with this outfit? What type of car will fill my needs as a consumer? If you are unsure of which answer would be the best to choose, it may help to use process of elimination. We use "filtering" all the time on sites such as eBay® or Craigslist® to eliminate the ads that are not right for us. We can do the same thing on an exam. Process of elimination is crossing out the answer choices we know for sure are wrong and leaving the ones that might be correct. It may help to cover up the incorrect answer choices with a piece of paper, although if the exam is computer-based, you may have to use your hand or mentally cross out the incorrect answer choices. Covering incorrect choices is a psychological act that alleviates stress due to the brain being exposed to a smaller amount of information. Choosing between two answer choices is much easier than choosing between four or five, and you have a better chance of selecting the correct answer if you have less to focus on.

4. Sticking to the World of the Question

When we are attempting to answer questions, our minds will often wander away from the question and what it is asking. We begin to see answer choices that are true in the real world instead of true in the world of the question. It may be helpful to think of each test question as its own little world. This world may be different from ours. This world may know as a truth that the chicken came before the egg or may assert that two plus two equals five. Remember that, no matter what hypothetical nonsense may be in the question, assume it to be true. If the question states that the chicken came before the egg, then choose your answer based on that truth. Sticking to the world of the question means placing all of our biases and assumptions aside and relying on the question to guide us to the correct answer. If we are simply looking for answers that are correct based on our own judgment, then we may choose incorrectly. Remember an answer that is true does not necessarily answer the question.

5. Key Words

If you come across a complex test question that you have to read over and over again, try pulling out some key words from the question in order to understand what exactly it is asking. Key words may be words that surround the question, such as *main idea, analogous, parallel, resembles, structured,* or *defines.* The question may be asking for the main idea, or it may be asking you to define something. Deconstructing the sentence may also be helpful in making the question simpler before trying to answer it. This means taking the sentence apart and obtaining meaning in pieces, or separating the question from the foundation of the question. For example, let's look at this question:

> Given the author's description of the content of paleontology in the first paragraph, which of the following is most parallel to what it taught?

The question asks which one of the answers most *parallels* the following information: The *description* of paleontology in the first paragraph. The first step would be to see *how* paleontology is described in the first paragraph. Then, we would find an answer choice that parallels that description. The question seems complex at first, but after we deconstruct it, the answer becomes much more attainable.

6. Subtle Negatives

Negative words in question stems will be words such as *not, but, neither,* or *except.* Test writers often use these words in order to trick unsuspecting test takers into selecting the wrong answer—or, at least, to test their reading comprehension of the question. Many exams will feature the negative words in all caps (*which of the following is NOT an example*), but some questions will add the negative word seamlessly into the sentence. The following is an example of a subtle negative used in a question stem:

> According to the passage, which of the following is *not* considered to be an example of paleontology?

If we rush through the exam, we might skip that tiny word, *not,* inside the question, and choose an answer that is opposite of the correct choice. Again, it's important to read the question fully, and double check for any words that may negate the statement in any way.

7. Spotting the Hedges

The word "hedging" refers to language that remains vague or avoids absolute terminology. Absolute terminology consists of words like *always, never, all, every, just, only, none,* and *must.* Hedging refers to words like *seem, tend, might, most, some, sometimes, perhaps, possibly, probability,* and *often.* In some cases, we want to choose answer choices that use hedging and avoid answer choices that use absolute terminology. Of course, this always depends on what subject you are being tested on. Humanities subjects like history and literature will contain hedging, because those subjects often do not have absolute answers. However, science and math may contain absolutes that are necessary for the question to be answered. It's important to pay attention to what subject you are on and adjust your response accordingly.

8. Restating to Understand

Every now and then we come across questions that we don't understand. The language may be too complex, or the question is structured in a way that is meant to confuse the test taker. When you come across a question like this, it may be worth your time to rewrite or restate the question in your own words in order to understand it better. For example, let's look at the following complicated question:

> Which of the following words, if substituted for the word *parochial* in the first paragraph, would LEAST change the meaning of the sentence?

Let's restate the question in order to understand it better. We know that they want the word *parochial* replaced. We also know that this new word would "least" or "not" change the meaning of the sentence. Now let's try the sentence again:

> Which word could we replace with *parochial,* and it would not change the meaning?

Restating it this way, we see that the question is asking for a synonym. Now, let's restate the question so we can answer it better:

> Which word is a synonym for the word *parochial*?

Before we even look at the answer choices, we have a simpler, restated version of a complicated question. Remember that, if you have paper, you can always rewrite the simpler version of the question so as not to forget it.

9. Guessing

When is it okay to guess on an exam? This question depends on the test format of the particular exam you're taking. On some tests, answer choices that are answered incorrectly are penalized. If you know that you are penalized for wrong answer choices, avoid guessing on the test question. If you can narrow the question down to fifty percent by process of elimination, then perhaps it may be worth it to guess between two answer choices. But if you are unsure of the correct answer choice among three or four answers, it may help to leave the question unanswered. Likewise, if the exam you are taking does *not* penalize for wrong answer choices, answer the questions first you know to be true, then go back through and mark an answer choice, even if you do not know the correct answer. This way, you will at least have a one in four chance of getting the answer correct. It may also be helpful to do some research on the exam you plan to take in order to understand how the questions are graded.

10. Avoiding Patterns

One popular myth in grade school relating to standardized testing is that test writers will often put multiple-choice answers in patterns. A runoff example of this kind of thinking is that the most common answer choice is "C," with "B" following close behind. Or, some will advocate certain made-up word patterns that simply do not exist. Test writers do not arrange their correct answer choices in any kind of pattern; their choices are randomized. There may even be times where the correct answer choice will be the same letter for two or three questions in a row, but we have no way of knowing when or if this might happen. Instead of trying to figure out what choice the test writer probably set as being correct, focus on what the *best answer choice* would be out of the answers you are presented with. Use the tips above, general knowledge, and reading comprehension skills in order to best answer the question, rather than looking for patterns that do not exist.

FREE DVD OFFER

Achieving a high score on your exam depends not only on understanding the content, but also on understanding how to apply your knowledge and your command of test taking strategies. **Because your success is our primary goal, we offer a FREE Study Tips DVD. It provides top-notch test taking strategies to help you optimize your testing experience.**

Our simple request in exchange for the strategy-packed DVD is that you email us your feedback about our study guide.

To receive your **FREE Study Tips DVD**, email freedvd@apexprep.com. Please put "FREE DVD" in the subject line and put the following in the email:

    a. The name of the study guide you purchased.

    b. Your rating of the study guide on a scale of 1-5, with 5 being the highest score.

    c. Any thoughts or feedback about your study guide.

    d. Your first and last name and your mailing address, so we know where to send your free DVD!

Introduction to the HiSET

Function of the Test

The High School Equivalency Test (HiSET) is a high school equivalency (HSE) exam designed for youth or adults who are ready to obtain their high school diploma. The HSE credential is a portable credential that is recognized by federal programs as students navigate their way through applying for student aid, loans, work-study funds, or showing their HSE credential to programs such as the U.S. Military or the U.S. Department of Labor Job Corps Program for acceptance.

Note that the HiSET is 7 hours long, but the test can be broken up so that examinees are able to take the test over a period of time rather than all at once. Because of this format, many test takers who begin the exam do not complete it. In 2016, 78,410 people took the HiSET exam, 72.9% completed the exam, and 65.4% passed the exam. Check the HiSET's website for eligibility requirements for information on minimum age, school enrollment status, residency, taking a practice test, and identification, as these will vary based on the state.

Test Administration

Test takers can schedule to take the HiSET exam through the website, by phone, or at a test center. Since the HiSET is made up of five subtests, test takers can schedule the tests all at once or one at a time, depending on the particular state. The subtests can be taken in any order. The administration of the HiSET exam is flexible. Test takers should review requirements to take the HiSET for their state, find a test center in that state, set up a HiSET account through the website, and then schedule the appointment.

The HiSET exam subtests can be taken up to three times per one calendar year. Additional ETS fees are not required for retesting with 12 months, but there may be extra fees through the testing center. Retaking any subtest after 12 months will require an ETS fee plus the testing center fee. Keep in mind that although ETS has standards for retesting, each state may have their own requirements for taking a retest, so contact your test center to get accurate information on retesting.

Test takers with disabilities may visit the HiSET ETS website to learn more about the process of requesting accommodations. The accommodation process can take up to six weeks, so submit your request before scheduling to take the HiSET exam.

Test Format

On the day of the exam, test takers should bring a valid ID, payment for the test center (if applicable), and proof of documents for state-required pretests (if applicable). The testing environment does not allow cell phones, calculators, pens, pencils, or papers. The testing center should provide everything you need. A locker with a locker key may be provided so test takers can store their belongings in a secure area.

The HiSET exam consists of 5 separate subtests that are taken over 7 hours. The subtests can be taken all at once or spread out over a period of time. The following table shows information on the subtest areas.

Subtest	Time	Number of Questions
Language Arts – Reading	65 minutes	50
Language Arts – Writing	120 minutes	60 1 essay
Mathematics	90 minutes	55
Science	80 minutes	60
Social Studies	70 minutes	60

Scoring

Computer-delivered HiSET exams will be given an unofficial test score at the test center. Paper-delivered exams or scores for the Writing test will not receive unofficial scores. Official scores will be delivered within three business days for multiple-choice tests and within five business days for the Writing subtest. HiSET scores will be sent through the HiSET account on the website. Test centers also have access to test scores. Test scores are automatically sent to the state that will issue the high school equivalency credential. Additionally, you may choose to send your HiSET scores through your online account to a college, university, employer, or the military.

Passing criteria for the HiSET is scoring at least an 8 out of 20 on each of the subtests, scoring at least a 2 out of 6 on the essay, and achieving a total score of at least 45 out of 100 on all 5 subtests. The score report should let you know whether or not you have passed the HiSET.

Language Arts: Reading
Comprehension

Understanding Explicit Details

Information explicitly stated in a passage leaves the reader no room for confusion. Information explicitly stated in the passage can be identified and used as text evidence. Test takers should consider if the information is an author's opinion or an objective fact, and whether the information contains bias or stereotypes. Also important to consider is the following question: Within the information stated, which words are directly stated and what words leave room for a connotative interpretation?

Facts and Opinions

A **fact** is a statement that is true empirically or an event that has actually occurred in reality and can be proven or supported by evidence; it is generally objective. In contrast, an **opinion** is subjective, representing something that someone believes rather than something that exists in the absolute. People's individual understandings, feelings, and perspectives contribute to variations in opinion. Though facts are typically objective in nature, in some instances, a statement of fact may be both factual and yet also subjective. For example, emotions are individual subjective experiences. If an individual says that they feel happy or sad, the feeling is subjective, but the statement is factual; hence, it is a subjective fact. In contrast, if one person tells another that the other is feeling happy or sad—whether this is true or not— that is an assumption or an opinion.

Biases

Biases usually occur when someone allows their personal preferences or ideologies to interfere with what should be an objective decision. In personal situations, someone is biased towards someone else if they favor them in an unfair way. In academic writing, being biased in your sources means leaving out objective information that would turn the argument one way or the other. The evidence of bias in academic writing makes the text less credible, so be sure to present all viewpoints when writing, not just your own so to avoid coming off as biased. Being objective when presenting information or dealing with people usually allows the person to gain more credibility.

Stereotypes

Stereotypes are preconceived notions that place a particular rule or characteristics on an entire group of people. Stereotypes are usually offensive to the group they refer to or to allies of that group, and often have negative connotations. The reinforcement of stereotypes isn't always obvious. Sometimes stereotypes can be very subtle and are still widely used in order for people to understand categories within the world. For example, saying that women are more emotional and intuitive than men is a stereotype, although this is still an assumption used by many in order to understand the differences between one another.

Summaries

An important skill is the ability to read a complex text and then reduce its length and complexity by focusing on the key events and details. A **summary** is a shortened version of the original text, written by the reader in their own words. The summary should be shorter than the original text, and it must be thoughtfully formed to include critical points from the original text.

In order to effectively summarize a complex text, it's necessary to understand the original source and identify the major points covered. It may be helpful to outline the original text to get the big picture and avoid getting bogged down in the minor details. For example, a summary wouldn't include a statistic from the original source unless it was the major focus of the text. It's also important for readers to use their own words, yet retain the original meaning of the passage. The key to a good summary is emphasizing the main idea without changing the focus of the original information.

The more complex a text, the more difficult it is to summarize. Readers must evaluate all points from the original source and then filter out what they feel are the less necessary details. Only the essential ideas should remain. The summary often mirrors the original text's organizational structure. For example, in a problem-solution text structure, the author typically presents readers with a problem and then develops solutions through the course of the text. An effective summary would likely retain this general structure, rephrasing the problem and then reporting the most useful solutions.

Paraphrasing is somewhat similar to summarizing. It calls for the reader to take a small part of the passage and list or describe its main points. Paraphrasing is more than rewording the original passage, though. Like summary, it should be written in the reader's own words, while still retaining the meaning of the original source. The main difference between summarizing and paraphrasing is that a summary would be appropriate for a much larger text, while paraphrase might focus on just a few lines of text. Effective paraphrasing will indicate an understanding of the original source, yet still help the reader expand on their interpretation. A paraphrase should neither add new information nor remove essential facts that change the meaning of the source.

Meaning of Words and Phrases as They are Used in the Text

Readers can often figure out what unfamiliar words mean without interrupting their reading to look them up in dictionaries by examining context. **Context** includes the other words or sentences in a passage. One common context clue is the root word and any affixes (prefixes/suffixes). Another common context clue is a synonym or definition included in the sentence. Sometimes both exist in the same sentence. Here's an example:

> Scientists who study birds are *ornithologists*.

Many readers may not know the word *ornithologist*. However, the example contains a definition (scientists who study birds). The reader may also have the ability to analyze the suffix (*-logy*, meaning the study of) and root (*ornitho-*, meaning bird).

Another common context clue is a sentence that shows differences. Here's an example:

> Birds *incubate* their eggs outside of their bodies, unlike mammals.

Some readers may be unfamiliar with the word *incubate*. However, since we know that "unlike mammals," birds incubate their eggs outside of their bodies, we can infer that *incubate* has something to do with keeping eggs warm outside the body until they are hatched.

In addition to analyzing the etymology of a word's root and affixes and extrapolating word meaning from sentences that contrast an unknown word with an antonym, readers can also determine word meanings from sentence context clues based on logic. Here's an example:

> Birds are always looking out for predators that could attack their young.

The reader who is unfamiliar with the word *predator* could determine from the context of the sentence that predators usually prey upon baby birds and possibly other young animals. Readers might also use the context clue of etymology here, as *predator* and *prey* have the same root.

When readers encounter an unfamiliar word in text, they can use the surrounding context—the overall subject matter, specific chapter/section topic, and especially the immediate sentence context to gain an understanding of what is being read. Among others, one category of context clues is grammar. For example, the position of a word in a sentence and its relationship to the other words can help the reader establish whether the unfamiliar word is a verb, a noun, an adjective, an adverb, etc. This narrows down the possible meanings of the word to one part of speech. However, this may be insufficient. In a sentence that many birds *migrate* twice yearly, the reader can determine the word is a verb, and probably does not mean eat or drink; but it could mean travel, mate, lay eggs, hatch, or molt.

Some words can have a number of different meanings depending on how they are used. For example, the word *fly* has a different meaning in each of the following sentences:

- "His trousers have a fly on them."
- "He swatted the fly on his trousers."
- "Those are some fly trousers."
- "They went fly fishing."
- "She hates to fly."
- "If humans were meant to fly, they would have wings."

As strategies, readers can try substituting a familiar word for an unfamiliar one and see whether it makes sense in the sentence. They can also identify other words in a sentence, offering clues to an unfamiliar word's meaning.

Word Choice, Style, and Tone

Words can be very powerful. When written words are used with the intent to make an argument or support a position, the words used—and the way in which they are arranged—can have a dramatic effect on the readers. Clichés, colloquialisms, run-on sentences, and misused words are all examples of ways that word choice can negatively affect writing quality. Unless the writer carefully considers word choice, a written work stands to lose credibility.

If a writer's overall intent is to provide a clear meaning on a subject, he or she must consider not only the exact words to use, but also their placement, repetition, and suitability. Academic writing should be intentional and clear, and it should be devoid of awkward or vague descriptions that can easily lead to misunderstandings. When readers find themselves reading and rereading just to gain a clear understanding of the writer's intent, there may be an issue with word choice. Although the words used in academic writing are different from those used in a casual conversation, they shouldn't necessarily be overly academic either. It may be relevant to employ key words that are associated with the subject but struggling to inject these words into a paper just to sound academic may defeat the purpose. If the message cannot be clearly understood the first time, word choice may be the culprit.

Word choice also conveys the author's attitude and sets a tone. Although each word in a sentence carries a specific **denotation**, it might also carry positive or negative **connotations**—and it is the connotations that set the tone and convey the author's attitude. Consider the following similar sentences:

> It was the same old routine that happens every Saturday morning—eat, exercise, chores.

The Saturday morning routine went off without a hitch—eat, exercise, chores.

The first sentence carries a negative connotation with the author's "same old routine" word choice. The feelings and attitudes associated with this phrase suggest that the author is bored or annoyed at the Saturday morning routine. Although the second sentence carries the same topic—explaining the Saturday morning routine—the choice to use the expression "without a hitch" conveys a positive or cheery attitude.

An author's writing style can likewise be greatly affected by word choice. When writing for an academic audience, for example, it is necessary for the author to consider how to convey the message by carefully considering word choice. If the author interchanges between third-person formal writing and second-person informal writing, the author's writing quality and credibility are at risk. Formal writing involves complex sentences, an objective viewpoint, and the use of full words as opposed to the use of a subjective viewpoint, contractions, and first- or second-person usage commonly found in informal writing.

Content validity, the author's ability to support the argument, and the audience's ability to comprehend the written work are all affected by the author's word choice.

Analyzing Word Parts

By learning some of the etymologies of words and their parts, readers can break new words down into components and analyze their combined meanings. For example, the root word *soph* is Greek for wise or knowledge. Knowing this informs the meanings of English words including *sophomore, sophisticated,* and *philosophy*. Those who also know that *phil* is Greek for love will realize that *philosophy* means the love of knowledge. They can then extend this knowledge of *phil* to understand *philanthropist* (one who loves people), *bibliophile* (book lover), *philharmonic* (loving harmony), *hydrophilic* (water-loving), and so on. In addition, *phob-* derives from the Greek *phobos,* meaning fear. This informs all words ending with it as meaning fear of various things: *acrophobia* (fear of heights), *arachnophobia* (fear of spiders), *claustrophobia* (fear of enclosed spaces), *ergophobia* (fear of work), and *hydrophobia* (fear of water), among others.

Some English word origins from other languages, like ancient Greek, are found in large numbers and varieties of English words. An advantage of the shared ancestry of these words is that once readers recognize the meanings of some Greek words or word roots, they can determine or at least get an idea of what many different English words mean. As an example, the Greek word *métron* means to measure, a measure, or something used to measure; the English word meter derives from it. Knowing this informs many other English words, including *altimeter, barometer, diameter, hexameter, isometric,* and *metric*. While readers must know the meanings of the other parts of these words to decipher their meaning fully, they already have an idea that they are all related in some way to measures or measuring.

While all English words ultimately derive from a proto-language known as Indo-European, many of them historically came into the English vocabulary, later, from sources like the ancient Greeks' language, the Latin used throughout Europe and much of the Middle East during the reign of the Roman Empire, and the Anglo-Saxon languages used by England's early tribes. In addition to classic revivals and native foundations, by the Renaissance era, other influences included French, German, Italian, and Spanish. Today we can often discern English word meanings by knowing common roots and affixes, particularly from Greek and Latin.

The following is a list of common prefixes and their meanings:

Prefix	Definition	Examples
a-	without	atheist, agnostic

ad-	to, toward	advance
ante-	before	antecedent, antedate
anti-	opposing	antipathy, antidote
auto-	self	autonomy, autobiography
bene-	well, good	benefit, benefactor
bi-	two	bisect, biennial
bio-	life	biology, biosphere
chron-	time	chronometer, synchronize
circum-	around	circumspect, circumference
com-	with, together	commotion, complicate
contra-	against, opposing	contradict, contravene
cred-	belief, trust	credible, credit
de-	from	depart
dem-	people	demographics, democracy
dis-	away, off, down, not	dissent, disappear
equi-	equal, equally	equivalent
ex-	former, out of	extract
for-	away, off, from	forget, forswear
fore-	before, previous	foretell, forefathers
homo-	same, equal	homogenized
hyper-	excessive, over	hypercritical, hypertension
in-	in, into	intrude, invade
inter-	among, between	intercede, interrupt
mal-	bad, poorly, not	malfunction
micr-	small	microbe, microscope
mis-	bad, poorly, not	misspell, misfire
mono-	one, single	monogamy, monologue
mor-	die, death	mortality, mortuary
neo-	new	neolithic, neoconservative
non-	not	nonentity, nonsense
omni-	all, everywhere	omniscient
over-	above	overbearing
pan-	all, entire	panorama, pandemonium
para-	beside, beyond	parallel, paradox
phil-	love, affection	philosophy, philanthropic
poly-	many	polymorphous, polygamous
pre-	before, previous	prevent, preclude
prim-	first, early	primitive, primary
pro-	forward, in place of	propel, pronoun
re-	back, backward, again	revoke, recur
sub-	under, beneath	subjugate, substitute
super-	above, extra	supersede, supernumerary
trans-	across, beyond, over	transact, transport
ultra-	beyond, excessively	ultramodern, ultrasonic, ultraviolet
un-	not, reverse of	unhappy, unlock

vis-	to see	visage, visible

The following is a list of common suffixes and their meanings:

Suffix	Definition	Examples
-able	likely, able to	capable, tolerable
-ance	act, condition	acceptance, vigilance
-ard	one that does excessively	drunkard, wizard
-ation	action, state	occupation, starvation
-cy	state, condition	accuracy, captaincy
-er	one who does	teacher
-esce	become, grow, continue	convalesce, acquiesce
-esque	in the style of, like	picturesque, grotesque
-ess	feminine	waitress, lioness
-ful	full of, marked by	thankful, zestful
-ible	able, fit	edible, possible, divisible
-ion	action, result, state	union, fusion
-ish	suggesting, like	churlish, childish
-ism	act, manner, doctrine	barbarism, socialism
-ist	doer, believer	monopolist, socialist
-ition	action, result, state,	sedition, expedition
-ity	quality, condition	acidity, civility
-ize	cause to be, treat with	sterilize, mechanize, criticize
-less	lacking, without	hopeless, countless
-like	like, similar	childlike, dreamlike
-ly	like, of the nature of	friendly, positively
-ment	means, result, action	refreshment, disappointment
-ness	quality, state	greatness, tallness
-or	doer, office, action	juror, elevator, honor
-ous	marked by, given to	religious, riotous
-some	apt to, showing	tiresome, lonesome
-th	act, state, quality	warmth, width
-ty	quality, state	enmity, activity

Inference and Interpretation

Making Inferences from the Text

Making an **inference** from a selection means to make an educated guess from the passage read. Inferences should be conclusions based off of sound evidence and reasoning. When multiple-choice test questions ask about the logical conclusion that can be drawn from reading text, the test-taker must identify which choice will unavoidably lead to that conclusion. In order to eliminate the incorrect choices, the test-taker should come up with a hypothetical situation wherein an answer choice is true, but the conclusion is not true. Here is an example with three answer choices:

> Fred purchased the newest PC available on the market. Therefore, he purchased the most expensive PC in the computer store.
>
> What can one assume for this conclusion to follow logically?
>
> a. Fred enjoys purchasing expensive items.
> b. PCs are some of the most expensive personal technology products available.
> c. The newest PC is the most expensive one.

The premise of the text is the first sentence: *Fred purchased the newest PC.* The conclusion is the second sentence: *Fred purchased the most expensive PC.* Recent release and price are two different factors; the difference between them is the logical gap. To eliminate the gap, one must equate whatever new information the conclusion introduces with the pertinent information the premise has stated. This example simplifies the process by having only one of each: one must equate product recency with product price. Therefore, a possible bridge to the logical gap could be a sentence stating that the newest PCs always cost the most.

Drawing Conclusions Not Explicitly Present in the Text

Authors describe settings, characters, character emotions, and events. Readers must infer to understand the text fully. Inferring enables readers to figure out meanings of unfamiliar words, make predictions about upcoming text, draw conclusions, and reflect on reading. Readers can infer about text before, during, and after reading. In everyday life, we use sensory information to infer. Readers can do the same with text. When authors do not answer all readers' questions, readers must infer by looking at illustrations, considering characters' behaviors, and asking questions during reading. Taking clues from text and connecting that text to prior knowledge helps to draw conclusions. Readers can infer word meanings, settings, reasons for occurrences, character emotions, pronoun referents, author messages, and answers to questions unstated in text.

Making inferences and drawing conclusions involve skills that are quite similar: both require readers to fill in information the author has omitted. Authors may omit information as a technique for inducing readers to discover the outcomes themselves; or they may consider certain information unimportant; or they may assume their reading audience already knows certain information. To make an inference or draw a conclusion about text, readers should observe all facts and arguments the author has presented and consider what they already know from their own personal experiences. Students taking multiple-choice reading tests that refer to text passages can determine correct and incorrect choices based on the information in the passage. For example, from a text passage describing an individual's signs of anxiety while unloading groceries and nervously clutching their wallet at a grocery store checkout, readers can infer or conclude that the individual may not have enough money to pay for everything.

Inferring the Traits, Feelings, and Motives of Characters

When authors create characters for fictional works, the audience is not presented with a list of character traits, feelings, or motives the characters experience throughout the narrative. Instead, readers must infer the traits, feelings, and motives of characters in each text. Readers may infer these characteristics through watching the behavior of the characters as well as the dialogue they engage in to determine what they may be thinking or feeling. Part of character analysis is researching the character and their most important moments in the text.

Traits

Character **traits** are the details that make up a character's *personality*. Traits are a distinguishing quality belonging to a person, like a feature or an attribute. Traits are usually described as the way a person is on the inside: someone can be *weird, arrogant, funny, caring, clever, creative, heroic, lovable, popular, sensitive, delicate, demanding, disorganized,* or *timid,* among many other attributes. Let's look at an example of a character trait below:

> When Christopher came in through the classroom door, the whole atmosphere changed. More than half the class was smiling, and a few students motioned him over because they had saved a seat for him. Christopher smiled back, slung his backpack over his shoulder, and made his way over to his friends in the far corner.

The audience may be asked to find the particular traits of the character, Christopher, in the selection above. Traits are the attributes that make up a person's personality or ego, so let's think of some words that describe Christopher's personality. The word *popular* immediately comes to mind, because Christopher's presence has the ability to change the atmosphere of the classroom to a friendly, welcoming one. He is also being motioned over by his friends who have saved a seat for him, which means he is a desired member of his friend group. *Confident* is another word that would describe Christopher's personality because he smiles and makes his way over to the group without hesitation.

Feelings

Character **feelings** are the *emotions* characters feel in reaction to another person or event. Feelings can be positive or negative depending on the situation, but feelings are always the emotions *felt by* the character and can usually be seen on the outside. For example, if a character is feeling cheerful, they are experiencing the emotion of happiness, and they will have a smile on their face or a sense of peace surrounding them. For an audience to tell what the character is feeling, we have to rely on either imagery or dialogue. We may be able to tell how a character feels by the way they look; likewise, we can tell what a character may be feeling by the way they speak to another character:

> Beth's face drew into a frown, and her bottom lip began to quiver, "I thought you were my friend, but apparently I was wrong. Friends support each other through everything, and you left right when I needed you the most."

From the imagery of Beth's face frowning and quivering, and the dialogue expressing her views on a betrayed friendship, we can infer that Beth is feeling *disappointed, hurt,* or *unloved.* There are many ways to express a single emotion, so be sure that you've glanced at all key words before you decide on which one fits a character best.

Motives

Character motives are important tools for the author to use; motives are what make a character complex and give them dimension. The word **motive** means a reason for doing something, usually hidden or not apparent. It's difficult to tell exactly what a character's motives are; authors will never announce a character's motives explicitly. Again, subtle imagery, dialogue, and behavior will open a door to what may be motivating the character in any given scene. Another technique to understand character motivation is their background: Where did they grow up? What were their life experiences? What was their family like? All of these questions will lead a reader to decide why the character behaves the way that they do.

Interpreting Information in Different Formats

Line Graphs

Line graphs are useful for visually representing data that *vary continuously over time*, like an individual student's test scores. The horizontal or x-axis shows dates/times; the vertical or y-axis shows point values. A dot is plotted on the point where each horizontal date line intersects each vertical number line, and then these dots are connected, forming a line. Line graphs show whether changes in values over time exhibit trends like ascending, descending, flat, or more variable, like going up and down at different times. For example, suppose a student's scores on the same type of reading test were 75% in October, 80% in November, 78% in December, 82% in January, 85% in February, 88% in March, and 90% in April. A line graph of these scores would look like this.

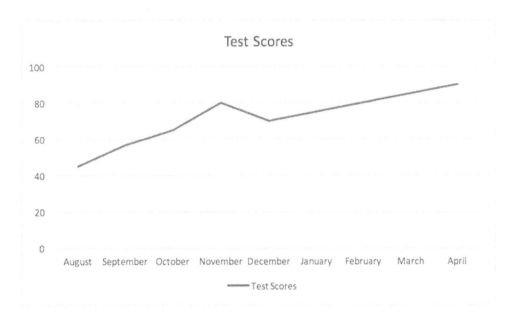

Bar Graphs

Bar graphs feature equally spaced, horizontal or vertical rectangular bars representing *numerical values*. They can show change over time as line graphs do, but unlike line graphs, bar graphs can also show differences and similarities among values at a single point in time. Bar graphs are also helpful for visually representing data from different categories, especially when the horizontal axis displays some value that is not numerical, like various countries with inches of annual rainfall.

The following is a bar graph that compares different classes and how many books they read:

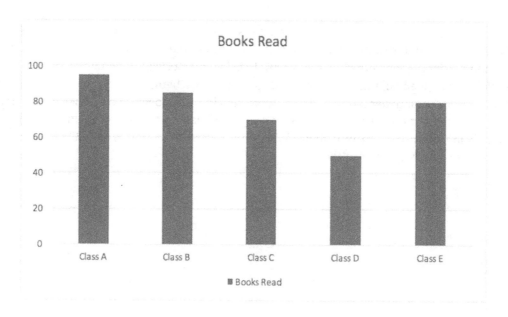

Pie Charts

Pie charts, also called circle graphs, are good for representing *percentages or proportions of a whole quantity* because they represent the whole as a circle or "pie," with the various proportion values shown as "slices" or wedges of the pie. This gives viewers a clear idea of how much of a total each item occupies. To calculate central angles to make each portion the correct size, multiply each percentage by 3.6 (= 360/100). For example, biologists may have information that 60% of Americans have brown eyes, 20% have hazel eyes, 15% have blue eyes, and 5% have green eyes. A pie chart of these distributions would look like this:

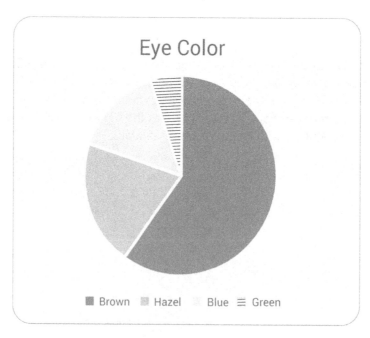

Line Plots

Rather than showing trends or changes over time like line graphs, **line plots** show the *frequency with which a value occurs in a group*. Line plots are used for visually representing data sets that total 50 or fewer values. They make visible features like gaps between some data points, clusters of certain numbers/number ranges, and outliers (data points with significantly smaller or larger values than others). For example, the age ranges in a class of nursing students might appear like this in a line plot:

```
xxxxxxxxx    xxxxx    xx       x       xxx      xx     x
_____
   18         23       28      33      38        43    48
```

Pictograms

Magazines, newspapers, and other similar publications designed for consumption by the general public often use pictograms to represent data. **Pictograms** feature *icons or symbols* that look like the category of data being counted; for example, little silhouettes shaped like human beings are commonly used to represent people. If the data involve large numbers, like populations, one person symbol might represent ten people, one thousand people, one million people, etc. For smaller values, such as how many individuals out of ten fit a given description, one symbol might equal one person. Male and female silhouettes are used to differentiate gender and child shapes for children. Clock symbols are used to represent amounts of time, such as a given number of hours; calendar pages might depict months; suns and moons could show days and nights; hourglasses might represent minutes. While pictogram symbols are easily recognizable and appealing to general viewers, one disadvantage is that it is difficult to display partial symbols for in-between quantities.

Interpreting Nonliteral Language

Authors of a text use language with multiple levels of meaning for many different reasons. When the meaning of a text calls for directness, literal language should be used to provide clarity to the reader. Figurative language can be used when the author wants to produce an emotional effect in the reader or facilitate a deeper understanding of a word or passage. For example, if someone wanted to write a set of instructions on how to use a computer, they would write in literal language. However, if someone wanted to comment on the social implications of banning immigration, they might want to use a wide range of figurative language to highlight an empathetic response. It is important to keep in mind, too, that a single text can have a mixture of both literal and figurative language.

Literal Language

Literal language uses words in accordance with their actual definition. Many informational texts employ literal language because it is straightforward and precise. Documents such as instructions, proposals, technical documents, and workplace documents use literal language for the majority of their writing, so there is no confusion or complexity of meaning for readers to decipher. The information is best communicated through clear and precise language. The following are brief examples of literal language:

- I cook with olive oil.
- There are 365 days in a year.
- My grandma's name is Barbara.
- Yesterday we had some scattered thunderstorms.
- World War II began in 1939.
- Blue whales are the largest species of whale.

Figurative Language

Not meant to be taken literal, **figurative language** is useful when the author of a text wants to produce an emotional effect in the reader or add a heightened complexity to the meaning of the text. Figurative language is used more heavily in texts such as literary fiction, poetry, critical theory, and speeches. Figurative language goes beyond literal language, allowing readers to form associations they wouldn't normally form with literal language. Using language in a figurative sense appeals to the imagination of the reader. It is important to remember that words themselves are signifiers of objects and ideas and not the objects and ideas themselves. Figurative language can highlight this detachment by creating multiple associations but also points to the fact that language is fluid and capable of creating a world full of linguistic possibilities. Figurative language, it can be argued, is the heart of communication even outside of fiction and poetry. People connect through humor, metaphors, cultural allusions, puns, and symbolism in their everyday rhetoric. The following are terms associated with figurative language:

Simile

A **simile** is a comparison of two things using *like*, *than*, or *as*. A simile usually takes objects that have no apparent connection, such as a mind and an orchid, and compares them:

> His mind was as complex and rare as a field of ghost orchids.

Similes encourage new, fresh perspectives on objects or ideas that wouldn't otherwise occur. Similes are different than metaphors. Metaphors do not use *like*, *than*, or *as*. So, a metaphor from the above example would be:

> His mind was a field of ghost orchids.

Similes highlight the comparison by focusing on the figurative side of the language, elucidating more the author's intent: a field of ghost orchids is something complex and rare, like the mind of a genius. With the metaphor, however, we get a beautiful yet somewhat equivocal comparison.

Metaphor

A popular use of figurative language, **metaphors** compare objects or ideas directly, asserting that something *is* a certain thing, even if it isn't. The following is an example of a metaphor used by author Virginia Woolf:

> Books are the mirrors of the soul.

Metaphors have a *vehicle* and a *tenor*. The tenor is "books" and the vehicle is "mirrors of the soul." That is, the tenor is what is meant to be described, and the vehicle is that which carries the weight of the comparison. In this metaphor, perhaps the author means to say that written language (books) reflect a person's most inner thoughts and desires.

There are also **dead metaphors**, which means that the phrases have been so overused to the point where the figurative meaning becomes literal, like the phrase "What you're saying is crystal clear." The phrase compares "what's being said" to something "crystal clear." However, since the latter part of the phrase is in such popular use, the meaning seems literal ("I understand what you're saying") even when it's not.

Finally, an **extended metaphor** is a metaphor that goes on for several paragraphs, or even an entire text. John Keats' poem "On First Looking into Chapman's Homer" begins, "Much have I travell'd in the realms of gold," and goes on to explain the first time he hears Chapman's translation of Homer's writing. We see the extended metaphor begin in the first line. Keats is comparing travelling into "realms of gold" and exploration of new lands to the act of hearing a certain kind of literature for the first time. The extended metaphor goes on until the end of the poem where Keats stands "Silent, upon a peak in Darien," having heard the end of Chapman's translation. Keats has gained insight into new lands (new text) and is the richer for it.

The following are brief definitions and examples of popular figurative language:

Onomatopoeia: A word that, when spoken, imitates the sound to which it refers. Ex: "We heard a loud *boom* while driving to the beach yesterday."

Personification: When human characteristics are given to animals, inanimate objects, or abstractions. An example would be in William Wordsworth's poem "Daffodils" where he sees a "crowd . . . / of golden daffodils . . . / Fluttering and dancing in the breeze." Dancing is usually a characteristic attributed solely to humans, but Wordsworth personifies the daffodils here as a crowd of people dancing.

Hyperbole: A hyperbole is an exaggeration. Ex: "I'm so tired I could sleep for centuries."

Pun: Puns are used in popular culture to invoke humor by exploiting the meanings of words. They can also be used in literature to give hints of meaning in unexpected places. One example of a pun is when Mercutio is giving his monologue after he is stabbed by Tybalt in "Romeo and Juliet" and says, "look for me tomorrow and you will find me a grave man."

Imagery: This is a collection of images given to the reader by the author. If a text is rich in imagery, it is easier for the reader to imagine themselves in the author's world. One example of a poem that relies on imagery is William Carlos Williams' "The Red Wheelbarrow":

> so much depends
> upon
>
> a red wheel
> barrow
>
> glazed with rain
> water
>
> beside the white
> chickens

The starkness of the imagery and the placement of the words in the poem, to some readers, throws the poem into a meditative state where, indeed, the world of this poem is made up solely of images of a purely simple life. This poem tells a story in sixteen words by using imagery.

Symbolism: A symbol is used to represent an idea or belief system. For example, poets in Western civilization have been using the symbol of a rose for hundreds of years to represent love. In Japan, poets have used the firefly to symbolize passionate love, and sometimes even spirits of those who have died. Symbols can also express powerful political commentary and can be used in propaganda.

Analysis

Topic, Main Idea, and Supporting Details

The **topic** of a text is the general subject matter. Text topics can usually be expressed in one word or a few words at most. Concerning a text's topic, readers should ask themselves what point the author is trying to make with said topic. This point is the **main idea** or *primary purpose* of the text; it is the one thing the author wants readers to know concerning the topic. Once the author has established the main idea, they will support the main idea by filling in the text with supporting details. *Supporting details* are evidence that support the main idea and include personal testimonies, examples, or statistics.

One analogy for these components and their relationships is that a text is like a well-designed house. The topic is the roof, covering all rooms. The main idea is the frame. The supporting details are the various rooms. To identify the topic of a text, readers can ask themselves what or who the author is writing about in the paragraph. To locate the main idea, readers can ask themselves what one idea the author wants readers to know about the main topic. To identify supporting details, readers can put the main idea into question form and ask, "what does the author use to prove or explain their main idea?"

Let's look at an example. An author is writing an essay about the Amazon rainforest and trying to convince the audience that more funding should go into protecting the area from deforestation. The author makes the argument stronger by including evidence of the benefits of the rainforest: it provides habitats to a variety of species, it provides much of the earth's oxygen which in turn cleans the atmosphere, and it is the home to medicinal plants that may be the answer to some of the world's deadliest diseases. Here is an outline of the essay looking at the topic, main idea, and supporting details:

- *Topic*: Amazon rainforest
- *Main Idea*: The Amazon rainforest should receive more funding in order to protect it from deforestation.
- *Supporting Details*:
     1. It provides habitats to a variety of species
     2. It provides much of the earth's oxygen which in turn cleans the atmosphere
     3. It is home to medicinal plants that may be the answer to some of the world's deadliest diseases.

Notice that the topic of the essay is listed in a few key words: "Amazon rainforest." The main idea tells us what about the topic is important: the topic should be funded in order to prevent deforestation. Finally, the supporting details are what the author relies on to convince the audience to act or to believe in the truth of the main idea.

Theme of a Text

The **theme** of a text is the central idea the author communicates. Whereas the topic of a passage of text may be concrete in nature, by contrast *the theme is always conceptual*. For example, while the topic of Mark Twain's novel *The Adventures of Huckleberry Finn* might be described as something like the coming-of-age experiences of a poor, illiterate, functionally orphaned boy around and on the Mississippi River in 19th-century Missouri, one theme of the book might be that human beings are corrupted by society. Another theme might be that slavery and "civilized" society itself are hypocritical. Whereas the main idea in a text is the most important single point that the author wants to make, the theme is the concept or view around which the author centers the text.

Themes are underlying meanings in literature. For example, if a story's main idea is a character succeeding against all odds, the theme is overcoming obstacles. If a story's main idea is one character wanting what another character has, the theme is jealousy. If a story's main idea is a character doing something they were afraid to do, the theme is courage. Themes differ from topics in that a topic is a subject matter; a theme is the author's opinion about it. For example, a work could have a topic of war and a theme that war is a curse. Authors present themes through characters' feelings, thoughts, experiences, dialogue, plot actions, and events. Themes function as "glue" holding other essential story elements together. They offer readers insights into characters' experiences, the author's philosophy, and how the world works.

Author's Purpose

Authors may have many **purposes** for writing a specific text. Their purposes may be to try and convince readers to agree with their position on a subject, to impart information, or to entertain. Other writers are motivated to write from a desire to express their own feelings. Authors' purposes are their reasons for writing something. A single author may have one overriding purpose for writing or multiple reasons. An author may explicitly state their intention in the text, or the reader may need to infer that intention. Those who read reflectively benefit from identifying the purpose because it enables them to analyze information in the text. By knowing why the author wrote the text, readers can glean ideas for how to approach it. The following is a list of questions readers can ask in order to discern an author's purpose for writing a text:

- From the title of the text, why do you think the author wrote it?
- Was the purpose of the text to give information to readers?
- Did the author want to describe an event, issue, or individual?
- Was it written to express emotions and thoughts?
- Did the author want to convince readers to consider a particular issue?
- Was the author primarily motivated to write the text to entertain?
- Why do you think the author wrote this text from a certain point of view?
- What is your response to the text as a reader?
- Did the author state their purpose for writing it?

Students should *read to interpret* information rather than simply content themselves with roles as text consumers. Being able to identify an author's purpose efficiently improves reading comprehension, develops critical thinking, and makes students more likely to consider issues in depth before accepting writer viewpoints. Authors of fiction frequently write to entertain readers. Another purpose for writing fiction is making a political statement; for example, Jonathan Swift wrote "A Modest Proposal" (1729) as a political satire. Another purpose for writing fiction as well as nonfiction is to persuade readers to take some action or further a particular cause. Fiction authors and poets both frequently write to evoke certain moods; for example, Edgar Allan Poe wrote novels, short stories, and poems that evoke moods of gloom, guilt, terror, and dread. Another purpose of poets is evoking certain emotions: love is popular, as in Shakespeare's sonnets and numerous others. In "The Waste Land" (1922), T.S. Eliot evokes society's alienation, disaffection, sterility, and fragmentation.

Authors seldom directly state their purposes in texts. Some students may be confronted with nonfiction texts such as biographies, histories, magazine and newspaper articles, and instruction manuals, among others. To identify the purpose in nonfiction texts, students can ask the following questions:

- Is the author trying to teach something?
- Is the author trying to persuade the reader?
- Is the author imparting factual information only?
- Is this a reliable source?

- Does the author have some kind of hidden agenda?

To apply the author's purpose in nonfictional passages, students can also analyze sentence structure, word choice, and transitions to answer the aforementioned questions and to make inferences. For example, authors wanting to convince readers to view a topic negatively often choose words with negative connotations.

Narrative Writing

Narrative writing tells a story. The most prominent examples of narrative writing are fictional novels. Here are some examples:

- Mark Twain's *The Adventures of Tom Sawyer* and *The Adventures of Huckleberry Finn*
- Victor Hugo's *Les Misérables*
- Charles Dickens' *Great Expectations, David Copperfield,* and *A Tale of Two Cities*
- Jane Austen's *Northanger Abbey, Mansfield Park, Pride and Prejudice, Sense and Sensibility,* and *Emma*
- Toni Morrison's Beloved, *The Bluest Eye,* and *Song of Solomon*
- Gabriel García Márquez's *One Hundred Years of Solitude* and *Love in the Time of Cholera*

Some nonfiction works are also written in narrative form. For example, some authors choose a narrative style to convey factual information about a topic, such as a specific animal, country, geographic region, and scientific or natural phenomenon.

Since narrative is the type of writing that tells a story, it must be told by someone, who is the narrator. The **narrator** may be a fictional character telling the story from their own viewpoint. This narrator uses the first person (*I, me, my, mine* and *we, us, our,* and *ours*) point of view. The narrator may simply be the author; for example, when Louisa May Alcott writes "Dear reader" in *Little Women,* she (the author) addresses us as readers. In this case, the novel is typically told in third person, referring to the characters as *he, she, they,* or *them.* Another more common technique is the omniscient narrator; i.e. the story is told by an unidentified individual who sees and knows everything about the events and characters—not only their externalized actions, but also their internalized feelings and thoughts. Second person, i.e. writing the story by addressing readers as "you" throughout, is less frequently used.

Expository Writing

Expository writing is also known as informational writing. Its purpose is not to tell a story as in narrative writing, to paint a picture as in descriptive writing, nor to persuade readers to agree with something as in argumentative writing. Rather, its point is to communicate information to the reader. As such, the point of view of the author will necessarily be more objective. Whereas other types of writing appeal to the reader's emotions, appeal to the reader's reason by using logic, or use subjective descriptions to sway the reader's opinion or thinking, expository writing seeks to do none of these but simply to provide facts, evidence, observations, and objective descriptions of the subject matter. Some examples of expository writing include research reports, journal articles, articles and books about historical events or periods, academic subject textbooks, news articles and other factual journalistic reports, essays, how-to articles, and user instruction manuals.

Technical Writing

Technical writing is similar to expository writing in that it is factual, objective, and intended to provide information to the reader. Indeed, it may even be considered a subcategory of expository writing. However, technical writing differs from expository writing in that (1) it is specific to a particular field, discipline, or subject; and (2) it uses the specific technical terminology that belongs only to that area.

Writing that uses technical terms is intended only for an audience familiar with those terms. A primary example of technical writing today is writing related to computer programming and usage.

Persuasive Writing
Persuasive writing is intended to persuade the reader to agree with the author's position. It is also known as argumentative writing. Some writers may be responding to other writers' arguments, in which case they make reference to those authors or text and then disagree with them. However, another common technique is for the author to anticipate opposing viewpoints in general, both from other authors and from the author's own readers. The author brings up these opposing viewpoints, and then refutes them before they can even be raised, strengthening the author's argument. Writers persuade readers by appealing to their reason, which Aristotle called **logos**; appealing to emotion, which Aristotle called **pathos**; or appealing to readers based on the author's character and credibility, which Aristotle called **ethos**.

Analyzing Individuals, Events, and Ideas

In order to better comprehend more complex texts, readers strive to draw connections between ideas or events. Authors often have a main idea or argument that is supported by ideas, facts, or expert opinion. These relationships that are built into writing can take on several different forms.

The relationships between people, events, and ideas may be clearly stated, or the reader might have to infer the relationships based on clues in the passage. To infer means to arrive at a conclusion based on evidence, clues, or facts.

People might be related through connections like family or friendship, or through events that link them directly or indirectly. In the passage, relationships may be described in background information or dialogue, or they may be implied through interactions between the characters.

Events and ideas in a passage can be related through sequence or through cause and effect. To relate events and ideas through sequence means to show what happened first and what happened next. Or a sequence could be ordered in another way, such as alphabetically, or geographically. Ideas and events can also be related through a comparison the author makes, showing the similarities and differences.

Sequence
When ideas are related through sequence, as in a series of events, the author typically uses signal words like *after, and then, while,* and *before*.

Cause and Effect
In cause and effect relationships, the cause is an event or circumstance that occurs before and is directly responsible for the effect. Signal words such as *because, due to,* and *as a result of* indicate a cause and effect relationship. When the relationship is implied, the reader must use clues in the passage to infer that the effect resulted from the cause.

Compare and Contrast
Ideas are often connected through comparisons of their similarities to one another using the signal words such as *like* or *and*. Ideas can also be connected through a contrast of their differences using signal words such as *but* or *however*.

Recognizing Aspects of Style, Structure, Mood, or Tone

While it may seem impossible to know exactly what an author feels toward their subject, there are clues to indicate the emotion, or lack thereof, of the author. Clues like *word choice* or *style* will alert readers to the author's attitude.

Some possible words that name the author's attitude are listed below:

- Admiring
- Angry
- Critical
- Defensive
- Enthusiastic
- Humorous
- Moralizing
- Neutral
- Objective
- Patriotic
- Persuasive
- Playful
- Sentimental
- Serious
- Supportive
- Sympathetic
- Unsupportive

An author's **tone** is the author's attitude toward their subject and is usually indicated by word choice. If an author's attitude toward their subject is one of disdain, the author will show the subject in a negative light, using deflating words or words that are negatively charged. If an author's attitude toward their subject is one of praise, the author will use agreeable words and show the subject in a positive light. If an author takes a neutral tone towards their subject, their words will be neutral as well, and they probably will show all sides of their subject, not just the negative or positive side.

Style is another indication of the author's attitude and includes aspects such as sentence structure, type of language, and formatting. **Sentence structure** is how a sentence is put together. Sometimes, short, choppy sentences will indicate a certain tone given the surrounding context, while longer sentences may serve to create a buffer to avoid being too harsh or may be used to explain additional information. Style may also include formal or informal language. Using **formal language** to talk about a subject may indicate a level of respect. Using **informal language** may be used to create an atmosphere of friendliness or familiarity with a subject. Again, it depends on the surrounding context whether or not language is used in a negative or positive way. Style may also include **formatting**, such as determining the length of paragraphs or figuring out how to address the reader at the very beginning of the text.

The following is a passage from *The Florentine Painters of the Renaissance* by Bernhard Berenson. Following the passage is a question stem regarding the author's attitude toward their subject:

> Let us look now at an even greater triumph of movement than the Nudes, Pollaiuolo's "Hercules Strangling Antæus." As you realise the suction of Hercules' grip on the earth, the swelling of his calves with the pressure that falls on them, the violent throwing back of his chest, the stifling force of his embrace; as you realise the supreme effort of Antæus, with one hand crushing down upon

the head and the other tearing at the arm of Hercules, you feel as if a fountain of energy had sprung up under your feet and were playing through your veins. I cannot refrain from mentioning still another masterpiece, this time not only of movement, but of tactile values and personal beauty as well—Pollaiuolo's "David" at Berlin. The young warrior has sped his stone, cut off the giant's head, and now he strides over it, his graceful, slender figure still vibrating with the rapidity of his triumph, expectant, as if fearing the ease of it. What lightness, what buoyancy we feel as we realise the movement of this wonderful youth!

Which one of the following best captures the author's attitude toward the paintings depicted in the passage?
    a. Neutrality towards the subject in this passage.
    b. Disdain for the violence found in the paintings.
    c. Excitement for the physical beauty found within the paintings.
    d. Passion for the movement and energy of the paintings.
    e. Seriousness for the level of artistry the paintings hold.

Choice *D* is the best answer. We know that the author feels positively about the subject because of the word choice. Berenson uses words and phrases like "supreme," "fountain of energy," "graceful," "figure still vibrating," "lightness," "buoyancy," and "wonderful youth." Notice, also, the exclamation mark at the end of the paragraph. These words and style depict an author full of passion, especially for the movement and energy found within the paintings.

Choice *A* is incorrect because the author is biased towards the subject due to the energy he writes with—he calls the movement in the paintings "wonderful" and by the other word choices and phrases, readers can tell that this is not an objective analysis of these paintings. Choice *B* is incorrect because, although the author does mention the "violence" in the stance of Hercules, he does not exude disdain towards this. Choice *C* is incorrect. There is excitement in the author's tone, and some of this excitement is directed towards the paintings' physical beauty. However, this is not the *best* answer choice. Choice *D* is more accurate when stating the passion is for the movement and energy of the paintings, of which physical beauty is included. Finally, Choice *E* is incorrect. The tone is partly serious, but we see the author getting carried away with enthusiasm for the beauty of the paintings towards the middle and especially the end of the passage.

Text structure is the way in which the author organizes and presents textual information so readers can follow and comprehend it. One kind of text structure is **sequence**. This means the author arranges the text in a logical order from beginning to middle to end. There are three types of sequences:

- **Chronological**: ordering events in time from earliest to latest

- **Spatial**: describing objects, people, or spaces according to their relationships to one another in space

- **Order of Importance**: addressing topics, characters, or ideas according to how important they are, from either least important to most important

Chronological sequence is the most common sequential text structure. Readers can identify sequential structure by looking for words that signal it, like *first, earlier, meanwhile, next, then, later, finally;* and specific times and dates the author includes as chronological references.

Problem-Solution Text Structure

The **problem-solution text** structure organizes textual information by presenting readers with a problem and then developing its solution throughout the course of the text. The author may present a variety of alternatives as possible solutions, eliminating each as they are found unsuccessful, or gradually leading up to the ultimate solution. For example, in fiction, an author might write a murder mystery novel and have the character(s) solve it through investigating various clues or character alibis until the killer is identified. In nonfiction, an author writing an essay or book on a real-world problem might discuss various alternatives and explain their disadvantages or why they would not work before identifying the best solution. For scientific research, an author reporting and discussing scientific experiment results would explain why various alternatives failed or succeeded.

Comparison-Contrast Text Structure

Comparison identifies similarities between two or more things. **Contrast** identifies differences between two or more things. Authors typically employ both to illustrate relationships between things by highlighting their commonalities and deviations. For example, a writer might compare Windows and Linux as operating systems, and contrast Linux as free and open-source vs. Windows as proprietary. When writing an essay, sometimes it is useful to create an image of the two objects or events you are comparing or contrasting. Venn diagrams are useful because they show the differences as well as the similarities between two things. Once you've seen the similarities and differences on paper, it might be helpful to create an outline of the essay with both comparison and contrast. Every outline will look different, because every two or more things will have a different number of comparisons and contrasts. Say you are trying to compare and contrast carrots with sweet potatoes. Here is an example of a compare/contrast outline using those topics:

- *Introduction:* Talk about why you are comparing and contrasting carrots and sweet potatoes. Give the thesis statement.
- *Body paragraph 1:* Sweet potatoes and carrots are both root vegetables (similarity)
- *Body paragraph 2:* Sweet potatoes and carrots are both orange (similarity)
- *Body paragraph 3:* Sweet potatoes and carrots have different nutritional components (difference)
- *Conclusion:* Restate the purpose of your comparison/contrast essay.

Of course, if there is only one similarity between your topics and two differences, you will want to rearrange your outline. Always tailor your essay to what works best with your topic.

Descriptive Text Structure

Description can be both a type of text structure and a type of text. Some texts are descriptive throughout entire books. For example, a book may describe the geography of a certain country, state, or region, or tell readers all about dolphins by describing many of their characteristics. Many other texts are not descriptive throughout but use descriptive passages within the overall text. The following are a few examples of descriptive text:

- When the author describes a character in a novel
- When the author sets the scene for an event by describing the setting
- When a biographer describes the personality and behaviors of a real-life individual
- When a historian describes the details of a particular battle within a book about a specific war
- When a travel writer describes the climate, people, foods, and/or customs of a certain place

A hallmark of description is using sensory details, painting a vivid picture so readers can imagine it almost as if they were experiencing it personally.

Cause and Effect Text Structure

When using **cause** and **effect** to extrapolate meaning from text, readers must determine the cause when the author only communicates effects. For example, if a description of a child eating an ice cream cone includes details like beads of sweat forming on the child's face and the ice cream dripping down her hand faster than she can lick it off, the reader can infer or conclude it must be hot outside. A useful technique for making such decisions is wording them in "If...then" form, e.g. "If the child is perspiring and the ice cream melting, it may be a hot day." Cause and effect text structures explain why certain events or actions resulted in particular outcomes. For example, an author might describe America's historical large flocks of dodo birds, the fact that gunshots did not startle/frighten dodos, and that because dodos did not flee, settlers killed whole flocks in one hunting session, explaining how the dodo was hunted into extinction.

Informational Text Structure

The structure in **informational texts** depends again on the genre. For example, a newspaper article may start by stating an exciting event that happened and then move on to talk about that event in chronological order, known as sequence or order structure. Many informational texts also use cause and effect structure, which describes an event and then identifies reasons for why that event occurred. Some essays may write about their subjects by way of comparison and contrast, which is a structure that compares two things or contrasts them to highlight their differences. Other documents, such as proposals, will have a problem to solution structure, where the document highlights some kind of problem and then offers a solution toward the end.

Evaluating Arguments and Use of Literary Devices

Arguments

When authors write text for the purpose of persuading others to agree with them, they assume a position with the subject matter about which they are writing. Rather than presenting information objectively, the author treats the subject matter subjectively so that the information presented supports his or her position. In their argumentation, the author presents information that refutes or weakens opposing positions. Another technique authors use in persuasive writing is to anticipate arguments against the position. When students learn to read subjectively, they gain experience with the concept of persuasion in writing and learn to identify positions taken by authors. This enhances their reading comprehension and develops their skills for identifying pro and con arguments and biases.

There are five main parts of the classical argument that writers employ in a well-designed stance:

- **Introduction**: In the introduction to a classical argument, the author establishes goodwill and rapport with the reading audience, warms up the readers, and states the thesis or general theme of the argument.

- **Narration**: In the narration portion, the author gives a summary of pertinent background information, informs the readers of anything they need to know regarding the circumstances and environment surrounding and/or stimulating the argument, and establishes what is at risk or the stakes in the issue or topic. Literature reviews are common examples of narrations in academic writing.

- **Confirmation**: The confirmation states all claims supporting the thesis and furnishes evidence for each claim, arranging this material in logical order—e.g. from most obvious to most subtle or strongest to weakest.

- **Refutation and Concession:** The refutation and concession discuss opposing views and anticipate reader objections without weakening the thesis, yet permitting as many oppositions as possible.

- **Summation:** The summation strengthens the argument while summarizing it, supplying a strong conclusion and showing readers the superiority of the author's solution.

Introduction

A classical argument's **introduction** must pique reader interest, get readers to perceive the author as a writer, and establish the author's position. Shocking statistics, new ways of restating issues, or quotations or anecdotes focusing the text can pique reader interest. Personal statements, parallel instances, or analogies can also begin introductions—so can bold thesis statements if the author believes readers will agree. Word choice is also important for establishing author image with readers.

The introduction should typically narrow down to a clear, sound thesis statement. If readers cannot locate one sentence in the introduction explicitly stating the writer's position or the point they support, the writer probably has not refined the introduction sufficiently.

Narration and Confirmation

The narration part of a classical argument should create a context for the argument by explaining the issue to which the argument is responding, and by supplying any background information that influences the issue. Readers should understand the issues, alternatives, and stakes in the argument by the end of the narration to enable them to evaluate the author's claims equitably. The confirmation part of the classical argument enables the author to explain why they believe in the argument's thesis. The author builds a chain of reasoning by developing several individual supporting claims and explaining why that evidence supports each claim and also supports the overall thesis of the argument.

Refutation and Concession and Summation

The classical argument is the model for argumentative/persuasive writing, so authors often use it to establish, promote, and defend their positions. In the refutation aspect of the refutation and concession part of the argument, authors disarm reader opposition by anticipating and answering their possible objections, persuading them to accept the author's viewpoint. In the concession aspect, authors can concede those opposing viewpoints with which they agree. This can avoid weakening the author's thesis while establishing reader respect and goodwill for the author: all refutation and no concession can antagonize readers who disagree with the author's position. In the conclusion part of the classical argument, a less skilled writer might simply summarize or restate the thesis and related claims; however, this does not provide the argument with either momentum or closure. More skilled authors revisit the issues and the narration part of the argument, reminding readers of what is at stake.

Literary Devices

When the main idea is not explicitly stated, it is more difficult to identify. Implicit main ideas must be deduced or decoded by gathering together the facts, arguments, and images hinted at in the text and drawing information from these clues. Some historical speeches, poems, songs, and political cartoons have implicit main ideas. They are likely to incorporate indirect literary devices to express their main points. The following are examples of literary devices:

Irony: There are three types of irony. **Verbal irony** is when a person states one thing and means the opposite. For example, a person is probably using irony when they say, "I can't wait to study for this exam next week." **Dramatic irony** occurs in a narrative and happens when the audience knows something that the characters do not. In the modern TV series *Hannibal*, we as an audience know that Hannibal Lecter is a serial killer, but most of the main characters do not. This is dramatic irony. Finally, **situational irony** is when one expects something to happen, and the opposite occurs. For example, we can say that a fire station burning down would be an instance of situational irony.

Juxtaposition: Juxtaposition is placing two objects side by side for comparison. In literature, this might look like placing two characters side by side for contrasting effect, like God and Satan in Milton's "Paradise Lost."

Paradox: A paradox is a statement that is self-contradictory but will be found nonetheless true. One example of a paradoxical phrase is when Socrates said "I know one thing; that I know nothing." Seemingly, if Socrates knew nothing, he wouldn't know that he knew nothing. However, it is one thing he knows: that true wisdom begins with casting all presuppositions one has about the world aside.

Allusion: An allusion is a reference to a character or event that happened in the past. An example of a poem littered with allusions is T.S. Eliot's "The Waste Land." An example of a biblical allusion manifests when the poet says, "I will show you fear in a handful of dust," creating an ominous tone from Genesis 3:19 "For you are dust, and to dust you shall return."

Flashback: A flashback in a story or passage is a way of showing what events happened previously that were not depicted in the story. Events from the past are shown out of sequence to give additional information or context to the story. In the Harry Potter novels, many events which took place before Harry was even born are revealed through flashbacks that provide insight into some of the other characters and their motivations.

Foreshadowing: Foreshadowing is a suggestion or hint by the author, of events that have not yet happened. Foreshadowing can be used to create tension or raise questions about what might happen later in a story. Foreshadowing can be indicated by a tone, mood, or even the weather; for example, in *Great Expectations,* tumultuous weather in the novel foreshadows the dramatic changes coming in Pip's life.

Synthesis and Generalization

Drawing Conclusions and Making Generalizations

Some of the conclusion question types will ask what test takers can "infer," "imply," or "conclude" from the given information. Other language used in conclusion questions might consist of the following:

- Must also be true
- Provide the most support for
- Which one of the following conclusions
- Most strongly supported by
- Properly inferred

Making inferences and drawing conclusions involve skills that are quite similar: both require readers to fill in information the test writer has omitted. To make an inference or draw a conclusion about the text, test takers should observe all facts and arguments the test writer has presented. The best way to understand ways to drawing well-supported conclusions is by practice. Let's take a look at the following example:

Nutritionist: More and more bodybuilders each year turn to whey protein as a source for their supplement intake to repair muscle tissue after working out. More and more studies are showing that using whey as a source of protein is linked to prostate cancer in men. Bodybuilders who use whey protein may consider switching to a plant-based protein source in order to avoid developing the negative effects that come with whey protein consumption.

Which of the following most accurately expresses the conclusion of the nutritionist's argument?

  a. Whey protein is an excellent way to repair muscles after a workout.
  b. Bodybuilders should switch from whey to a plant-based protein.
  c. Whey protein causes every single instance of prostate cancer in men.
  d. We still don't know the causes of prostate cancer in men.
  e. It's possible that bodybuilding may cause prostate cancer.

The correct answer choice is *B:* bodybuilders should switch from whey to a plant-based protein. We can gather this from the entirety of the passage, as it begins with what kind of protein bodybuilders consume, the dangers of that protein, and what kind of protein to switch to. Choice *A* is incorrect; this is the opposite of what the passage states. When reading through answer choices, it's important to look for choices that include the words "every," "always," or "all." In many instances, absolute answer choices will not be the correct answer. This example is shown in Choice *C;* the passage does not state that whey protein causes "every single instance" of prostate cancer in men, only that it is *linked* to prostate cancer in men. Choice *D* is incorrect; although the nutritionist doesn't list all the causes of prostate cancer in men, the nutritionist does not conclude that we don't know the causes of prostate cancer in men either. Finally, Choice *E* is incorrect. This answer choice makes a jump from bodybuilding to prostate cancer, which is incorrect. The passage states that bodybuilders consume more whey protein, which is linked to cancer, not that bodybuilding *itself* causes cancer.

The key to drawing well-supported conclusions is to read the question stem in its entirety a few times over and then paraphrase the passage in your own words. Once you do this, you will get an idea of the passage's conclusion before you are confused by all the different answer choices. Remember that drawing a conclusion is different than making an assumption. With drawing a conclusion, we are relying solely on the passage for facts to come to our conclusion. Making an assumption goes beyond the facts of the passage, so be careful of answer choices depicting assumptions instead of passage-based conclusions.

Making Predictions

Before and during reading, readers can apply the strategy of making **predictions** about what they think may happen next. For example, what plot and character developments will occur in fiction? What points will the author discuss in nonfiction? Making predictions about portions of text they have not yet read prepares readers mentally, and also gives them a purpose for reading. To inform and make predictions about text, the reader can do the following:

- Consider the title of the text and what it implies
- Look at the cover of the book
- Look at any illustrations or diagrams for additional visual information
- Analyze the structure of the text
- Apply outside experience and knowledge to the text

Readers may adjust their predictions as they read. Reader predictions may or may not come true in text.

Compare and Contrast Points of View

When reading, you will notice three main **points of view** from which a story is written. Each point of view allows the author to tell their story from a certain perspective. The three points of view are first person, second person, and third person. It's important to note that specific pronouns will be used for their respective points of view; it's additionally worth noting that not every point of view is appropriate to use depending on the writing style and audience for which the text is being written.

First Person

First person uses pronouns such as *I, me, we, my, us, and our.* Some writers naturally find it easier to tell stories from their own points of view, so writing in the first person offers advantages for them. The first-person voice is better for interpreting the world from a single viewpoint, and for enabling reader immersion in one protagonist's experiences. However, others find it difficult to use the first-person narrative voice. Its disadvantages can include overlooking the emotions of characters, forgetting to include description, producing stilted writing, using too many sentence structures involving "I did . . .", and not devoting enough attention to the story's "here-and-now" immediacy.

When a writer tells a story using the first person, readers can identify this by the use of first person pronouns, like *I, me, we, us,* etc. However, first person narratives can be told by different people or from different points of view. For example, some authors write in the first person point of view to tell the story from the main character's viewpoint, as Charles Dickens did in his novels *David Copperfield* and *Great Expectations.* Some authors write in the first person point of view as a fictional character in the story, but not necessarily the main character. For example, F. Scott Fitzgerald wrote *The Great Gatsby* as narrated by Nick Carraway, a character in the story, about the main characters, Jay Gatsby and Daisy Buchanan. Other authors write in the first person, but as the omniscient narrator—an often unnamed person who knows all of the characters' inner thoughts and feelings. Writing in first person as oneself is more common in nonfiction.

Second Person

Narrative written in the **second person** addresses someone else as "you." In novels and other fictional works, the second person is the narrative voice most seldom used. The primary reason for this is that it often reads in an awkward manner, which prevents readers from being drawn into the fictional world of the novel. The second person is more often used in informational text, especially in how-to manuals, guides, and other instructions.

Third Person

The **third person** narrative is probably the most prevalent voice used in fictional literature. While some authors tell stories from the point of view and in the voice of a fictional character using the first person, it is a more common practice to describe the actions, thoughts, and feelings of fictional characters in the third person using *he, him, she, her, they, them,* etc.

Although plot and character development are both necessary and possible when writing narrative from a first person point of view, they are also more difficult, particularly for new writers and those who find it unnatural or uncomfortable to write from that perspective. Therefore, writing experts advise beginning writers to start out writing in the third person. A big advantage of third-person narration is that the writer can describe the thoughts, feelings, and motivations of every character in a story, which is not possible for the first-person narrator. Third person narrative can impart information to readers that the characters do not know. On the other hand, beginning writers often regard using the third person point of view as more difficult because they must write about the feelings and thoughts of every character, rather than only about those of the protagonist.

Analyzing Two or More Texts

Texts with Similar Themes
Throughout time, humans have told stories with similar themes. Some themes are universal across time, space, and culture. These include themes of the individual as a hero, conflicts of the individual against nature, the individual against society, change vs. tradition, the circle of life, coming-of-age, and the complexities of love. Themes involving war and peace have featured prominently in diverse works, like Homer's *Iliad*, Tolstoy's *War and Peace* (1869), Stephen Crane's *The Red Badge of Courage* (1895), Hemingway's *A Farewell to Arms* (1929), and Margaret Mitchell's *Gone with the Wind* (1936). Another universal literary theme is that of the *quest*. These appear in folklore from countries and cultures worldwide, including the Gilgamesh Epic, Arthurian legend's Holy Grail quest, Virgil's *Aeneid*, Homer's *Odyssey*, and the *Argonautica*. Cervantes' *Don Quixote* is a parody of chivalric quests. J.R.R. Tolkien's *The Lord of the Rings* trilogy (1954) also features a quest.

Similar themes across cultures often occur in countries that share a border or are otherwise geographically close together. For example, a folklore story of a rabbit in the moon using a mortar and pestle is shared among China, Japan, Korea, and Thailand—making medicine in China, making rice cakes in Japan and Korea, and hulling rice in Thailand. Another instance is when cultures are more distant geographically, but their languages are related. For example, East Turkestan's Uighurs and people in Turkey share tales of folk hero Effendi Nasreddin Hodja. Another instance, which may either be called cultural diffusion or simply reflect commonalities in the human imagination, involves shared themes among geographically and linguistically different cultures: both Cameroon's and Greece's folklore tell of centaurs; Cameroon, India, Malaysia, Thailand, and Japan, of mermaids; Brazil, Peru, China, Japan, Malaysia, Indonesia, and Cameroon, of underwater civilizations; and China, Japan, Thailand, Vietnam, Malaysia, Brazil, and Peru, of shape-shifters.

Two prevalent literary themes are love and friendship, which can end happily, sadly, or both. William Shakespeare's *Romeo and Juliet*, Emily Brontë's *Wuthering Heights*, Leo Tolstoy's *Anna Karenina*, and both *Pride and Prejudice* and *Sense and Sensibility* by Jane Austen are famous examples. Another theme recurring in popular literature is of revenge, an old theme in dramatic literature, e.g. Elizabethans Thomas Kyd's *The Spanish Tragedy* and Thomas Middleton's *The Revenger's Tragedy*. Some more well-known instances include Shakespeare's tragedies *Hamlet* and *Macbeth*, Alexandre Dumas' *The Count of Monte Cristo*, John Grisham's *A Time to Kill*, and Stieg Larsson's *The Girl Who Kicked the Hornet's Nest*.

Comparing Informational Texts that Address the Same Topic
Informational texts about the same topic can include vastly different pieces of information, perceptions, or opinions. When comparing two or more informational texts that address the same topic, the first comparison to make is to determine and understand the source of each text. Based on the source, the authors may hold rather different intentions for writing the text. For example, an author of an academic textbook is likely to present information that is evidence based, instructional, and/or reviewed by peers with expert-level credentialing. A journalist who writes about the same topic may present similar information yet include a personal editorial opinion, such as the application of the information in the real world. These two authors are writing on the same topic but presenting to largely different readerships; therefore, their method of information sharing is likely to be different. Understanding the audience for whom the author is writing, the purpose of writing their text, and the author's own credentials can be useful components of a comparison analysis.

Additionally, the point of view from which an informational text is written can also be useful in understanding the different values between two pieces of text that address the same topic. For example, the first-person account of a historical event from someone who experienced it directly will likely present different information, a different perspective, and evoke different emotions from the reader than the recount of the same event by an objective researcher who is simply sharing facts about the event. In addition to focusing on where the two accounts differ, readers should note similarities between the two passages (such as factual information or similar feelings that are expressed by both authors).

Practice Questions

The next six questions are based on the following passage from The Life, Crime, and Capture of John Wilkes Booth *by George Alfred Townsend.*

The box in which the President sat consisted of two boxes turned into one, the middle partition being removed, as on all occasions when a state party visited the theater. The box was on a level with the dress circle; about twelve feet above the stage. There were two entrances—the door nearest to the wall having been closed and locked; the door nearest the balustrades of the dress circle, and at right angles with it, being open and left open, after the visitors had entered. The interior was carpeted, lined with crimson paper, and furnished with a sofa covered with crimson velvet, three arm chairs similarly covered, and six cane-bottomed chairs. Festoons of flags hung before the front of the box against a background of lace.

President Lincoln took one of the arm-chairs and seated himself in the front of the box, in the angle nearest the audience, where, partially screened from observation, he had the best view of what was transpiring on the stage. Mrs. Lincoln sat next to him, and Miss Harris in the opposite angle nearest the stage. Major Rathbone sat just behind Mrs. Lincoln and Miss Harris. These four were the only persons in the box.

The play proceeded, although "Our American Cousin," without Mr. Sothern, has, since that gentleman's departure from this country, been justly esteemed a very dull affair. The audience at Ford's, including Mrs. Lincoln, seemed to enjoy it very much. The worthy wife of the President leaned forward, her hand upon her husband's knee, watching every scene in the drama with amused attention. Even across the President's face at intervals swept a smile, robbing it of its habitual sadness.

About the beginning of the second act, the mare, standing in the stable in the rear of the theater, was disturbed in the midst of her meal by the entrance of the young man who had quitted her in the afternoon. It is presumed that she was saddled and bridled with exquisite care.

Having completed these preparations, Mr. Booth entered the theater by the stage door; summoned one of the scene shifters, Mr. John Spangler, emerged through the same door with that individual, leaving the door open, and left the mare in his hands to be held until he (Booth) should return. Booth who was even more fashionably and richly dressed than usual, walked thence around to the front of the theater, and went in. Ascending to the dress circle, he stood for a little time gazing around upon the audience and occasionally upon the stage in his usual graceful manner. He was subsequently observed by Mr. Ford, the proprietor of the theater, to be slowly elbowing his way through the crowd that packed the rear of the dress circle toward the right side, at the extremity of which was the box where Mr. and Mrs. Lincoln and their companions were seated. Mr. Ford casually noticed this as a slightly extraordinary symptom of interest on the part of an actor so familiar with the routine of the theater and the play.

1. Which of the following best describes the author's attitude toward the events leading up to the assassination of President Lincoln?
   a. Excitement due to the setting and its people.
   b. Sadness due to the death of a beloved president.
   c. Anger because of the impending violence.
   d. Neutrality due to the style of the report.

2. What does the author mean by the last sentence in the passage?

    a. Mr. Ford was suspicious of Booth and assumed he was making his way to Mr. Lincoln's box.

    b. Mr. Ford assumed Booth's movement throughout the theater was due to being familiar with the theater.

    c. Mr. Ford thought that Booth was making his way to the theater lounge to find his companions.

    d. Mr. Ford thought that Booth was elbowing his way to the dressing room to get ready for the play.

3. Given the author's description of the play "Our American Cousin," which one of the following is most analogous to Mr. Sothern's departure from the theater?

    a. A ballet dancer who leaves the New York City Ballet just before they go on to their final performance.

    b. A basketball player leaves an NBA team and the next year they make it to the championship but lose.

    c. A lead singer leaves their band to begin a solo career, and the band drops in sales by 50 percent on their next album.

    d. A movie actor who dies in the middle of making a movie and the movie is made anyway by actors who resemble the deceased.

4. Based on the organizational structure of the passage, which of the following texts most closely relates?

    a. A chronological account in a fiction novel of a woman and a man meeting for the first time.

    b. A cause-and-effect text ruminating on the causes of global warming.

    c. An autobiography that begins with the subject's death and culminates to his birth.

    d. A text focusing on finding a solution to the problem of the Higgs boson particle.

5. Which of the following words, if substituted for the word *festoons* in the first paragraph, would LEAST change the meaning of the sentence?

    a. Feathers

    b. Armies

    c. Adornments

    d. Buckets

6. What is the primary purpose of the passage?

    a. To persuade the audience that John Wilkes Booth killed Abraham Lincoln.

    b. To inform the audience of the setting wherein Lincoln was shot.

    c. To narrate the bravery of Lincoln and his last days as President.

    d. To recount in detail the events that led up to Abraham Lincoln's death.

The next seven questions are based on the following passage from The Story of Germ Life *by Herbert William Conn.*

The first and most universal change effected in milk is its souring. So universal is this phenomenon that it is generally regarded as an inevitable change which can not be avoided, and, as already pointed out, has in the past been regarded as a normal property of milk. To-day, however, the phenomenon is well understood. It is due to the action of certain of the milk bacteria upon the milk sugar which converts it into lactic acid, and this acid gives the sour taste and curdles the milk. After this acid is produced in small quantity its presence proves deleterious to the growth of the bacteria, and further bacterial growth is checked. After souring, therefore, the milk for some time does not ordinarily undergo any further changes.

Milk souring has been commonly regarded as a single phenomenon, alike in all cases. When it was first studied by bacteriologists it was thought to be due in all cases to a single species of micro-organism which was discovered to be commonly present and named Bacillus acidi lactici. This bacterium has certainly the power of souring milk rapidly, and is found to be very common in dairies in Europe. As soon as bacteriologists turned their attention more closely to the subject it was found that the spontaneous souring of milk was not always caused by the same species of bacterium. Instead of finding this Bacillus acidi lactici always present, they found that quite a number of different species of bacteria have the power of souring milk, and are found in different specimens of soured milk. The number of species of bacteria which have been found to sour milk has increased until something over a hundred are known to have this power. These different species do not affect the milk in the same way. All produce some acid, but they differ in the kind and the amount of acid, and especially in the other changes which are affected at the same time that the milk is soured, so that the resulting soured milk is quite variable. In spite of this variety, however, the most recent work tends to show that the majority of cases of spontaneous souring of milk are produced by bacteria which, though somewhat variable, probably constitute a single species, and are identical with the Bacillus acidi lactici. This species, found common in the dairies of Europe, according to recent investigations occurs in this country as well. We may say, then, that while there are many species of bacteria infesting the dairy which can sour the milk, there is one which is more common and more universally found than others, and this is the ordinary cause of milk souring.

When we study more carefully the effect upon the milk of the different species of bacteria found in the dairy, we find that there is a great variety of changes which they produce when they are allowed to grow in milk. The dairyman experiences many troubles with his milk. It sometimes curdles without becoming acid. Sometimes it becomes bitter, or acquires an unpleasant "tainted" taste, or, again, a "soapy" taste. Occasionally a dairyman finds his milk becoming slimy, instead of souring and curdling in the normal fashion. At such times, after a number of hours, the milk becomes so slimy that it can be drawn into long threads. Such an infection proves very troublesome, for many a time it persists in spite of all attempts made to remedy it. Again, in other cases the milk will turn blue, acquiring about the time it becomes sour a beautiful sky-blue colour. Or it may become red, or occasionally yellow. All of these troubles the dairyman owes to the presence in his milk of unusual species of bacteria which grow there abundantly.

7. The word *deleterious* in the first paragraph can be best interpreted as referring to which one of the following?
   a. Amicable
   b. Smoldering
   c. Luminous
   d. Ruinous

8. Which of the following best explains how the passage is organized?
   a. The author begins by presenting the effects of a phenomenon, then explains the process of this phenomenon, and then ends by giving the history of the study of this phenomenon.
   b. The author begins by explaining a process or phenomenon, then gives the history of the study of this phenomenon, this ends by presenting the effects of this phenomenon.
   c. The author begins by giving the history of the study of a certain phenomenon, then explains the process of this phenomenon, then ends by presenting the effects of this phenomenon.
   d. The author begins by giving a broad definition of a subject, then presents more specific cases of the subject, then ends by contrasting two different viewpoints on the subject.

9. What is the primary purpose of the passage?
   a. To inform the reader of the phenomenon, investigation, and consequences of milk souring.
   b. To persuade the reader that milk souring is due to Bacillus acidi lactici, found commonly in the dairies of Europe.
   c. To describe the accounts and findings of researchers studying the phenomenon of milk souring.
   d. To discount the former researchers' opinions on milk souring and bring light to new investigations.

10. What does the author say about the ordinary cause of milk souring?
   a. Milk souring is caused mostly by a species of bacteria called Bacillus acidi lactici, although former research asserted that it was caused by a variety of bacteria.
   b. The ordinary cause of milk souring is unknown to current researchers, although former researchers thought it was due to a species of bacteria called Bacillus acidi lactici.
   c. Milk souring is caused mostly by a species of bacteria identical to that of Bacillus acidi lactici, although there are a variety of other bacteria that cause milk souring as well.
   d. The ordinary cause of milk souring will sometimes curdle without becoming acidic, though sometimes it will turn colors other than white, or have strange smells or tastes.

11. The author of the passage would most likely agree most with which of the following?
   a. Milk researchers in the past have been incompetent and have sent us on a wild goose chase when determining what causes milk souring.
   b. Dairymen are considered more expert in the field of milk souring than milk researchers.
   c. The study of milk souring has improved throughout the years, as we now understand more of what causes milk souring and what happens afterward.
   d. Any type of bacteria will turn milk sour, so it's best to keep milk in an airtight container while it is being used.

12. Given the author's account of the consequences of milk souring, which of the following is most closely analogous to the author's description of what happens after milk becomes slimy?
   a. The chemical change that occurs when a firework explodes.
   b. A rainstorm that overwaters a succulent plant.
   c. Mercury inside of a thermometer that leaks out.
   d. A child who swallows flea medication.

13. What type of paragraph would most likely come after the third?
    a. A paragraph depicting the general effects of bacteria on milk.
    b. A paragraph explaining a broad history of what researchers have found in regard to milk souring.
    c. A paragraph outlining the properties of milk souring and the way in which it occurs.
    d. A paragraph showing the ways bacteria infiltrate milk and ways to avoid this infiltration.

The following passage is taken from Chapter 6 of Sense and Sensibility, by Jane Austen:

The first part of their journey was performed in too melancholy a disposition to be otherwise than tedious and unpleasant. But as they drew toward the end of it, their interest in the appearance of a country which they were to inhabit overcame their dejection, and a view of Barton Valley as they entered it gave them cheerfulness. It was a pleasant fertile spot, well wooded, and rich in pasture. After winding along it for more than a mile, they reached their own house. A small green court was the whole of its demesne in front; and a neat wicket gate admitted them into it.

As a house, Barton Cottage, though small, was comfortable and compact; but as a cottage it was defective, for the building was regular, the roof was tiled, the window shutters were not painted green, nor were the walls covered with honeysuckles. A narrow passage led directly through the house into the garden behind. On each side of the entrance was a sitting room, about sixteen feet square; and beyond them were the offices and the stairs. Four bed-rooms and two garrets formed the rest of the house. It had not been built many years and was in good repair. In comparison of Norland, it was poor and small indeed!—but the tears which recollection called forth as they entered the house were soon dried away. They were cheered by the joy of the servants on their arrival, and each for the sake of the others resolved to appear happy. It was very early in September; the season was fine, and from first seeing the place under the advantage of good weather, they received an impression in its favour which was of material service in recommending it to their lasting approbation.

The situation of the house was good. High hills rose immediately behind, and at no great distance on each side; some of which were open downs, the others cultivated and woody. The village of Barton was chiefly on one of these hills, and formed a pleasant view from the cottage windows. The prospect in front was more extensive; it commanded the whole of the valley, and reached into the country beyond. The hills which surrounded the cottage terminated the valley in that direction; under another name, and in another course, it branched out again between two of the steepest of them.

With the size and furniture of the house Mrs. Dashwood was upon the whole well satisfied; for though her former style of life rendered many additions to the latter indispensable, yet to add and improve was a delight to her; and she had at this time ready money enough to supply all that was wanted of greater elegance to the apartments. "As for the house itself, to be sure," said she, "it is too small for our family, but we will make ourselves tolerably comfortable for the present, as it is too late in the year for improvements. Perhaps in the spring, if I have plenty of money, as I dare say I shall, we may think about building. These parlors are both too small for such parties of our friends as I hope to see often collected here; and I have some thoughts of throwing the passage into one of them with perhaps a part of the other, and so leave the remainder of that other for an entrance; this, with a new drawing room which may be easily added, and a bed-chamber and garret above, will make it a very snug little cottage. I could wish the stairs were handsome. But one must not expect every thing; though I suppose it would be no difficult matter to widen them. I shall see how much I am before-hand with the world in the spring, and we will plan our improvements accordingly."

In the mean time, till all these alterations could be made from the savings of an income of five hundred a-year by a woman who never saved in her life, they were wise enough to be contented with the house as it was; and each of them was busy in arranging their particular concerns, and endeavoring, by placing around them books and other possessions, to form themselves a home. Marianne's pianoforte was unpacked and properly disposed of; and Elinor's drawings were affixed to the walls of their sitting room.

In such employments as these they were interrupted soon after breakfast the next day by the entrance of their landlord, who called to welcome them to Barton, and to offer them every accommodation from his own house and garden in which theirs might at present be deficient. Sir John Middleton was a good looking man about forty. He had formerly visited at Stanhill, but it was too long for his young cousins to remember him. His countenance was thoroughly good-humoured; and his manners were as friendly as the style of his letter. Their arrival seemed to afford him real satisfaction, and their comfort to be an object of real solicitude to him. He said much of his earnest desire of their living in the most sociable terms with his family, and pressed them so cordially to dine at Barton Park every day till they were better settled at home, that, though his entreaties were carried to a point of perseverance beyond civility, they could not give offence. His kindness was not confined to words; for within an hour after he left them, a large basket full of garden stuff and fruit arrived from the park, which was followed before the end of the day by a present of game.

14. What is the point of view in this passage?
    a. Third-person omniscient
    b. Second-person
    c. First-person
    d. Third-person objective

15. Which of the following events occurred first?
    a. Sir John Middleton stopped by for a visit.
    b. The servants joyfully cheered for the family.
    c. Mrs. Dashwood discussed improvements to the cottage.
    d. Elinor hung her drawings up in the sitting room.

16. Over the course of the passage, the Dashwoods' attitude shifts. Which statement best describes that shift?
    a. From appreciation of the family's former life of privilege to disdain for the family's new landlord
    b. From confidence in the power of the family's wealth to doubt in the family's ability to survive
    c. From melancholy about leaving Norland to excitement about reaching Barton Cottage in the English countryside
    d. From cheerfulness about the family's expedition to anxiety about the upkeep of such a big home

17. Which of the following is a theme of this passage?
    a. All-conquering love
    b. Power of wealth
    c. Wisdom of experience
    d. Reality vs. expectations

18. At the start of paragraph five, the narrator says, "till all these alterations could be made from the savings of an income of five hundred a-year by a woman who never saved in her life, they were wise enough to be contented with the house as it was." What does the narrator mean?
    a. The family is going through a transition phase.
    b. Mrs. Dashwood needs to obtain meaningful employment.
    c. The family is going through a growth phase.
    d. The Dashwood children need to be concerned about the future.

19. What is the relationship between the new landlord and the Dashwoods?
    a. He is a former social acquaintance.
    b. He is one of their cousins.
    c. He is Mrs. Dashwood's father.
    d. He is a long-time friend of the family.

20. Why does the narrator describe the generosity of Sir John Middleton?
    a. To identify one of many positive traits that a landlord should possess.
    b. To explain how a landlord should conduct himself in order to be successful.
    c. To illustrate how his kindness eased the family's adaptation to their new home and circumstances.
    d. To demonstrate that he did not need to be cold and businesslike all of the time.

The next seven questions are from Rhetoric and Poetry in the Renaissance: A Study of Rhetorical Terms in English Renaissance Literary Criticism *by DL Clark*

To the Greeks and Romans rhetoric meant the theory of oratory. As a pedagogical mechanism it endeavored to teach students to persuade an audience. The content of rhetoric included all that the ancients had learned to be of value in persuasive public speech. It taught how to work up a case by drawing valid inferences from sound evidence, how to organize this material in the most persuasive order, how to compose in clear and harmonious sentences. Thus to the Greeks and Romans rhetoric was defined by its function of discovering means to persuasion and was taught in the schools as something that every free-born man could and should learn.

In both these respects the ancients felt that poetics, the theory of poetry, was different from rhetoric. As the critical theorists believed that the poets were inspired, they endeavored less to teach men to be poets than to point out the excellences which the poets had attained. Although these critics generally, with the exceptions of Aristotle and Eratosthenes, believed the greatest value of poetry to be in the teaching of morality, no one of them endeavored to define poetry, as they did rhetoric, by its purpose. To Aristotle, and centuries later to Plutarch, the distinguishing mark of poetry was imitation. Not until the renaissance did critics define poetry as an art of imitation endeavoring to inculcate morality . . .

The same essential difference between classical rhetoric and poetics appears in the content of classical poetics. Whereas classical rhetoric deals with speeches which might be delivered to convict or acquit a defendant in the law court, or to secure a certain action by the deliberative assembly, or to adorn an occasion, classical poetic deals with lyric, epic, and drama. It is a commonplace that classical literary critics paid little attention to the lyric. It is less frequently realized that they devoted almost as little space to discussion of metrics. By far the greater bulk of classical treatises on poetics is devoted to characterization and to the technique of plot construction, involving as it does narrative and dramatic unity and movement as distinct from logical unity and movement.

21. What does the author say about one way in which the purpose of poetry changed for later philosophers?

    a. The author says that at first, poetry was not defined by its purpose but was valued for its ability to be used to teach morality. Later, some philosophers would define poetry by its ability to instill morality. Finally, during the renaissance, poetry was believed to be an imitative art, but was not necessarily believed to instill morality in its readers.

    b. The author says that the classical understanding of poetry dealt with its ability to be used to teach morality. Later, philosophers would define poetry by its ability to imitate life. Finally, during the renaissance, poetry was believed to be an imitative art that instilled morality in its readers.

    c. The author says that at first, poetry was thought to be an imitation of reality, then later philosophers valued poetry more for its ability to instill morality.

    d. The author says that the classical understanding of poetry was that it dealt with the search for truth through its content; later, the purpose of poetry would be through its entertainment.

22. What does the author of the passage say about classical literary critics in relation to poetics?

    a. That rhetoric was more valued than poetry because rhetoric had a definitive purpose to persuade an audience, and poetry's wavering purpose made it harder for critics to teach.

    b. That although most poetry was written as lyric, epic, or drama, the critics were most focused on the techniques of lyric and epic and their performance of musicality and structure.

    c. That although most poetry was written as lyric, epic, or drama, the critics were most focused on the techniques of the epic and drama and their performance of structure and character.

    d. That the study of poetics was more pleasurable than the study of rhetoric due to its ability to assuage its audience, and the critics therefore focused on what poets did to create that effect.

23. What is the primary purpose of this passage?

    a. To contemplate the differences between classical rhetoric and poetry and to consider their purposes in a particular culture.

    b. To inform the readers of the changes in poetic critical theory throughout the years and to contrast those changes to the solidity of rhetoric.

    c. To educate the audience on rhetoric by explaining the historical implications of using rhetoric in the education system.

    d. To convince the audience that poetics is a subset of rhetoric as viewed by the Greek and Roman culture.

24. The word *inculcate* in paragraph two can be best interpreted as referring to which one of the following?

    a. Imbibe
    b. Instill
    c. Implode
    d. Inquire

25. Which of the following most closely resembles the way in which the passage is structured?
   a. The first paragraph presents us with an issue. The second paragraph offers a solution to the problem. The third paragraph summarizes the first two paragraphs.
   b. The first paragraph presents us with definitions and examples of a particular subject. The second paragraph presents a second subject in the same way. The third paragraph offers a contrast of the two subjects.
   c. The first paragraph presents an inquiry. The second paragraph explains the details of that inquiry. The last paragraph offers a solution.
   d. The first paragraph presents us with two subjects alongside definitions and examples. The second paragraph presents us with a comparison of the two subjects. The third paragraph presents us with a contrast of the two subjects.

26. Given the author's description of the content of rhetoric in the first paragraph, which one of the following is most analogous to what it taught? (The sentence is shown below.)

It taught how to work up a case by drawing valid inferences from sound evidence, how to organize this material in the most persuasive order, how to compose in clear and harmonious sentences.

   a. As a musician, they taught me that the end product of the music is everything—what I did to get there was irrelevant, whether it was my ability to read music or the reliance on my intuition to compose.
   b. As a detective, they taught me that time meant everything when dealing with a new case, that the simplest explanation is usually the right one, and that documentation is extremely important to credibility.
   c. As a writer, they taught me the most important thing about writing was consistently showing up to the page every single day, no matter where my muse was.
   d. As a football player, they taught me how to understand the logistics of the game, how my placement on the field affected the rest of the team, and how to run and throw with a mixture of finesse and strength.

27. Which of the following words, if substituted for the word *treatises* in paragraph two, would LEAST change the meaning of the sentence?
   a. Commentary
   b. Encyclopedias
   c. Sermons
   d. Anthems

The next four questions are based on the following passage. It is from Oregon, Washington, and Alaska. Sights and Scenes for the Tourist, *written by E.L. Lomax in 1890:*

Portland is a very beautiful city of 60,000 inhabitants and situated on the Willamette river twelve miles from its junction with the Columbia. It is perhaps true of many of the growing cities of the West, that they do not offer the same social advantages as the older cities of the East. But this is principally the case as to what may be called boom cities, where the larger part of the population is of that floating class which follows in the line of temporary growth for the purposes of speculation, and in no sense applies to those centers of trade whose prosperity is based on the solid foundation of legitimate business. As the metropolis of a vast section of country, having broad agricultural valleys filled with improved farms, surrounded by mountains rich in mineral wealth, and boundless forests of as fine timber as the world produces, the cause of Portland's growth and prosperity is the trade which it has as the center of collection and distribution of this great wealth of natural resources, and it has attracted, not the boomer and speculator, who find their profits in the wild excitement of the boom, but the merchant, manufacturer, and investor, who seek the surer if slower channels of legitimate business and investment. These have come from the East, most of them within the last few years. They came as seeking a better and wider field to engage in the same occupations they had followed in their Eastern homes, and bringing with them all the love of polite life which they had acquired there, have established here a new society, equaling in all respects that which they left behind. Here are as fine churches, as complete a system of schools, as fine residences, as great a love of music and art, as can be found at any city of the East of equal size.

But while Portland may justly claim to be the peer of any city of its size in the United States in all that pertains to social life, in the attractions of beauty of location and surroundings it stands without its peer. The work of art is but the copy of nature. What the residents of other cities see but in the copy, or must travel half the world over to see in the original, the resident of Portland has at its very door.

The city is situate on a gently-sloping ground, with, on the one side, the river, and on the other a range of hills, which, within easy walking distance, rise to an elevation of a thousand feet above the river, affording a most picturesque building site. From the very streets of the thickly settled portion of the city, the Cascade Mountains, with the snow-capped peaks of Hood, Adams, St. Helens, and Rainier, are in plain view.

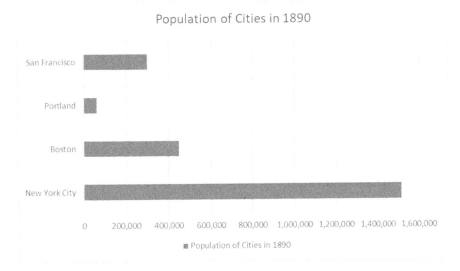

Population of Cities in 1890

28. What is a characteristic of a "boom city," as indicated by the passage?
	a. A city that is built on solid business foundation of mineral wealth and farming.
	b. An area of land on the west coast that quickly becomes populated by residents from the east coast.
	c. A city that, due to the hot weather and dry climate, catches fire frequently, resulting in a devastating population drop.
	d. A city whose population is made up of people who seek quick fortunes rather than building a solid business foundation.

29. By stating that "they do not offer the same social advantages as the older cities of the East" in the first paragraph, the author most likely intends to suggest that:
	a. Inhabitants who reside in older cities in the East are much more social than inhabitants who reside in newer cities in the West because of background and experience.
	b. Cities in the West have no culture compared to the East because the culture in the East comes from European influence.
	c. Cities in the East are older than cities in the West, and older cities always have better culture than newer cities.
	d. Since cities in the West are newly established, it takes them a longer time to develop cultural roots and societal functions than those cities that are already established in the East.

30. Based on the information at the end of paragraph 1, what would the author say of Portland?
	a. It has twice as much culture as the cities in the East.
	b. It has as much culture as the cities in the East.
	c. It doesn't have as much culture as cities in the East.
	d. It doesn't have as much culture as cities in the West.

31. How many more citizens did San Francisco have than Portland in 1890?
	a. Approximately 240,000
	b. Approximately 500,000
	c. Approximately 1,000,000
	d. Approximately 1,500,000

The following four questions are based on the excerpt from Variation of Animals and Plants by Charles Darwin.

> Peach (Amygdalus persica).—In the last chapter I gave two cases of a peach-almond and a double-flowered almond which suddenly produced fruit closely resembling true peaches. I have also given many cases of peach-trees producing buds, which, when developed into branches, have yielded nectarines. We have seen that no less than six named and several unnamed varieties of the peach have thus produced several varieties of nectarine. I have shown that it is highly improbable that all these peach-trees, some of which are old varieties, and have been propagated by the million, are hybrids from the peach and nectarine, and that it is opposed to all analogy to attribute the occasional production of nectarines on peach-trees to the direct action of pollen from some neighbouring nectarine-tree. Several of the cases are highly remarkable, because, firstly, the fruit thus produced has sometimes been in part a nectarine and in part a peach; secondly, because nectarines thus suddenly produced have reproduced themselves by seed; and thirdly, because nectarines are produced from peach-trees from seed as well as from buds. The seed of the nectarine, on the other hand, occasionally produces peaches; and we have seen in one instance that a nectarine-tree yielded peaches by bud-variation. As the peach is certainly the oldest or primary variety, the production of peaches from nectarines, either by seeds or buds, may

perhaps be considered as a case of reversion. Certain trees have also been described as indifferently bearing peaches or nectarines, and this may be considered as bud-variation carried to an extreme degree.

The grosse mignonne peach at Montreuil produced "from a sporting branch" the grosse mignonne tardive, "a most excellent variety," which ripens its fruit a fortnight later than the parent tree, and is equally good. (11/2. "Gardener's Chronicle" 1854 pg. 821.) This same peach has likewise produced by bud-variation the early grosse mignonne. Hunt's large tawny nectarine "originated from Hunt's small tawny nectarine, but not through seminal reproduction." (11/3. Lindley "Guide to Orchard" as quoted in "Gardener's Chronicle" 1852 pg. 821.)

Plums

Mr. Knight states that a tree of the yellow magnum bonum plum, forty years old, which had always borne ordinary fruit, produced a branch which yielded red magnum bonums. (11/4 "Transact. Hort. Soc." Volume 2 pg. 160.) Mr. Rivers, a Sawbridgeworth, informs me (January 1863) that a single tree out of 400 or 500 trees of the Early Prolific plum, which is a purple kind, descended from an old French variety bearing purple fruit, produced when about ten years old bright yellow plums; these differed in no respect except colour from those on the other trees, but were unlike any other known kind of yellow plum (11/5. See also "Gardener's Chronicle" 1863 pg. 27).

32. Which statement is not a detail from the passage?
    a. At least six named varieties of the peach have produced several varieties of nectarine.
    b. It is not probable that all of the peach-trees mentioned are hybrids from the peach and nectarine.
    c. An unremarkable case is the fact that nectarines are produced from peach-trees from seed as well as from buds.
    d. The production of peaches from nectarines might be considered a case of reversion.

33. What is the meaning of the word "propagated" in the first paragraph of this passage?
    a. Multiplied
    b. Diminished
    c. Watered
    d. Uprooted

34. Which of the following most closely reveals the author's tone in this passage?
    a. Enthusiastic
    b. Objective
    c. Critical
    d. Desperate

35. Which of the following is an accurate paraphrasing of the following phrase?

> Certain trees have also been described as indifferently bearing peaches or nectarines, and this may be considered as bud-variation carried to an extreme degree.

a. Some trees are described as bearing peaches and some trees have been described as bearing nectarines, but individually the buds are extreme examples of variation.
b. One way in which bud-variation is said to be carried to an extreme degree is when specific trees have been shown to casually produce peaches or nectarines.
c. Certain trees are indifferent to bud-variation, as recently shown in the trees that produce both peaches and nectarines in the same season.
d. Nectarines and peaches are known to have cross-variation in their buds, which indifferently bears other sorts of fruit to an extreme degree.

The following is entitled Architecture and Democracy *by Claude Bragdon. The next five questions are based on the following passage:*

The world war represents not the triumph, but the birth of democracy. The true ideal of democracy—the rule of a people by the *demos*, or group soul—is a thing unrealized. How then is it possible to consider or discuss an architecture of democracy—the shadow of a shade? It is not possible to do so with any degree of finality, but by an intention of consciousness upon this juxtaposition of ideas—architecture and democracy—signs of the times may yield new meanings, relations may emerge between things apparently unrelated, and the future, always existent in every present moment, may be evoked by that strange magic which resides in the human mind.

Architecture, at its worst as at its best, reflects always a true image of the thing that produced it; a building is revealing even though it is false, just as the face of a liar tells the thing his words endeavor to conceal. This being so, let us make such architecture as is ours declare to us our true estate.

The architecture of the United States, from the period of the Civil War, up to the beginning of the present crisis, everywhere reflects a struggle to be free of a vicious and depraved form of feudalism, grown strong under the very ægis of democracy. The qualities that made feudalism endeared and enduring; qualities written in beauty on the cathedral cities of mediaeval Europe—faith, worship, loyalty, magnanimity—were either vanished or banished from this pseudo-democratic, aridly scientific feudalism, leaving an inheritance of strife and tyranny—a strife grown mean, a tyranny grown prudent, but full of sinister power the weight of which we have by no means ceased to feel.

Power, strangely mingled with timidity; ingenuity, frequently misdirected; ugliness, the result of a false ideal of beauty—these in general characterize the architecture of our immediate past; an architecture "without ancestry or hope of posterity," an architecture devoid of coherence or conviction; willing to lie, willing to steal. What impression such a city as Chicago or Pittsburgh might have made upon some denizen of those cathedral-crowned feudal cities of the past we do not know. He would certainly have been amazed at its giant energy, and probably revolted at its grimy dreariness. We are wont to pity the mediaeval man for the dirt he lived in, even while smoke greys our sky and dirt permeates the very air we breathe: we think of castles as grim and cathedrals as dim, but they were beautiful and gay with color compared with the grim, dim canyons of our city streets.

36. Which of the following does the author NOT consider to be a characteristic of modern architecture?
    a. Power, strangely mingled with timidity
    b. Ugliness as a result of the false ideal of beauty
    c. Giant energy with grimy and dreariness
    d. Cathedral-crowned grim and dim castles

37. By stating that "Architecture, at its worst as at its best, reflects always a true image of the thing that produced it," the author most likely intends to suggest that:
    a. People always create buildings to look like themselves.
    b. Architecture gets more grim, drab, and depressing as the years go by.
    c. Architecture reflects—in shape, color, and form—the attitude of the society which built it.
    d. Modern architecture is a lot like democracy because it is uniform yet made up of more pieces than your traditional architecture.

38. The author refers to "mediaeval man" in the fourth paragraph in order to:
    a. Make the audience look at feudalism with a sense of nostalgia and desire.
    b. Make the audience feel gratitude for modern comforts such as architecture.
    c. Make the audience realize the irony produced from pitying him.
    d. Make the audience look back at feudalism as a time which was dark and dreary.

39. Based on the discussion in paragraph 3, which of the following would be considered architecture from medieval Europe?
    a. Canyon
    b. Castle
    c. Factory
    d. Skyscraper

40. The author's attitude toward modern architecture can best be characterized as:
    a. Narcissistic
    b. Aggrieved
    c. Virtuous
    d. Sarcastic

Questions 41–44 are based on the excerpt from A Christmas Carol *by Charles Dickens:*

> Meanwhile the fog and darkness thickened so, that people ran about with flaring links, proffering their services to go before horses in carriages, and conduct them on their way. The ancient tower of a church, whose gruff old bell was always peeping slyly down at Scrooge out of a Gothic window in the wall, became invisible, and struck the hours and quarters in the clouds, with tremulous vibrations afterwards as if its teeth were chattering in its frozen head up there. The cold became intense. In the main street, at the corner of the court, some labourers were repairing the gas-pipes, and had lighted a great fire in a brazier, round which a party of ragged men and boys were gathered: warming their hands and winking their eyes before the blaze in rapture. The water-plug being left in solitude, its overflowings sullenly congealed, and turned to misanthropic ice. The brightness of the shops where holly sprigs and berries crackled in the lamp heat of the windows, made pale faces ruddy as they passed. Poulterers' and grocers' trades became a splendid joke; a glorious pageant, with which it was next to impossible to believe that such dull principles as bargain and sale had anything to do. The Lord Mayor, in the stronghold of the mighty Mansion House, gave orders to his fifty cooks and butlers to keep Christmas as a Lord Mayor's household should; and even the little tailor, whom he had fined five shillings on the

previous Monday for being drunk and bloodthirsty in the streets, stirred up to-morrow's pudding in his garret, while his lean wife and the baby sallied out to buy the beef.

Foggier yet, and colder. Piercing, searching, biting cold. If the good Saint Dunstan had but nipped the Evil Spirit's nose with a touch of such weather as that, instead of using his familiar weapons, then indeed he would have roared to lusty purpose. The owner of one scant young nose, gnawed and mumbled by the hungry cold as bones are gnawed by dogs, stopped down at Scrooge's keyhole to regale him with a Christmas carol: but at the first sound of

"God bless you, merry gentleman! May nothing you dismay"

Scrooge seized the ruler with such energy of action, that the singer fled in terror, leaving the keyhole to the fog and even more congenial frost.

41. In the context in which it appears, *congealed* most nearly means which of the following?
   a. Burst
   b. Loosened
   c. Shrank
   d. Thickened

42. Which of the following can NOT be inferred from the passage?
   a. The season of this narrative is in the winter time.
   b. The majority of the narrative is located in a bustling city street.
   c. This passage takes place during the night time.
   d. The Lord Mayor is a wealthy person within the narrative.

43. According to the passage, which of the following regarding the poulterers and grocers is true?
   a. They were so poor in the quality of their products that customers saw them as a joke.
   b. They put on a pageant in the streets every year for Christmas to entice their customers.
   c. They did not believe in Christmas, so they refused to participate in the town parade.
   d. They set their shops up to be entertaining public spectacles rather than a dull trade exchange.

44. The author's depiction of the scene in the last few paragraphs does all EXCEPT which of the following?
   a. Offer an allusion to religious affiliation in England.
   b. Attempt to evoke empathy for the character of Scrooge.
   c. Provide a palpable experience through the use of imagery and diction.
   d. Depict Scrooge as an uncaring, terrifying character to his fellows.

Questions 45–48 are based on the book On the Trail *by Lina Beard and Adelia Belle Beard:*

For any journey, by rail or by boat, one has a general idea of the direction to be taken, the character of the land or water to be crossed, and of what one will find at the end. So it should be in striking the trail. Learn all you can about the path you are to follow. Whether it is plain or obscure, wet or dry; where it leads; and its length, measured more by time than by actual miles. A smooth, even trail of five miles will not consume the time and strength that must be expended upon a trail of half that length which leads over uneven ground, varied by bogs and obstructed by rocks and fallen trees, or a trail that is all up-hill climbing. If you are a novice and accustomed to walking only over smooth and level ground, you must allow more time for covering the distance than an experienced person would require and must count upon the expenditure of more strength, because your feet are not trained to the wilderness paths with their pitfalls and traps for the unwary, and every nerve and muscle will be strained to secure a safe foothold amid the tangled roots, on the slippery, moss-covered logs, over precipitous rocks that lie in your path. It will take time to pick your way over boggy places where the water oozes up through the thin, loamy soil as through a sponge; and experience alone will teach you which hummock of grass or moss will make a safe stepping-place and will not sink beneath your weight and soak your feet with hidden water. Do not scorn to learn all you can about the trail you are to take . . . It is not that you hesitate to encounter difficulties, but that you may prepare for them. In unknown regions take a responsible guide with you, unless the trail is short, easily followed, and a frequented one. Do not go alone through lonely places; and, being on the trail, keep it and try no explorations of your own, at least not until you are quite familiar with the country and the ways of the wild.

Blazing the Trail

A woodsman usually blazes his trail by chipping with his axe the trees he passes, leaving white scars on their trunks, and to follow such a trail you stand at your first tree until you see the blaze on the next, then go that and look for the one farther on; going in this way from tree to tree you keep the trail though it may, underfoot, be overgrown and indistinguishable.

If you must make a trail of your own, blaze it as you go by bending down and breaking branches of trees, underbrush, and bushes. Let the broken branches be on the side of bush or tree in the direction you are going, but bent down away from that side, or toward the bush, so that the lighter underside of the leaves will show and make a plain trail. Make these signs conspicuous and close together, for in returning, a dozen feet without the broken branch will sometimes confuse you, especially as everything has a different look when seen from the opposite side. By this same token it is a wise precaution to look back frequently as you go and impress the homeward-bound landmarks on your memory. If in your wanderings you have branched off and made ineffectual or blind trails which lead nowhere, and, in returning to camp, you are led astray by one of them, do not leave the false trail and strike out to make a new one, but turn back and follow the false trail to its beginning, for it must lead to the true trail again. Don't lose sight of your broken branches.

Adelia B. Beard

45. What part of the text is the girl most likely emulating in the image?
   a. Building a trap
   b. Setting up camp
   c. Blazing the trail
   d. Picking berries to eat

46. According to the passage, what does the author say about unknown regions?
   a. You should try and explore unknown regions in order to learn the land better.
   b. Unless the trail is short or frequented, you should take a responsible guide with you.
   c. All unknown regions will contain pitfalls, traps, and boggy places.
   d. It's better to travel unknown regions by rail rather than by foot.

47. Which statement is NOT a detail from the passage?
   a. Learning about the trail beforehand is imperative.
   b. Time will differ depending on the land.
   c. Once you are familiar with the outdoors you can go places on your own.
   d. Be careful of wild animals on the trail you are on.

48. In the last paragraph, which of the following does the author suggest when being led astray by a false trail?
   a. Bend down and break the branches off trees, underbrush, and bushes.
   b. Ignore the false trail and strike out to make a new one.
   c. Follow the false trail back to its beginning so that you can rediscover the real trail.
   d. Make the signs conspicuous so that you won't be confused when you turn around.

Questions 49–50 are based on the following passage, which is a preface for Poems by Alexander Pushkin *by Ivan Panin:*

I do not believe there are as many as five examples of deviation from the literalness of the text. Once only, I believe, have I transposed two lines for convenience of translation; the other deviations are (*if* they are such) a substitution of an *and* for a comma in order to make now and then the reading of a line musical. With these exceptions, I have sacrificed *everything* to faithfulness of rendering. My object was to make Pushkin himself, without a prompter, speak to English readers. To make him thus speak in a foreign tongue was indeed to place him at a disadvantage; and music and rhythm and harmony are indeed fine things, but truth is finer still. I wished to present not what Pushkin would have said, or [Pg 10] should have said, if he had written in English, but what he does say in Russian. That, stripped from all ornament of his wonderful melody and grace of form, as he is in a translation, he still, even in the hard English tongue, soothes and stirs, is in itself a sign that through the individual soul of Pushkin sings that universal soul whose strains appeal forever to man, in whatever clime, under whatever sky.

I ask, therefore, no forgiveness, no indulgence even, from the reader for the crudeness and even harshness of the translation, which, I dare say, will be found in abundance by those who *look* for something to blame. Nothing of the kind is necessary. I have done the only thing there was to be done. Nothing more *could* be done (I mean by me, of course), and if critics still demand more, they must settle it not with me, but with the Lord Almighty, who in his grim, yet arch way, long before critics appeared on the stage, hath ordained that it shall be impossible for a thing to be and not to be at the same time.

I have therefore tried neither for measure nor for rhyme. What I have done was this: I first translated each line word for word, and then by reading it aloud let mine ear arrange for me the words in such a way as to make some kind of rhythm. Where this could be done, I was indeed glad; where this could not be done, I was not sorry. It is idle to regret the impossible.

49. From clues in this passage, what type of work is the author doing?
   a. Translation work
   b. Criticism
   c. Historical validity
   d. Writing a biography

50. Where would you most likely find this passage in a text?
   a. Appendix
   b. Table of contents
   c. First chapter
   d. Preface

Answer Explanations

1. D: Neutrality due to the style of the report. The report is mostly objective; we see very little language that entails any strong emotion whatsoever. The story is told almost as an objective documentation of a sequence of actions—we see the president sitting in his box with his wife, their enjoyment of the show, Booth's walk through the crowd to the box, and Ford's consideration of Booth's movements. There is perhaps a small amount of bias when the author mentions the president's "worthy wife." However, the word choice and style show no signs of excitement, sadness, anger, or apprehension from the author's perspective, so the best answer is Choice *D*.

2. B: Mr. Ford assumed Booth's movement throughout the theater was due to being familiar with the theater. Choice *A* is incorrect; although Booth does eventually make his way to Lincoln's box, Mr. Ford does not make this distinction in this part of the passage. Choice *C* is incorrect; although the passage mentions "companions," it mentions Lincoln's companions rather than Booth's companions. Choice *D* is incorrect; the passage mentions "dress circle," which means the first level of the theater, but this is different from a "dressing room."

3. C: A lead singer leaves their band to begin a solo career, and the band drops in sales by 50 percent on their next album. The original source of the analogy displays someone significant to an event who leaves, and then the event becomes the worst for it. We see Mr. Sothern leaving the theater company, and then the play becoming a "very dull affair." Choice *A* depicts a dancer who backs out of an event before the final performance, so this is incorrect. Choice *B* shows a basketball player leaving an event, and then the team makes it to the championship but then loses. This choice could be a contestant for the right answer; however, we don't know if the team has become the worst for his departure or the better for it. We simply do not have enough information here. Choice *D* is incorrect. The actor departs an event, but there is no assessment of the quality of the movie. It simply states what actors filled in instead.

4. A: A chronological account in a fiction novel of a woman and a man meeting for the first time. It's tempting to mark Choice *A* wrong because the genres are different. Choice *A* is a fiction text, and the original passage is not a fictional account. However, the question stem asks specifically for organizational structure. Choice *A* is a chronological structure just like the passage, so this is the correct answer. The passage does not have a cause and effect or problem/solution structure, making Choices *B* and *D* incorrect. Choice *C* is tempting because it mentions an autobiography; however, the structure of this text starts at the end and works its way toward the beginning, which is the opposite structure of the original passage.

5. C: The word *adornments* would LEAST change the meaning of the sentence because it's the most closely related word to *festoons*. The other choices don't make sense in the context of the sentence. *Feathers* of flags, *armies* of flags, and *buckets* of flags are not as accurate as the word *adornments* of flags. The passage also talks about other décor in the setting, so the word adornments fits right in with the context of the paragraph.

6. D: To recount in detail the events that led up to Abraham Lincoln's death. Choice *A* is incorrect; the author makes no claims and uses no rhetoric of persuasion towards the audience. Choice *B* is incorrect, though it's a tempting choice; the passage depicts the setting in exorbitant detail, but the setting itself is not the primary purpose of the passage. Choice *C* is incorrect; one could argue this is a narrative, and the passage is about Lincoln's last few hours, but this isn't the *best* choice. The best choice recounts the details that leads up to Lincoln's death.

7. D: The word *deleterious* can be best interpreted as referring to the word *ruinous*. The first paragraph attempts to explain the process of milk souring, so the "acid" would probably prove "ruinous" to the growth of bacteria and cause souring. Choice *A, amicable*, means friendly, so this does not make sense in context. Choice *B, smoldering*, means to boil or simmer, so this is also incorrect. Choice *C,* luminous, has positive connotations and doesn't make sense in the context of the passage. Luminous means shining or brilliant.

8. B: The author begins by explaining a process or phenomenon, then gives the history of the study of this phenomenon, this ends by presenting the effects of this phenomenon. The author explains the process of souring in the first paragraph by informing the reader that "it is due to the action of certain of the milk bacteria upon the milk sugar which converts it into lactic acid, and this acid gives the sour taste and curdles the milk." In paragraph two, we see how the phenomenon of milk souring was viewed when it was "first studied," and then we proceed to gain insight into "recent investigations" toward the end of the paragraph. Finally, the passage ends by presenting the effects of the phenomenon of milk souring. We see the milk curdling, becoming bitter, tasting soapy, turning blue, or becoming thread-like. All of the other answer choices are incorrect.

9: A: To inform the reader of the phenomenon, investigation, and consequences of milk souring. Choice *B* is incorrect because the passage states that Bacillus acidi lactici is not the only cause of milk souring. Choice *C* is incorrect because, although the author mentions the findings of researchers, the main purpose of the text does not seek to describe their accounts and findings, as we are not even told the names of any of the researchers. Choice *D* is tricky. We do see the author present us with new findings in contrast to the first cases studied by researchers. However, this information is only in the second paragraph, so it is not the primary purpose of the *entire passage*.

10. C: Milk souring is caused mostly by a species of bacteria identical to that of Bacillus acidi lactici although there are a variety of other bacteria that cause milk souring as well. Choice *A* is incorrect because it contradicts the assertion that the souring is still caused by a variety of bacteria. Choice *B* is incorrect because the ordinary cause of milk souring *is known* to current researchers. Choice *D* is incorrect because this names mostly the effects of milk souring, not the cause.

11. C: The study of milk souring has improved throughout the years, as we now understand more of what causes milk souring and what happens afterward. None of the choices here are explicitly stated, so we have to rely on our ability to make inferences. Choice *A* is incorrect because there is no indication from the author that milk researchers in the past have been incompetent—only that recent research has done a better job of studying the phenomenon of milk souring. Choice *B* is incorrect because the author refers to dairymen in relation to the effects of milk souring and their "troubles" surrounding milk souring and does not compare them to milk researchers. Choice *D* is incorrect because we are told in the second paragraph that only certain types of bacteria are able to sour milk. Choice *C* is the best answer choice here because although the author does not directly state that the study of milk souring has improved, we can see this might be true due to the comparison of old studies to newer studies, and the fact that the newer studies are being used as a reference in the passage.

12. A: The chemical change that occurs when a firework explodes. The author tells us that after milk becomes slimy, "it persists in spite of all attempts made to remedy it," which means the milk has gone through a chemical change. It has changed its state from milk to sour milk by changing its odor, color, and material. After a firework explodes, there is nothing one can do to change the substance of a firework back to its original form—the original substance is turned into sound and light. Choice *B* is incorrect because, although the rain overwatered the plant, it's possible that the plant is able to recover from this. Choice *C* is incorrect because although Mercury leaking out may be dangerous, the actual substance itself stays the same and does not alter into something else. Choice *D* is incorrect; this situation is not analogous to the alteration of a substance.

13. D: A paragraph showing the ways bacteria infiltrate milk and ways to avoid this infiltration. Choices *A*, *B*, and *C* are incorrect because these are already represented in the third, second, and first paragraphs. Choice *D* is the best answer because it follows a sort of problem/solution structure in writing.

14. A: The point of view of the narrator of the passage can best be described as third-person omniscient. The narrator refers to the characters in the story by third-person pronouns: *he, it,* and *they*. The narrator also comes across as "all-knowing" (omniscient) by relating the information and feelings about all the characters (instead of just those of a single character). Second-person point of view would incorporate the second-person pronoun, *you*. First-person point of view would incorporate first-person pronouns: *I* and *we*. Finally, the third-person objective point of view would also refer to the characters in the story by third-person pronouns *he, it,* and *they*. However, the narrator would stay detached, only telling the story and not expressing what the characters feel and think.

15. B: The chronological order of these four events in the passage is as follows:

1. The servants joyfully cheered for the family.
2. Mrs. Dashwood discussed improvements to the cottage.
3. Elinor hung her drawings up in the sitting room.
4. Sir John Middleton stopped by for a visit.

16. C: In the course of this passage, the Dashwoods' attitude shifts. The narrator initially describes the family's melancholy disposition at the start of their journey from Norland. However, the narrator then describes their spirits beginning to lift and becoming more upbeat as they make their way through the scenic English countryside to their new home at Barton Cottage. Although the remaining answer choices may contain some partial truths (appreciation for the family's former privileged life and fear about the family's expedition), they do not accurately depict the narrator's theme throughout the entire passage. For example, the family did not express disdain for their new landlord or doubt their future ability to survive.

17. D: Reality versus expectations is a theme of this passage. Although Mrs. Dashwood expects her new financial circumstances to be trying, and the family expects to miss their former place of residence for quite some time, the Dashwoods seem to begin to adapt well with the help of Sir John Middleton, their generous new landlord. A theme of love conquers all is typically used in literature when a character overcomes an obstacle due to his or her love for someone. A theme of power of wealth is used in literature to show that money either accomplishes things or is the root of evil. Finally, a theme of wisdom of experience is typically used in literature to show that improved judgment comes with age.

18. A: The statement at the start of paragraph five in this passage signifies the family is going through a transition. The Dashwoods find themselves in a time of great transition as they learn to accept their family's demotion in social standing and their reduced income. The statement was not meant to signify that Mrs. Dashwood needs to obtain meaningful employment, that the family is going through a growth phase, or that the Dashwood children need to be concerned about the future.

19. B: The new landlord is a cousin to the Dashwoods. This accurately depicts their relationship. The new landlord was not a former social acquaintance, Mrs. Dashwood's father, or a long-time friend of the family.

20. C: The narrator describes the generosity of Sir John Middleton to explain how his kindness eased the family's adaptation to their new home and circumstances. He played a very important role in making the Dashwoods feel comfortable moving to Barton Cottage. The narrator did not describe Sir John Middleton's generosity to identify one of the many positive traits that a landlord should possess, to explain how a landlord should conduct himself in order to be successful, or to demonstrate that he did not need to be cold and businesslike all the time.

21. B: The author says that the classical understanding of poetry dealt with its ability to be used to teach morality. Later, philosophers would define poetry by its ability to imitate life. Finally, during the renaissance, poetry was believed to be an imitative art that instilled morality in its readers. The rest of the answer choices are mixed together from this explanation in the passage. Poetry was never mentioned for use in entertainment, which makes Choice *D* incorrect. Choices *A* and *C* are incorrect for mixing up the chronological order.

22. C: That although most poetry was written as lyric, epic, or drama, the critics were most focused on the techniques of the epic and drama and their performance of structure and character. This is the best answer choice as portrayed by paragraph three. Choice *A* is incorrect because nowhere in the passage does it say rhetoric was more valued than poetry, although it did seem to have a more definitive purpose than poetry. Choice *B* is incorrect; this almost mirrors Choice *A*, but the critics were *not* focused on the lyric, as the passage indicates. Choice *D* is incorrect because the passage does not mention that the study of poetics was more pleasurable than the study of rhetoric.

23. A: To contemplate the differences between classical rhetoric and poetry and to consider their purposes in a particular culture. Choice *B* is incorrect; although changes in poetics throughout the years is mentioned, this is not the main idea of the passage. Choice *C* is incorrect; although this is partly true, that rhetoric within the education system is mentioned, the subject of poetics is left out of this answer choice. Choice *D* is incorrect; the passage makes no mention of poetics being a subset of rhetoric.

24. B: The correct answer choice is Choice *B*, instill. Choice *A*, imbibe, means to drink heavily, so this choice is incorrect. Choice *C*, implode, means to collapse inward, and does not make sense in this context. Choice *D*, inquire, means to investigate. This option is better than the other options, but it is not as accurate as *instill*.

25. B: The first paragraph presents us with definitions and examples of a particular subject. The second paragraph presents a second subject in the same way. The third paragraph offers a contrast of the two subjects. In the passage, we see the first paragraph defining rhetoric and offering examples of how the Greeks and Romans taught this subject. In paragraph two we see poetics being defined along with examples of its dynamic definition. In the third paragraph, we see the contrast between rhetoric and poetry characterized through how each of these were studied in a classical context.

26. D: The best answer is Choice *D*. As a football player, they taught me how to understand the logistics of the game, how my placement on the field affected the rest of the team, and how to run and throw with a mixture of finesse and strength. The content of rhetoric in the passage . . . "taught how to work up a case by drawing valid inferences from sound evidence, how to organize this material in the most persuasive order, and how to compose in clear and harmonious sentences." What we have here is three general principles: 1) it taught me how to understand logic and reason (drawing inferences parallels to understanding the logistics of the game), 2) taught me how to understand structure and organization (organization of material parallels to organization on the field) and 3) it taught me how to make the end product beautiful (how to compose in harmonious sentences parallels to how to run with finesse and strength). Each part parallels by logic, organization, then style.

27. A: Treatises is most closely related to the word *commentary*. Choice *B* does not make sense because thesauruses and encyclopedias are not written about one single subject. Choice *C* is incorrect; sermons are usually given by religious leaders as advice or teachings. Choice *D* is incorrect; anthems are songs and do not fit within the context of this sentence.

28. D: A city whose population is made up of people who seek quick fortunes rather than building a solid business foundation. Choice *A* is a characteristic of Portland, but not that of a boom city. Choice *B* is close—a boom city is one that becomes quickly populated, but it is not necessarily always populated by residents from the east coast. Choice *C* is incorrect because a boom city is not one that catches fire frequently, but one made up of people who are looking to make quick fortunes from the resources provided on the land.

29. D: Choice *D* is the best answer because of the surrounding context. We can see that the fact that Portland is a "boom city" means that the "floating class" go through, a group of people who only have temporary roots put down. This would cause the main focus of the city to be on employment and industry, rather than society and culture. Choice *A* is incorrect, as we are not told about the inhabitants being social or antisocial. Choice *B* is incorrect because the text does not talk about the culture in the East regarding European influence. Finally, Choice *C* is incorrect; this is an assumption that has no evidence in the text to back it up.

30. B: The author would say that it has as much culture as the cities in the East. The author says that Portland has "as fine churches, as complete a system of schools, as fine residences, as great a love of music and art, as can be found at any city of the East of equal size," which proves that the culture is similar in this particular city to the cities in the East.

31. A: Approximately 240,000. We know from the image that San Francisco has around 300,000 inhabitants at this time. From the text (and from the graph) we can see that Portland has 60,000 inhabitants. Subtract these two numbers to come up with 240,000.

32. C: This question requires close attention to the passage. Choice *A* can be found where the passage says "no less than six named and several unnamed varieties of the peach have thus produced several varieties of nectarine, so this choice is incorrect. Choice *B* can be found where the passage says "it is highly improbable that all these peach-trees . . . are hybrids from the peach and nectarine." Choice *D* is incorrect because we see in the passage that "the production of peaches from nectarines, either by seeds or buds, may perhaps be considered as a case of reversion." Choice *C* is the correct answer because the word "unremarkable" should be changed to "remarkable" in order for it to be consistent with the details of the passage.

33. A: The word *multiplied* is synonymous with the word *propagated*, making Choice *A* correct. Choice *B* is incorrect because *diminished* means to decrease or recede and is the opposite of *propagated*. Choice *C* is incorrect; *watered* is closed, because it pertains to the growth of trees, but it is not exactly the same thing as *propagated*. Finally, Choice *D* is incorrect; *uprooted* could also pertain to trees, but this answer is incorrect.

34. B: The author's tone in this passage can be considered objective. An objective tone means that the author is open-minded and detached about the subject. Most scientific articles are objective. Choices *A, C,* and *D* are incorrect. The author is not very enthusiastic on the paper; the author is not critical, but rather interested in the topic. The author is not desperate in any way here.

35. B: Choice *B* is the correct answer because the meaning holds true even if the words have been switched out or rearranged some. Choice *A* is incorrect because it has trees either bearing peaches or nectarines, and the trees in the original phrase bear both. Choice *C* is incorrect because the statement does not say these trees are "indifferent to bud-variation," but that they have "indifferently [bore] peaches or nectarines." Choice *D* is incorrect; the statement may use some of the same words, but the meaning is skewed in this sentence.

36. D: The author does not consider modern architecture to be "cathedral-crowned grim and dim castles." The author is speaking of feudal architecture when they say this phrase. Choices *A, B,* and *C* are all mentioned in the text as characteristics of modern architecture, especially in the cities of Chicago and Pittsburgh.

37. C: Architecture reflects—in shape, color, and form—the attitude of the society which built it. Choice *A* is too specific and is taken too literal. It is not that the architecture represents the builders; it represents the builder's culture. Choice *B* is also incorrect; we do see the words "at its worst as at its best," but the meaning of this is skewed in Choice *B*. Choice *D* is incorrect; the statement does not suggest this analysis.

38. C: The author refers to the "mediaeval man" in order to make the audience realize the irony produced from pitying him. The author says that we pity him for the dirt he lived in; however, we inhale smoke and dirt from our own skies. Choices *A, B,* and *D* are incorrect, as they just miss the mark of the statement.

39. B: The text talks about castles and cathedrals as examples of architecture from medieval Europe. Factories and skyscrapers are considered pre-modern or modern architecture. Canyons are not buildings, so Choice *A* is incorrect.

40. B: The author's tone is best described as "aggrieved" because it is annoyed, offended, and disgruntled by the "power, strangely mingled with timidity," the "ingenuity, frequently misdirected" aspects of modern architecture. The author calls out its "giant energy" and "grimy and dreariness," along with its "grim, dim canyons of our city streets."

41. D: *Congealed* in this context most nearly means *thickened*, because we see liquid turning into ice. Choice *B, loosened,* is the opposite of the correct answer. Choices *A* and *C, burst* and *shrank,* are also incorrect.

42. C: Choice *A* is incorrect. We cannot infer that the passage takes place during the night time. While we do have a statement that says that the darkness thickened, this is the only evidence we have. The darkness could be thickening because it is foggy outside. We don't have enough proof to infer this otherwise. We *can* infer that the season of this narrative is in the winter time. Some of the evidence here is that "the cold became intense," and people were decorating their shops with "holly sprigs,"—a Christmas tradition. It also mentions that it's Christmastime at the end of the passage. Choice *B* is incorrect; we *can* infer that the narrative is located in a bustling city street by the actions in the story. People are running around trying to sell things, the atmosphere is busy, there is a church tolling the hours, etc. The scene switches to the Mayor's house at the end of the passage, but the answer says "majority," so this is still incorrect. Choice *D* is incorrect; we *can* infer that the Lord Mayor is wealthy—he lives in the "Mansion House" and has fifty cooks.

43. D: The passage tells us that the poulterers' and grocers' trades were "a glorious pageant, with which it was next to impossible to believe that such dull principles as bargain and sale had anything to do," which means they set up their shops to be entertaining public spectacles in order to increase sales. Choice *A* is incorrect; although the word "joke" is used, it is meant to be used as a source of amusement rather than something made in poor quality. Choice *B* is incorrect; that they put on a "pageant" is figurative for the public spectacle they made with their shops, not a literal play. Choice *C* is incorrect, as this is not mentioned anywhere in the passage.

44. B: The author, at least in the last few paragraphs, does not attempt to evoke empathy for the character of Scrooge. We see Scrooge lashing out at an innocent, cold boy, with no sign of affection or feeling for his harsh conditions. We see Choice *A* when the author talks about Saint Dunstan. We see Choice *C* providing a palpable experience and imaginable setting and character, especially with the "piercing, searching, biting cold," among other statements. Finally, we see Choice *D* when Scrooge chases the young boy away.

45. C: The section that is talked about in the text is blazing the trail, which is Choice *C*. The passage states that one must blaze the trail by "bending down and breaking branches of trees, underbrush, and bushes." The girl in the image is bending a branch in order to break it so that she can use it to "blaze the trail" so she won't get lost.

46. B: Choice *B* is the best answer here; the sentence states "In unknown regions take a responsible guide with you, unless the trail is short, easily followed, and a frequented one." Choice *A* is incorrect; the passage does not state that you should try and explore unknown regions. Choice *C* is incorrect; the passage talks about trails that contain pitfalls, traps, and boggy places, but it does not say that *all* unknown regions contain these things. Choice *D* is incorrect; the passage mentions "rail" and "boat" as means of transport at the beginning, but it does not suggest it is better to travel unknown regions by rail.

47. D: Choice *D* is correct; it may be real advice an experienced hiker would give to an inexperienced hiker. However, the question asks about details in the passage, and this is not in the passage. Choice *A* is incorrect; we do see the author encouraging the reader to learn about the trail beforehand . . . "wet or dry; where it leads; and its length." Choice *B* is also incorrect, because we do see the author telling us the time will lengthen with boggy or rugged places opposed to smooth places. Choice *C* is incorrect; at the end of the passage, the author tells us "do not go alone through lonely places . . . unless you are quite familiar with the country and the ways of the wild."

48. C: The best answer here is Choice *C:* "Follow the false trail back to its beginning so that you can rediscover the real trail." Choices *A* and *D* are represented in the text; but this is advice on how to blaze a trail, not what to do when being led astray by a false trail. Choice *B* is incorrect; this is the opposite of what the text suggests doing.

49. A: The author is doing translation work. We see this very clearly in the way the author talks about staying truthful to the original language of the text. The text also mentions "translation" towards the end. Criticism is taking an original work and analyzing it, making Choice *B* incorrect. The work is not being tested for historical validity, but being translated into the English language, making Choice *C* incorrect. The author is not writing a biography, as there is nothing in here about Pushkin himself, only his work, making Choice *D* incorrect.

50. D: You would most likely find this in the preface. A preface to a text usually explains what the author has done or aims to do with the work. An appendix is usually found at the end of a text and does not talk about what the author intends to do to the work, making Choice *A* incorrect. A table of contents does not contain prose, but bullet points listing chapters and sections found in the text, making Choice *B* incorrect. Choice *C* is incorrect; the first chapter would include the translation work (here, poetry), and not the author's intentions.

Writing

Organization of Ideas

Logical or Effective Opening, Transitional, and Closing Sentences

An **essay** is only successful if readers can understand its reasoning, follow its logic, determine its focus or thesis, and understand the included points that support the main idea. Readers must have confidence that the writer is presenting the full gamut of evidence and presenting the ideas without bias. If readers cannot follow the chain of reasoning or easily and decisively identify the thesis, they are likely to stop reading and walk away without being convinced of the point the author intended to support. Likewise, a reader will not be persuaded to adopt the author's opinion (if it is different from his or her own) if the essay is disjointed, if the ideas don't flow or follow logically, if conflicting points are argued such that the thesis is not consistent throughout, and if the language and writing style is not coherent. It is the writer's responsibility to guide readers through the essay and through its reasoning. The former involves producing writing that is clear, coherent, grammatically correct, and understandable.

Using appropriate and effective types of sentences can create structure within an essay. **Opening sentences** are often used to introduce topics for the essay or paragraph. They should be clear and also gain the interest of the reader. **Transitional sentences** that use transitional phrases such as "however" and "besides" can be used to join ideas within an essay so the reader can make connections between them. **Closing sentences** are used to provide a restatement or summary of the information provided in the paragraph or essay.

Sentence Fluency
Sentence fluency is achieved by varying the beginnings of sentences. Writers do this by starting most of their sentences with different words and phrases rather than monotonously repeating the same ones across multiple sentences. Another way writers can increase fluency is by varying the lengths of sentences. Since run-on sentences are incorrect, writers make sentences longer by also converting them from simple to compound, complex, and compound-complex sentences. The coordination and subordination involved in these also give the text more variation and interest, hence more fluency. Here are a few more ways writers can increase fluency:

- Varying the transitional language and conjunctions used makes sentences more fluent.
- Writing sentences with a variety of rhythms by using prepositional phrases.
- Varying sentence structure adds fluency.

Evaluating Relevance of Content

An essay, or part of an essay, meets its goal by communicating an idea clearly and effectively to the reader. Clear communication is achieved through logical organization of the information that supports each main point. Arguments that are well thought out and communicated in a suitable tone are the foundations of effective communication. An essay has unity if all the parts relate to the thesis sentence, and it has coherence if the reader can follow the author's thoughts as they progress. If the reader is motivated, the essay has succeeded.

An essay should have its own internal logic, and each idea and element in it should fit together. If something doesn't fit, chances are it is irrelevant information that will distract the reader.

Analyzing Organizational Structure

Sequence structure is the order of events in which a story or information is presented to the audience. Sometimes the text will be presented in chronological order, or sometimes it will be presented by displaying the most recent information first, then moving backwards in time. The sequence structure depends on the author, the context, and the audience. The structure of a text also depends on the genre in which the text is written. Is it literary fiction? Is it a magazine article? Is it instructions for how to complete a certain task? Different genres will have different purposes for switching up the sequence of their writing.

Narrative Structure

The structure presented in literary fiction is also known as **narrative structure**. Narrative structure is the foundation on which the text moves. The basic ways for moving the text along are in the plot and the setting. The plot is the sequence of events in the narrative that move the text forward through cause and effect. The setting of a story is the place or time period in which the story takes place. Narrative structure has two main categories: linear and nonlinear.

Linear Narrative

Linear narrative is a narrative told in chronological order. Traditional linear narratives will follow the plot diagram below depicting the narrative arc. The narrative arc consists of the exposition, conflict, rising action, climax, falling action, and resolution.

- **Exposition**: The exposition is in the beginning of a narrative and introduces the characters, setting, and background information of the story. The importance of the exposition lies in its framing of the upcoming narrative. Exposition literally means "a showing forth" in Latin.

- **Conflict**: The conflict, in a traditional narrative, is presented toward the beginning of the story after the audience becomes familiar with the characters and setting. The conflict is a single instance between characters, nature, or the self, in which the central character is forced to make a decision or move forward with some kind of action. The conflict presents something for the main character, or protagonist, to overcome.

- **Rising Action**: The rising action is the part of the story that leads into the climax. The rising action will feature the development of characters and plot, and creates the tension and suspense that eventually lead to the climax.

- **Climax**: The climax is the part of the story where the tension produced in the rising action comes to a culmination. The climax is the peak of the story. In a traditional structure, everything before the climax builds up to it, and everything after the climax falls from it. It is the height of the narrative, and is usually either the most exciting part of the story or is marked by some turning point in the character's journey.

- **Falling Action**: The falling action happens as a result of the climax. Characters continue to develop, although there is a wrapping up of loose ends here. The falling action leads to the resolution.

- **Resolution**: The resolution is where the story comes to an end and usually leaves the reader with the satisfaction of knowing what happened within the story and why. However, stories do not always end in this fashion. Sometimes readers can be confused or frustrated at the end from lack of information or the absence of a happy ending.

This is a chart that graphs the **linear narrative** previously discussed:

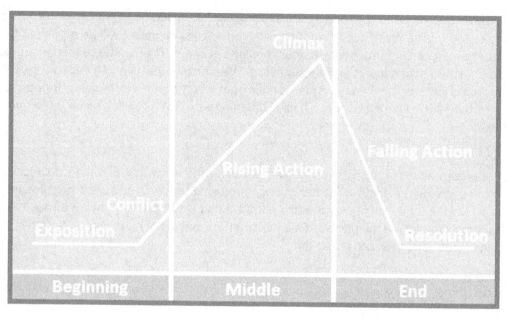

Nonlinear Narrative

A **nonlinear narrative** deviates from the traditional narrative in that it does not always follow the traditional plot structure of the narrative arc. Nonlinear narratives may include structures that are disjointed, circular, or disruptive, in the sense that they do not follow chronological order, but rather a nontraditional order of structure. *In medias res* is an example of a structure that predates the linear narrative. *In medias res* is Latin for "in the middle of things," which is how many ancient texts, especially epic poems, began their story, such as Homer's *Iliad*. Instead of having a clear exposition with a full development of characters, they would begin right in the middle of the action.

Modernist texts in the late nineteenth and early twentieth century are known for their experimentation with disjointed narratives, moving away from traditional linear narrative. Disjointed narratives are depicted in novels like *Catch 22*, where the author, Joseph Heller, structures the narrative based on free association of ideas rather than chronology. Another nonlinear narrative can be seen in the novel *Wuthering Heights*, written by Emily Bronte, which disrupts the chronological order by being told retrospectively after the first chapter. There seem to be two narratives in *Wuthering Heights* working at the same time: a present narrative as well as a past narrative. Authors employ disrupting narratives for various reasons; some use it for the purpose of creating situational irony for the readers, while some use it to create a certain effect in the reader, such as excitement, or even a feeling of discomfort or fear.

Sequence Structure in Technical Documents

The purpose of technical documents, such as instructions manuals, cookbooks, or "user-friendly" documents, is to provide information to users as clearly and efficiently as possible. In order to do this, the sequence structure in technical documents that should be used is one that is as straightforward as possible. This usually involves some kind of chronological order or a direct sequence of events. For example, someone who is reading an instruction manual on how to set up their Smart TV wants directions in a clear, simple, straightforward manner that does not leave the reader to guess at the proper sequence or lead to confusion.

Recognizing Logical Transitions

The use of transitional words and phrases that help link ideas with the appropriate connection will remove the guesswork for readers. Words like *thus, consequently, therefore, as such, however, besides*, and *accordingly* are helpful in this regard. It is also useful to include words that insinuate or shepherd readers toward a particular interpretation of statements that the writer wants the reader to adopt after reading the statement. Words like *fortunately, unfortunately*, and *thankfully* are examples of these types of qualifiers. The takeaway of a sentence, or it's emotional impact, can vary significantly by the tacking on of one of these words. As such, such words can be useful tools for a writer when crafting an argumentative piece to persuade readers to side with the writer's opinion. Consider the following example and notice how the meaning of the same sentence varies significantly when the sole qualifying word changes:

> Fortunately, her role as hall monitor has changed.

> Unfortunately, her role as hall monitor has changed.

While this sentence is short and it is difficult to get a sense of who "she" is and why her role as hall monitor has changed because there's no surrounding context, it is possible to see how the two sentences carry opposite meanings. In the first sentence, it's a good thing that her role has changed. Perhaps she was bad at her role as hall monitor, disliked it, or now has a better job. Although we don't know why it's a favorable change, we are left with the takeaway that this change is a positive thing. The opposite can be said for the second sentence. Here, we see that regardless of the reason she's no longer acting as hall monitor, it's a negative change.

Lastly, writers can include temporal words like *first, next*, and *finally* to help readers follow along in the list of evidence or points made. For example, if the writer asserts that there are four reasons why renewable energy sources are actually the most energy-efficient choices these days, he or she could use *firstly . . . , secondly . . . , thirdly . . . , lastly . . .*. Writers will increase the effectiveness of their arguments and the ease with which they can be followed by using language that helps guide the reader.

Language Facility

Phrases and Clauses

Conjunctions connect or coordinate words, phrases, clauses, or sentences together, typically as a way to demonstrate a relationship.

> Tony has a cat *and* a rabbit.

> Tony likes animals, *but* he is afraid of snakes.

Coordinating conjunctions join words or phrases that have equal rank or emphasis. There are seven coordinating conjunctions, all short words, that can be remembered by the mnemonic FANBOYS: *for, and, nor, but, or, yet, so*. They can join two words that are of the same part of speech (two verbs, two adjectives, two adverbs, or two nouns). They can also connect two phrases or two independent clauses.

Subordinating conjunctions help transition and connect two elements in the sentence, but in a way that diminishes the importance of the one it introduces, known as the dependent, or subordinate, clause. They include words like *because, since, unless, before, after, whereas, if*, and *while*.

Parallelism

Parallel structure in a sentence matches the forms of sentence components. Any sentence containing more than one description or phrase should keep them consistent in wording and form. Readers can easily follow writers' ideas when they are written in parallel structure, making it an important element of correct sentence construction. For example, this sentence lacks parallelism: "Our coach is a skilled manager, a clever strategist, and works hard." The first two phrases are parallel, but the third is not. Correction: "Our coach is a skilled manager, a clever strategist, and a hard worker." Now all three phrases match in form. Here is another example:

Fred intercepted the ball, escaped tacklers, and a touchdown was scored.

This is also non-parallel. Here is the sentence corrected:

Fred intercepted the ball, escaped tacklers, and scored a touchdown.

Modifier Placement

Modifiers are optional elements that can clarify or add details about a phrase or another element of a sentence. They are a dependent phrase and removing them usually does not change the grammatical correctness of the sentence; however, the meaning will be changed because, as their name implies, modifiers modify another element in the sentence. Consider the following:

Nico loves sardines.

Nico, who is three years old, loves sardines.

The first simple sentence is grammatically correct; however, we learn a lot more from the second sentence, which contains the modifier *who is three years old*. Sardines tend to be a food that young children don't like, so adding the modifier helps readers see why Nico loving sardines is noteworthy.

Beginning writers sometimes place modifiers incorrectly. Then, instead of enhancing comprehension and providing helpful description for the reader, the modifier causes more confusion. A **misplaced modifier** is located incorrectly in relation to the phrase or word it modifies. Consider the following sentence:

Because it is salty, Nico loves fish.

The modifier in this sentence is "because it is salty," and the noun it is intended to modify is "fish." However, due to the erroneous placement of the modifier next to the subject, Nico, the sentence is actually saying that Nico is salty.

Nico loves fish because it is salty.

The modifier is now adjacent to the appropriate noun, clarifying which of the two elements is salty.

Dangling modifiers are so named because they modify a phrase or word that is not clearly found in the sentence, making them rather unattached. They are not intended to modify the word or phrase they are placed next to. Consider the following:

Walking home from school, the sky opened and Bruce got drenched.

The modifier here, "walking home from school," should modify who was walking (Bruce). Instead, the noun immediately after the modifier is "the sky"—but the sky was not walking home from school. Although not always the case, dangling modifiers are often found at the beginning of a sentence.

Use of Sentences

For **fluent** composition, writers must use a variety of sentence types and structures, and also ensure that they smoothly flow together when they are read. To accomplish this, they must first be able to identify fluent writing when they read it. This includes being able to distinguish among simple, compound, complex, and compound-complex sentences in text; to observe variations among sentence types, lengths, and beginnings; and to notice figurative language and understand how it augments sentence length and imparts musicality. Once students/writers recognize superior fluency, they should revise their own writing to be more readable and fluent. They must be able to apply acquired skills to revisions before being able to apply them to new drafts.

One strategy for revising writing to increase its sentence fluency is **flipping sentences**. This involves rearranging the word order in a sentence without deleting, changing, or adding any words. For example, the student or other writer who has written the sentence, "We went bicycling on Saturday" can revise it to, "On Saturday, we went bicycling." Another technique is using appositives. An **appositive** is a phrase or word that renames or identifies another adjacent word or phrase. Writers can revise for sentence fluency by inserting main phrases/words from one shorter sentence into another shorter sentence, combining them into one longer sentence, e.g. from "My cat Peanut is a gray and brown tabby. He loves hunting rats." to "My cat Peanut, a gray and brown tabby, loves hunting rats." Revisions can also connect shorter sentences by using conjunctions and commas and removing repeated words: "Scott likes eggs. Scott is allergic to eggs" becomes "Scott likes eggs, but he is allergic to them."

One technique for revising writing to increase sentence fluency is "padding" short, simple sentences by adding phrases that provide more details specifying why, how, when, and/or where something took place. For example, a writer might have these two simple sentences: "I went to the market. I purchased a cake." To revise these, the writer can add the following informative dependent and independent clauses and prepositional phrases, respectively: "Before my mother woke up, I sneaked out of the house and went to the supermarket. As a birthday surprise, I purchased a cake for her." When revising sentences to make them longer, writers must also punctuate them correctly to change them from simple sentences to compound, complex, or compound-complex sentences.

Sentence Structure

All *complete sentences* contain the same two basic components: a subject and a predicate. The **subject** is who or what is doing the action or being described in the sentence. The **predicate** is everything else in the sentence; the predicate includes the **verb,** which describes the action the subject is doing or the condition of the subject. The predicate, therefore, describes what the subject does or is.

Although all sentences contain these same two basic elements, there are different ways that the subject and predicate can be combined. There are four general sentence structures used in the English language.

Simple sentences contain one subject and one verb, but still express a complete thought:

> The mouse ate cheese.

The subject is *the mouse.* The verb is *ate.*

A sentence can still be considered a simple sentence if it has a compound subject or compound verb. A **compound** noun or verb consists of more than one element. The following sentence contains a compound subject:

The mouse and the gerbil ate cheese.

When two or more simple sentences are joined together to form a single sentence with more than one subject-verb combinations, it is considered a **compound sentence:**

The mouse ate cheese, and *the boy built him a maze.*

This structure contains two **independent clauses**: (1) *the mouse ate cheese* and (2) *the boy built him a maze.* These two clauses are independent because they can stand on their own. However, they are combined with a comma and a coordinating conjunction *(and)* to form a compound sentence. Other coordinating conjunctions, such as *but* and *so* can also be used to form compound sentences.

Jenny read for two hours, *but* she did not finish her reading assignment.

The book was dense, *so* she was unable to understand it.

Complex sentences are formed from an independent clause and at least one dependent clause. Subordinating conjunctions, such as *although, because, unless, while, as soon as, since, if,* and *when,* are used to connect the dependent clause to the sentence.

Pablo bought a new bike helmet *because* his old one was cracked.

Unless you plan to renew your library books, they must be returned in two weeks.

The necessary punctuation in a complex sentence depends on the order of the clauses. When the dependent clause begins the sentence, a comma is used after it to separate it from the independent clause. However, when an independent clause precedes a dependent clause, a comma is not necessarily required.

Lastly, a **compound-complex sentence** consists of at least two independent clauses and at least one dependent clause:

Before you go home, please recycle your scrap paper, and stack your chair on your desk.

The first independent clause in the compound sentence structure includes a subordinating clause—*before you go home.* Therefore, the sentence structure is both complex and compound.

Idiomatic Usage

Idiomatic expressions are phrases or groups of words that have an established meaning when used together that's unrelated to the literal meanings of the individual words. For example, consider the following sentence that includes a common idiomatic phrase:

I know Phil is coming to visit this weekend because I heard it straight from the horse's mouth.

The speaker of this sentence did not consult a horse nor hear anything uttered from a horse in relation to Phil's visit. Instead, "straight from the horse's mouth" is an *idiom* that means it's the truth. As in the sentence in which it is used above, it often means whatever said should be taken as truth because it was spoken by a reliable source or by the person to which it pertains (in this case, Phil). The phrase is derived from the fact that sellers of horses at auctions would sometimes try to lie about the age of the horse. However, the size and shape of a horse's teeth can provide a fairly accurate estimate of the horse's true age. Therefore, essentially the truth regarding the horse's age come straight from their mouth. The idiomatic expression came to mean getting the truth in any situation.

Finding errors in idiomatic expressions can be difficult because it requires familiarity with the idiom. Because there are more than one thousand idioms in the English language, memorizing all of them is impractical. However, it is recommended to review the most common ones. There are many webpages dedicated to listing and explaining frequently used idioms.

Errors in the idiomatic expressions are typically one of two types. Either the idiomatic expression is stated improperly or it is used incorrectly. In the first type of issue, the prepositions used are often incorrect. Using the idiom from above, for example, it might say "straight *in* the horse's mouth" or "straight *with* the horse's mouth." In the second error type, the idiomatic expression is used improperly because the meaning it carries does not make sense in the context in which it is being used. Consider the following:

> He was looking straight from the horse's mouth when he complained about the phone his father bought him.

Here, the writer has confused the idiom "straight from the horse's mouth" with "looking a gift horse in the mouth," which means to find fault in a gift or favor.

Either type of error can be difficult to detect and correct without prior knowledge of the idiomatic phrase. Practicing using idioms and discussing and studying their origins can help students remember the meanings and precise wordings, which will help them identify and correct errors in their usage.

Expressing Ideas with Appropriate Style and Tone

Tone conveys the author's attitude toward the topic and the audience. The tone also reveals their level of confidence on the subject, and whether they intend to bring humor, emotion, or seriousness to the writing. Setting the tone ultimately determines how readers will receive the overall message.

In professional writing, it is imperative that authors maintain an appropriate and professional tone that strengthens the writing quality. Framing the writing ahead of time, determining the purpose of the paper, and even considering the author's own bias on the subject will help to set the appropriate tone. Why is the paper being written? What message does the author want to convey, and to whom? Why would the intended audience find this paper interesting, and how does the author expect the audience to react? These are just some of the questions that can help to frame a piece of writing and develop an appropriate tone. To help develop the tone before writing, it sometimes helps to consider the intended audience's perspective.

When an author wishes to convey a clear message but does not want to compromise credibility by writing subjectively, using emphasis can be very effective. This is done by introducing and stressing the important points of the subject in the opening sentence or at the beginning of specific paragraphs. Consider the following:

> Music soothes the soul and captures our hearts. The following is a study on how music has affected North American society during the past five decades.

The following is a study on how music has affected North American society during the past five decades. Music soothes the soul and captures our hearts.

Clearly, the author's attitude toward music is captured in the first sentence of the first example. When the sentences are rearranged in the second example, however, the emphasis is lost. With the subtle positioning of key words and phrases, an author can emphasize important points of the paper and set the tone.

Written tone can either capture readers' attention or turn them away. Considering the intended audience and how readers will perceive the message helps to develop the appropriate tone. In academic or professional writing, writers tend to employ a more serious tone, but serious does not mean dull or boring. Academic writing can engage readers by setting a tone that, although serious, also reveals a connection to readers. Authors who wish to connect with their audience on a more personal level might introduce their topic with a surprising or entertaining fact, or quote a famous individual who reinforces the paper's topic.

The style and tone of a piece of writing should match the intended audience and purpose. The **style** of a piece of writing refers to the way the writer has composed it. It entails the specific words chosen, the selection and arrangement of sentence structures, and the paragraph structure. The **tone** of a piece relates to the how the writer has conveyed his or her ideas. A tone, for example, might be serious, humorous, or somber. The tone of most formal pieces is impersonal or authoritative because the purpose is to inform and demonstrate a command of a topic or idea. Argumentative essays tend to be more emotional and passionate, as they usually attempt to evoke an emotional reaction in the reader.

Writers should strive to maintain consistency in the style and tone used in a given essay. For example, if the essay uses formal writing, it should be free from informal language, slang, and grammatical rule bending that is somewhat more permissible in conversational or informal writing. If the tone of a piece is authoritative and the writer is striving to assert himself or herself as an expert, the writer should not switch to a silly or playful tone. Writers can help ensure that style and tone remain consistent by maintaining the same "voice" in a given essay and remembering to write in a way that reflects the purpose and audience of the piece.

Nuances in the Meaning of Words with Similar Denotations

Denotation refers to a word's explicit definition, like that found in the dictionary. Denotation is often set in comparison to connotation. **Connotation** is the emotional, cultural, social, or personal implication associated with a word. Denotation is more of an objective definition, whereas connotation can be more subjective, although many connotative meanings of words are similar for certain cultures. The denotative meanings of words are usually based on facts, and the connotative meanings of words are usually based on emotion.

Here are some examples of words and their denotative and connotative meanings in Western culture:

Word	Denotative Meaning	Connotative Meaning
Home	A permanent place where one lives, usually as a member of a family.	A place of warmth; a place of familiarity; comforting; a place of safety and security. "Home" usually has a positive connotation.
Snake	A long reptile with no limbs and strong jaws that moves along the ground; some snakes have a poisonous bite.	An evil omen; a slithery creature (human or nonhuman) that is deceitful or unwelcome. "Snake" usually has a negative connotation.
Winter	A season of the year that is the coldest, usually from December to February in the northern hemisphere and from June to August in the southern hemisphere.	Circle of life, especially that of death and dying; cold or icy; dark and gloomy; hibernation, sleep, or rest. "Winter" can have a negative connotation, although many who have access to heat may enjoy the snowy season from their homes.

Writing Conventions

Verbs

A **verb** is a word or phrase that expresses action, feeling, or state of being. Verbs explain what their subject is *doing*. Three different types of verbs used in a sentence are action verbs, linking verbs, and helping verbs.

Action verbs show a physical or mental action. Some examples of action verbs are *play, type, jump, write, examine, study, invent, develop,* and *taste.* The following example uses an action verb:

> Kat *imagines* that she is a mermaid in the ocean.

The verb *imagines* explains what Kat is doing: she is imagining being a mermaid.

Linking verbs connect the subject to the predicate without expressing an action. The following sentence shows an example of a linking verb:

> The mango *tastes* sweet.

The verb *tastes* is a linking verb. The mango doesn't *do* the tasting, but the word *taste* links the mango to its predicate, sweet. Most linking verbs can also be used as action verbs, such as *smell, taste, look, seem, grow,* and *sound.* Saying something *is* something else is also an example of a linking verb. For example, if we were to say, "Peaches is a dog," the verb *is* would be a linking verb in this sentence, since it links the subject to its predicate.

Helping verbs are verbs that help the main verb in a sentence. Examples of helping verbs are *be, am, is, was, have, has, do, did, can, could, may, might, should,* and *must,* among others. The following are examples of helping verbs:

> Jessica *is* planning a trip to Hawaii.

> Brenda *does* not like camping.

> Xavier *should* go to the dance tonight.

Notice that after each of these helping verbs is the main verb of the sentence: *planning, like,* and *go.* Helping verbs usually show an aspect of time.

Pronouns

There are three **pronoun** cases: subjective case, objective case, and possessive case. Pronouns as subjects are pronouns that replace the subject of the sentence, such as *I, you, he, she, it, we, they* and *who.* Pronouns as objects replace the object of the sentence, such as *me, you, him, her, it, us, them,* and *whom.* Pronouns that show possession are *mine, yours, hers, its, ours, theirs,* and *whose.* The following are examples of different pronoun cases:

- *Subject pronoun:* *She* ate the cake for her birthday. *I* saw the movie.
- *Object pronoun:* You gave *me* the card last weekend. She gave the picture to *him.*
- *Possessive pronoun:* That bracelet you found yesterday is *mine. His* name was Casey.

Modifier Forms

Adjectives

Adjectives are descriptive words that modify nouns or pronouns. They may occur before or after the nouns or pronouns they modify in sentences. For example, in "This is a big house," *big* is an adjective modifying or describing the noun *house.* In "This house is big," the adjective is at the end of the sentence rather than preceding the noun it modifies.

A rule of punctuation that applies to adjectives is to separate a series of adjectives with commas. For example, "Their home was a large, rambling, old, white, two-story house." A comma should never separate the last adjective from the noun, though.

Adverbs

Whereas adjectives modify and describe nouns or pronouns, **adverbs** modify and describe adjectives, verbs, or other adverbs. Adverbs can be thought of as answers to questions in that they describe when, where, how, how often, how much, or to what extent.

Many (but not all) adjectives can be converted to adverbs by adding *–ly.* For example, in "She is a quick learner," *quick* is an adjective modifying *learner.* In "She learns quickly," *quickly* is an adverb modifying *learns.* One exception is *fast. Fast* is an adjective in "She is a fast learner." However, *–ly* is never added to the word *fast;* it retains the same form as an adverb in "She learns fast."

Agreement

Subject-Verb Agreement

Lack of **subject-verb agreement** is a very common grammatical error. One of the most common instances is when people use a series of nouns as a compound subject with a singular instead of a plural verb. Here is an example:

> Identifying the best books, locating the sellers with the lowest prices, and paying for them *is* difficult

instead of saying "*are* difficult." Additionally, when a sentence subject is compound, the verb is plural:

> He and his cousins *were* at the reunion.

However, if the conjunction connecting two or more singular nouns or pronouns is "or" or "nor," the verb must be singular to agree:

> That pen or another one like it is in the desk drawer.

If a compound subject includes both a singular noun and a plural one, and they are connected by "or" or "nor," the verb must agree with the subject closest to the verb: "Sally or her sisters go jogging daily"; but "Her sisters or Sally goes jogging daily."

Simply put, singular subjects require singular verbs and plural subjects require plural verbs. A common source of agreement errors is not identifying the sentence subject correctly. For example, people often write sentences incorrectly like, "The group of students *were* complaining about the test." The subject is not the plural "students" but the singular "group." Therefore, the correct sentence should read, "The group of students *was* complaining about the test." The converse also applies, for example, in this incorrect sentence: "The facts in that complicated court case *is* open to question." The subject of the sentence is not the singular "case" but the plural "facts." Hence the sentence would correctly be written: "The facts in that complicated court case *are* open to question." New writers should not be misled by the distance between the subject and verb, especially when another noun with a different number intervenes as in these examples. The verb must agree with the subject, not the noun closest to it.

Pronoun-Antecedent Agreement

Pronouns within a sentence must refer specifically to one noun, known as the **antecedent.** Sometimes, if there are multiple nouns within a sentence, it may be difficult to ascertain which noun belongs to the pronoun. It's important that the pronouns always clearly reference the nouns in the sentence so as not to confuse the reader. Here's an example of an unclear pronoun reference:

> After Catherine cut Libby's hair, David bought her some lunch.

The pronoun in the examples above is *her.* The pronoun could either be referring to *Catherine* or *Libby.* Here are some ways to write the above sentence with a clear pronoun reference:

> After Catherine cut Libby's hair, David bought Libby some lunch.

> David bought Libby some lunch after Catherine cut Libby's hair.

But many times, the pronoun will clearly refer to its antecedent, like the following:

> After David cut Catherine's hair, he bought her some lunch.

Sentence Fragments and Run-Ons

Incomplete Sentences

Four types of incomplete sentences are sentence fragments, run-on sentences, subject-verb and/or pronoun-antecedent disagreement, and non-parallel structure.

Sentence fragments are caused by absent subjects, absent verbs, or dangling/uncompleted dependent clauses. Every sentence must have a subject and a verb to be complete. An example of a fragment is "Raining all night long," because there is no subject present. "It was raining all night long" is one correction. Another example of a sentence fragment is the second part in "Many scientists think in unusual ways. Einstein, for instance." The second phrase is a fragment because it has no verb. One correction is "Many scientists, like Einstein, think in unusual ways." Finally, look for "cliffhanger" words like *if, when, because,* or *although* that introduce dependent clauses, which cannot stand alone without an independent clause. For example, to correct the sentence fragment "If you get home early," add an independent clause: "If you get home early, we can go dancing."

Run-On Sentences

A **run-on sentence** combines two or more complete sentences without punctuating them correctly or separating them. For example, a run-on sentence caused by a lack of punctuation is the following:

> There is a malfunction in the computer system however there is nobody available right now who knows how to troubleshoot it.

One correction is, "There is a malfunction in the computer system; however, there is nobody available right now who knows how to troubleshoot it." Another is, "There is a malfunction in the computer system. However, there is nobody available right now who knows how to troubleshoot it."

An example of a comma splice of two sentences is the following:

> Jim decided not to take the bus, he walked home.

Replacing the comma with a period or a semicolon corrects this. Commas that try and separate two independent clauses without a contraction are considered **comma splices**.

Capitalization and Punctuation

Rules of Capitalization

The first word of any document, and of each new sentence, is **capitalized**. Proper nouns, like names and adjectives derived from proper nouns, should also be capitalized. Here are some examples:

- Grand Canyon
- Pacific Palisades
- Golden Gate Bridge
- Freudian slip
- Shakespearian, Spenserian, or Petrarchan sonnet
- Irish song

Some exceptions are adjectives, originally derived from proper nouns, which through time and usage are no longer capitalized, like *quixotic, herculean*, or *draconian*. Capitals draw attention to specific instances of people, places, and things. Some categories that should be capitalized include the following:

- brand names
- companies
- weekdays
- months
- governmental divisions or agencies
- historical eras
- major historical events
- holidays
- institutions
- famous buildings
- ships and other manmade constructions
- natural and manmade landmarks
- territories
- nicknames
- epithets
- organizations
- planets
- nationalities
- tribes
- religions
- names of religious deities
- roads
- special occasions, like the Cannes Film Festival or the Olympic Games

Exceptions

Related to American government, capitalize the noun Congress but not the related adjective congressional. Capitalize the noun U.S. Constitution, but not the related adjective constitutional. Many experts advise leaving the adjectives federal and state in lowercase, as in federal regulations or state water board, and only capitalizing these when they are parts of official titles or names, like Federal Communications Commission or State Water Resources Control Board. While the names of the other planets in the solar system are capitalized as names, Earth is more often capitalized only when being described specifically as a planet, like Earth's orbit, but lowercase otherwise since it is used not only as a proper noun but also to mean *land, ground, soil*, etc.

Names of animal species or breeds are not capitalized unless they include a proper noun. Then, only the proper noun is capitalized. Antelope, black bear, and yellow-bellied sapsucker are not capitalized. However, Bengal tiger, German shepherd, Australian shepherd, French poodle, and Russian blue cat are capitalized.

Other than planets, celestial bodies like the sun, moon, and stars are not capitalized. Medical conditions like tuberculosis or diabetes are lowercase; again, exceptions are proper nouns, like Epstein-Barr syndrome, Alzheimer's disease, and Down syndrome. Seasons and related terms like winter solstice or autumnal equinox are lowercase. Plants, including fruits and vegetables, like poinsettia, celery, or avocados, are not capitalized unless they include proper names, like Douglas fir, Jerusalem artichoke, Damson plums, or Golden Delicious apples.

Titles and Names

When official titles precede names, they should be capitalized, except when there is a comma between the title and name. But if a title follows or replaces a name, it should not be capitalized. For example, "the president" without a name is not capitalized, as in "The president addressed Congress." But with a name it is capitalized, like "President Obama addressed Congress." Or, "Chair of the Board Janet Yellen was appointed by President Obama." One exception is that some publishers and writers nevertheless capitalize President, Queen, Pope, etc., when these are not accompanied by names to show respect for these high offices. However, many writers in America object to this practice for violating democratic principles of equality. Occupations before full names are not capitalized, like owner Mark Cuban, director Martin Scorsese, or coach Roger McDowell.

Some universal rules for capitalization in composition titles include capitalizing the following:

- The first and last words of the title
- Forms of the verb *to be* and all other verbs
- Pronouns
- The word *not*

Universal rules for NOT capitalizing include the articles *the, a,* or *an;* the conjunctions *and, or,* or *nor;* and the preposition *to,* or *to* as part of the infinitive form of a verb. The exception to all of these is UNLESS any of them is the first or last word in the title, in which case they are capitalized. Other words are subject to differences of opinion and differences among various stylebooks or methods. These include *as, but, if,* and *or,* which some capitalize and others do not. Some authorities say no preposition should ever be capitalized; some say prepositions five or more letters long should be capitalized. The *Associated Press Stylebook* advises capitalizing prepositions longer than three letters (like *about, across,* or *with*).

Ellipses

Ellipses (. . .) signal omitted text when quoting. Some writers also use them to show a thought trailing off, but this should not be overused outside of dialogue. An example of an ellipses would be if someone is quoting a phrase out of a professional source but wants to omit part of the phrase that isn't needed: "Dr. Skim's analysis of pollen inside the body is clearly a myth . . . that speaks to the environmental guilt of our society."

Commas

Commas separate words or phrases in a series of three or more. The Oxford comma is the last comma in a series. Many people omit this last comma, but many times it causes confusion. Here is an example:

> I love my sisters, the Queen of England and Madonna.

This example without the comma implies that the "Queen of England and Madonna" are the speaker's sisters. However, if the speaker was trying to say that they love their sisters, the Queen of England, as well as Madonna, there should be a comma after "Queen of England" to signify this.

Commas also separate two coordinate adjectives ("big, heavy dog") but not cumulative ones, which should be arranged in a particular order for them to make sense ("beautiful ancient ruins").

A comma ends the first of two independent clauses connected by conjunctions. Here is an example:

> I ate a bowl of tomato soup, and I was hungry very shortly after.

Here are some brief rules for commas:

- Commas follow introductory words like *however, furthermore, well, why,* and *actually,* among others.
- Commas go between city and state: Houston, Texas.
- If using a comma between a surname and Jr. or Sr. or a degree like M.D., also follow the whole name with a comma: "Martin Luther King, Jr., wrote that."
- A comma follows a dependent clause beginning a sentence: "Although she was very small, . . ."
- Nonessential modifying words/phrases/clauses are enclosed by commas: "Wendy, who is Peter's sister, closed the window."
- Commas introduce or interrupt direct quotations: "She said, 'I hate him.' 'Why,' I asked, 'do you hate him?'"

Semicolons

Semicolons are used to connect two independent clauses, but should never be used in the place of a comma. They can replace periods between two closely connected sentences: "Call back tomorrow; it can wait until then." When writing items in a series and one or more of them contains internal commas, separate them with semicolons, like the following:

People came from Springfield, Illinois; Alamo, Tennessee; Moscow, Idaho; and other locations.

Hyphens

Here are some rules concerning **hyphens**:

- Compound adjectives like state-of-the-art or off-campus are hyphenated.
- Original compound verbs and nouns are often hyphenated, like "throne-sat," "video-gamed," "no-meater."
- Adjectives ending in –*ly* are often hyphenated, like "family-owned" or "friendly-looking."
- "Five years old" is not hyphenated, but singular ages like "five-year-old" are.
- Hyphens can clarify. For example, in "stolen vehicle report," "stolen-vehicle report" clarifies that "stolen" modifies "vehicle," not "report."
- Compound numbers twenty-one through ninety-nine are spelled with hyphens.
- Prefixes before proper nouns/adjectives are hyphenated, like "mid-September" and "trans-Pacific."

Parentheses

Parentheses enclose information such as an aside or more clarifying information: "She ultimately replied (after deliberating for an hour) that she was undecided." They are also used to insert short, in-text definitions or acronyms: "His FBS (fasting blood sugar) was higher than normal." When parenthetical information ends the sentence, the period follows the parentheses: "We received new funds ($25,000)." Only put periods within parentheses if the whole sentence is inside them: "Look at this. (You'll be astonished.)" However, this can also be acceptable as a clause: "Look at this (you'll be astonished)." Although parentheses appear to be part of the sentence subject, they are not, and do not change subject-verb agreement: "Will (and his dog) was there."

Quotation Marks

Quotation marks are typically used when someone is quoting a direct word or phrase someone else writes or says. Additionally, quotation marks should be used for the titles of poems, short stories, songs, articles, chapters, and other shorter works. When quotations include punctuation, periods and commas should *always* be placed inside of the quotation marks.

When a quotation contains another quotation inside of it, the outer quotation should be enclosed in double quotation marks and the inner quotation should be enclosed in single quotation marks. For example: "Timmy was begging, 'Don't go! Don't leave!'" When using both double and single quotation marks, writers will find that many word-processing programs may automatically insert enough space between the single and double quotation marks to be visible for clearer reading. But if this is not the case, the writer should write/type them with enough space between to keep them from looking like three single quotation marks. Additionally, non-standard usages, terms used in an unusual fashion, and technical terms are often clarified by quotation marks. Here are some examples:

My "friend," Dr. Sims, has been micromanaging me again.

This way of extracting oil has been dubbed "fracking."

Apostrophes

One use of the **apostrophe** is followed by an *s* to indicate possession, like *Mrs. White's home* or *our neighbor's dog*. When using the *'s* after names or nouns that also end in the letter *s*, no single rule applies: some experts advise adding both the apostrophe and the *s*, like "the Jones's house," while others prefer using only the apostrophe and omitting the additional *s*, like "the Jones' house." The wisest expert advice is to pick one formula or the other and then apply it consistently. Newspapers and magazines often use *'s* after common nouns ending with *s*, but add only the apostrophe after proper nouns or names ending with *s*. One common error is to place the apostrophe before a name's final *s* instead of after it: "Ms. Hasting's book" is incorrect if the name is Ms. Hastings.

Plural nouns should not include apostrophes (e.g. "apostrophe's"). Exceptions are to clarify atypical plurals, like verbs used as nouns: "These are the do's and don'ts." Irregular plurals that do not end in *s* always take apostrophe-*s*, not *s*-apostrophe—a common error, as in "childrens' toys," which should be "children's toys." Compound nouns like mother-in-law, when they are singular and possessive, are followed by apostrophe-*s*, like "your mother-in-law's coat." When a compound noun is plural and possessive, the plural is formed before the apostrophe-*s*, like "your sisters-in-laws' coats." When two people named possess the same thing, use apostrophe-*s* after the second name only, like "Dennis and Pam's house."

Spelling

Homophones

Homophones are words that have different meanings and spellings, but sound the same. These can be confusing for English Language Learners (ELLs) and beginning students, but even native English-speaking adults can find them problematic unless informed by context. Whereas listeners must rely entirely on context to *differentiate* spoken homophone meanings, readers with good spelling knowledge have a distinct advantage since homophones are spelled differently. For instance, *their* means belonging to them; *there* indicates location; and *they're* is a contraction of *they are*; despite different meanings, they all sound the same. *Lacks* can be a plural noun or a present-tense, third-person singular verb; either way it refers to absence—*deficiencies* as a plural noun, and *is deficient in* as a verb. But *lax* is an adjective that means loose, slack, relaxed, uncontrolled, or negligent. These two spellings, derivations, and meanings are completely different. With speech, listeners cannot know spelling and must use context; but with print, readers with spelling knowledge can differentiate them with or without context.

Homonyms and Homographs

Homophones are words that sound the same in speech but have different spellings and meanings. For example, *to, too,* and *two* all sound alike, but have three different spellings and meanings. Homophones with different spellings are also called *heterographs*. **Homographs** are words that are spelled identically but have different meanings. If they also have different pronunciations, they are *heteronyms*. For instance, *tear* pronounced one way means a drop of liquid formed by the eye; pronounced another way, it means to rip. Homophones that are also homographs are **homonyms**. For example, *bark* can mean the outside of a tree or a dog's vocalization; both meanings have the same spelling. *Stalk* can mean a plant stem or to pursue and/or harass somebody; these are spelled and pronounced the same. *Rose* can mean a flower or the past tense of *rise*. Many non-linguists confuse things by using "homonym" to mean sets of words that are homophones but not homographs, and also those that are homographs but not homophones.

The word *row* can mean to use oars to propel a boat; a linear arrangement of objects or print; or an argument. It is pronounced the same with the first two meanings, but differently with the third. Because it is spelled identically regardless, all three meanings are homographs. However, the two meanings pronounced the same are homophones, whereas the one with the different pronunciation is a heteronym. By contrast, the word *read* means to peruse language, whereas the word *reed* refers to a marsh plant. Because these are pronounced the same way, they are homophones; because they are spelled differently, they are heterographs. Homonyms are both homophones and homographs—pronounced and spelled identically, but with different meanings. One distinction between homonyms is of those with separate, unrelated etymologies, called "true" homonyms, e.g. *skate* meaning a fish or *skate* meaning to glide over ice/water. Those with common origins are called polysemes or polysemous homonyms, e.g. the *mouth* of an animal/human or of a river.

Irregular Plurals
One type of *irregular English plural* involves words that are spelled the same whether they are singular or plural. These include *deer, fish, salmon, trout, sheep, moose, offspring, species, aircraft*, etc. The spelling rule for making these words plural is simple: they do not change. Another type of irregular English plurals does change from singular to plural form, but it does not take regular English *–s* or *–es* endings. Their irregular plural endings are largely derived from grammatical and spelling conventions in the other languages of their origins, like Latin, German, and vowel shifts and other linguistic mutations. Some examples of these words and their irregular plurals include *child* and *children; die* and *dice; foot* and *feet; goose* and *geese; louse* and *lice; man* and *men; mouse* and *mice; ox* and *oxen; person* and *people; tooth* and *teeth;* and *woman* and *women*.

Contractions
Contractions are formed by joining two words together, omitting one or more letters from one of the component words, and replacing the omitted words with an apostrophe. An obvious yet often forgotten rule for spelling contractions is to place the apostrophe where the letters were omitted; for example, spelling errors like *did'nt* for *didn't. Didn't* is a contraction of *did not.* Therefore, the apostrophe replaces the "o" that is omitted from the "not" component. Another common error is confusing contractions with **possessives** because both include apostrophes, e.g. spelling the possessive *its* as "it's," which is a contraction of "it is"; spelling the possessive *their* as "they're," a contraction of "they are"; spelling the possessive *whose* as "who's," a contraction of "who is"; or spelling the possessive *your* as "you're," a contraction of "you are."

Frequently Misspelled Words
One source of spelling errors is not knowing whether to drop the final letter *e* from a word when its form is changed by adding an ending to indicate the past tense or progressive participle of a verb, converting an adjective to an adverb, a noun to an adjective, etc. Some words retain the final *e* when another syllable is added; others lose it. For example, *true* becomes *truly, argue* becomes *arguing, come* becomes *coming, write* becomes *writing,* and *judge* becomes *judging.* In these examples, the final *e* is dropped before

adding the ending. But *severe* becomes *severely*, *complete* becomes *completely*, *sincere* becomes *sincerely*, *argue* becomes *argued*, and *care* becomes *careful*. In these instances, the final *e* is retained before adding the ending. Note that some words, like *argue* in these examples, drops the final *e* when the *–ing* ending is added to indicate the participial form; but the regular past tense ending of *–ed* makes it *argued*, in effect, replacing the final *e* so that *arguing* is spelled without an *e* but *argued* is spelled with one.

Some English words contain the vowel combination of *ei*, while some contain the reverse combination of *ie*. Many people confuse these. Some examples include these:

> *ceiling, conceive, leisure, receive, weird, their, either, foreign, sovereign, neither, neighbors, seize, forfeit, counterfeit, height, weight, protein,* and *freight*

Words with *ie* include *piece, believe, chief, field, friend, grief, relief, mischief, siege, niece, priest, fierce, pierce, achieve, retrieve, hygiene, science,* and *diesel.* A rule that also functions as a mnemonic device is "I before E except after C, or when sounded like A as in 'neighbor' or 'weigh'." However, it is obvious from the list above that many exceptions exist.

Many people often misspell certain words by confusing whether they have the vowel *a, e,* or *i,* frequently in the middle syllable of three-syllable words or beginning the last syllables that sound the same in different words. For example, in the following correctly-spelled words, the vowel in boldface is the one people typically get wrong by substituting one or either of the others for it:

> *cem**e**tery, quant**i**ties, benef**i**t, privil**e**ge, unpleas**a**nt, sep**a**rate, independ**e**nt, excell**e**nt, cat**e**gories, indispens**a**ble,* and *irrelev**a**nt*

The words with final syllables that sound the same when spoken but are spelled differently include *unpleasant, independent, excellent,* and *irrelevant.* Another source of misspelling is whether or not to double consonants when adding suffixes. For example, we double the last consonant before *–ed* and *–ing* endings in *controlled, beginning, forgetting, admitted, occurred, referred,* and *hopping;* but we do not double the last consonant before the suffix in *shining, poured, sweating, loving, hating, smiling,* and *hoping.*

One way in which people misspell certain words frequently is by failing to include letters that are silent. Some letters are articulated when pronounced correctly but elided in some people's speech, which then transfers to their writing. Another source of misspelling is the converse: people add extraneous letters. For example, some people omit the silent *u* in *g**u**arantee,* overlook the first *r* in *su**r**prise,* leave out the *z* in *reali**z**e,* fail to double the *m* in *reco**m**mend,* leave out the middle *i* from *aspirin,* and exclude the *p* from *tem**p**erature.* The converse error, adding extra letters, is common in words like *until* by adding a second *l* at the end; or by inserting a superfluous syllabic *a* or *e* in the middle of *athletic,* reproducing a common mispronunciation.

Using Reference Sources

Books as Resources
When a student has an assignment to research and write a paper, one of the first steps after determining the topic is to select research sources. The student may begin by conducting an Internet or library search of the topic, may refer to a reading list provided by the instructor, or may use an annotated bibliography of works related to the topic. To evaluate the worth of the book for the research paper, the student first considers the book title to get an idea of its content. Then, the student can scan the book's table of

contents for chapter titles and topics to get further ideas of their applicability to the topic. The student may also turn to the end of the book to look for an alphabetized index. Most academic textbooks and scholarly works have these; students can look up key topic terms to see how many are included and how many pages are devoted to them.

Journal Articles
Like books, journal articles are primary or secondary sources the student may need to use for researching any topic. To assess whether a journal article will be a useful source for a particular paper topic, a student can first get some idea about the content of the article by reading its title and subtitle, if any exists. Many journal articles, particularly scientific ones, include abstracts. These are brief summaries of the content. The student should read the abstract to get a more specific idea of whether the experiment, literature review, or other work documented is applicable to the paper topic. Students should also check the references at the end of the article, which today often contain links to related works for exploring the topic further.

Dictionaries, Encyclopedias, and Thesauruses
Dictionaries, encyclopedias, and thesauruses are reference books for looking up information alphabetically. Dictionaries are more exclusively focused on vocabulary words. They include each word's correct spelling, pronunciation, variants, part(s) of speech, definitions of one or more meanings, and examples used in a sentence. Some dictionaries provide illustrations of certain words when these inform the meaning. Some dictionaries also offer synonyms, antonyms, and related words under a word's entry. A thesaurus is a book of words that primarily focuses on synonyms and antonyms. They can be used to find alternate terms to use when writing essays or papers. Encyclopedias, like dictionaries, often provide word pronunciations and definitions. However, they have broader scopes: one can look up entire subjects in encyclopedias, not just words, and find comprehensive, detailed information about historical events, famous people, countries, disciplines of study, and many other things.

Card Catalogs
A card catalog is a means of organizing, classifying, and locating the large numbers of books found in libraries. Without being able to look up books in library card catalogs, it would be virtually impossible to find them on library shelves. Card catalogs may be on traditional paper cards filed in drawers, or electronic catalogs accessible online; some libraries combine both. Books are shelved by subject area; subjects are coded using formal classification systems—standardized sets of rules for identifying and labeling books by subject and author. These assign each book a call number: a code indicating the classification system, subject, author, and title. Call numbers also function as bookshelf "addresses" where books can be located. Most public libraries use the Dewey Decimal Classification System. Most university, college, and research libraries use the Library of Congress Classification. Nursing students will also encounter the National Institute of Health's National Library of Medicine Classification System, which major collections of health sciences publications utilize.

Databases
A database is a collection of digital information organized for easy access, updating, and management. Users can sort and search databases for information. One way of classifying databases is by content, i.e. full-text, numerical, bibliographical, or images. Another classification method used in computing is by organizational approach. The most common approach is a relational database, which is tabular and defines data so they can be accessed and reorganized in various ways. A distributed database can be reproduced or interspersed among different locations within a network. An object-oriented database is organized to be aligned with object classes and subclasses defining the data. Databases usually collect files like product inventories, catalogs, customer profiles, sales transactions, student bodies, and resources.

An associated set of application programs is a database management system or database manager. It enables users to specify which reports to generate, control access to reading and writing data, and analyze database usage. Structured Query Language (SQL) is a standard computer language for updating, querying, and otherwise interfacing with databases.

Text Features

Textbooks that are designed well employ varied text features for organizing their main ideas, illustrating central concepts, spotlighting significant details, and signaling evidence that supports the ideas and points conveyed. When a textbook uses these features in recurrent patterns that are predictable, it makes it easier for readers to locate information and come up with connections. When readers comprehend how to make use of text features, they will take less time and effort deciphering how the text is organized, leaving them more time and energy for focusing on the actual content in the text. Instructional activities can include not only previewing text through observing main text features, but moreover through examining and deconstructing the text and ascertaining how the text features can aid them in locating and applying text information for learning.

Included among various text features are a table of contents, headings, subheadings, an index, a glossary, a foreword, a preface, paragraphing spaces, bullet lists, footnotes, sidebars, diagrams, graphs, charts, pictures, illustrations, captions, italics, boldface, colors, and symbols. A **glossary** is a list of key vocabulary words and/or technical terminology and definitions. This helps readers recognize or learn specialized terms used in the text before reading it. A **foreword** is typically written by someone other than the text author and appears at the beginning to introduce, inform, recommend, and/or praise the work. A **preface** is often written by the author and also appears at the beginning, to introduce or explain something about the text, like new additions. A **sidebar** is a box with text and sometimes graphics at the left or right side of a page, typically focusing on a more specific issue, example, or aspect of the subject. **Footnotes** are additional comments/notes at the bottom of the page, signaled by superscript numbers in the text.

Practice Questions

Questions 1–5 are based on the following passage:

Nobody knew where she went. She wasn't in the living room, the dining room, or the kitchen. She wasn't underneath the staircase, and she certainly wasn't in her bedroom. The bathroom was thoroughly checked, and the closets were emptied. Where could she have possibly gone? Outside was dark, cold, and damp. Each person took a turn calling for her. <u>"Eva! Eva!"</u> (1) they'd each cry out, but not a sound was heard in return. The littlest sister in the house began to cry. "Where could my cat have gone?" Everyone just looked at <u>each other</u> (2) in utter silence. <u>Lulu was only five years old and brought this kitten home only six weeks ago and she was so fond of her.</u> (3) Every morning, Lulu would comb her fur, feed her, and play a cat-and-mouse game with her before she left for school. At night, Eva would curl up on the pillow next to Lulu and stay with her <u>across the night</u> (4). They were already becoming best friends. But this evening, right after dinner, Eva snuck outside when Lulu's brother came in from the yard. It was already dark out and Eva ran so quickly that nobody could see where she went.

It was really starting to feel hopeless when all of a sudden Lulu had an idea. She remembered how much Eva loved the cat-and-mouse game, so she decided to go outside and pretend that she was playing with Eva. Out she went. She curled up on the grass and began to play cat and mouse. Within seconds, Lulu could see shining eyes in the distance. The eyes got closer, and closer, and closer. Lulu continued to play <u>in order that</u> (5) she wouldn't scare the kitten away. In less than five minutes, her beautiful kitten had crawled right up to her lap as if asking to join in the game. Eva was back, and Lulu couldn't have been more thrilled! Gently, she scooped up her kitten, grabbed the cat-and-mouse game, and headed back inside, grinning from ear to ear. That night, Eva and Lulu snuggled right back down again, side by side, and had a very restful sleep.

1. Choose the best replacement punctuation.
    a. No change
    b. 'Eva! Eva!'
    c. 'Eva, Eva'!
    d. "Eva, Eva"!

2. Choose the best replacement word or phrase.
    a. No change
    b. one another
    c. themselves
    d. each one

3. Choose the best replacement sentence.
    a. No change
    b. Lulu brought this kitten home only six weeks ago and she was so fond of her
    c. Lulu brought this kitten home only six weeks ago, and she was so fond of her
    d. Lulu was only five years old, and brought this kitten home only six weeks ago, and she was so fond of her

4. Choose the best replacement phrase.
    a. No change
    b. below the night
    c. through the night
    d. amidst the night

5. Choose the best replacement word or phrase.
    a. No change
    b. so,
    c. so that
    d. so that,

Questions 6–10 are based on the following passage:

In our busy family, there are chores that each member must do every day. There are chores for our daughter, for me, and for my husband. There is always so much going on in our family and so many scheduled events every week that without a set list of <u>chores and</u> (6) without ensuring the chores are completed, our household would be utterly chaotic.

I work days, from home. My husband works evenings about a half-hour away, and our <u>13-year-old</u> (7) daughter, Niamh, attends a homeschooling program. She is an avid hockey player and the third-highest scorer on the team. She simply lives for hockey. Three times a week, she has hockey practice, and <u>once, approximately, a week</u> (8), she has a hockey game. Niamh is responsible for washing her jerseys and her hockey pants and making sure all her gear is properly stored after every practice and game in the garage. Occasionally she forgets and leaves it all in the middle of the dining room floor. Niamh must also clean the kitchen every evening after dinner, empty the dishwasher, and feed her two guinea pigs and her cat. Every Saturday, we also give Niamh an allowance for doing extra chores. She sweeps and washes our floors, cleans windows, vacuums, and tidies up the bathrooms.

As for me, I am responsible for walking and feeding the dogs, preparing the lunches and dinners, and keeping up with the laundry. My husband takes out the trash and recycling, keeps up with the yard work, and pays all the bills. Both my husband and <u>myself</u> (9) drive our daughter to friends' houses, school, the arena, the grocery store, medical appointments, and more.

We very rarely seem to get a break, but we are always happy and healthy, and that's what really matters. So, now that you know the breakdown of chores between our daughter, my husband, and me, who do you think has <u>more</u> (10) chores?

6. Choose the best replacement.
    a. No change
    b. chores, and,
    c. chores and,
    d. chores, and

7. Choose the best replacement.
    a. No change
    b. 13 year old
    c. 13-year old
    d. 13 year-old

8. Choose the best replacement.
    a. No change
    b. approximately once a week
    c. approximately bi-weekly
    d. once a week

9. Choose the best replacement.
    a. No change
    b. I
    c. me
    d. myself,

10. Choose the best replacement.
    a. No change
    b. the most
    c. lesser
    d. less

Questions 11–15 are based on the following passage:

Imagine yourself in a secluded room. There is no one to talk to. The room has a bed, a desk, and a computer, and its one window faces the yard. Every day, you must work in this isolated room, away from everyone. There isn't a sound heard <u>accept</u> (11) for the <u>birds outside</u> (12) and your own hands tapping on the computer keyboard. Imagine having to work in that somber environment every day. The only breaks you receive are to go to the bathroom or grab a quick bite to eat. It feels like solitary confinement, and you start to feel yourself slipping <u>ever-so</u> (13) slowly into a depressed state. What should you do? If you don't get the work done, you won't get paid.

Suddenly, you have an idea. You decide that at least two days a week, you can pack up and work at the library. Although it is quiet at the library, there are people coming and going. You will be able to hear the librarians helping students and visitors, and you can people-watch. Feeling optimistic, <u>the day suddenly makes you smile</u>. (14) When the morning arrives, you pack up your computer, notebooks, calendar, phone, and you head out the door. You arrive at the library just as the doors are opening. You find a quiet place to <u>set up and</u> (15) you settle down into your work day. At first, you are thrilled with the new scenery and you love the extra space you have to work with. But before long, you realize your work isn't getting done. You spent so much time watching the different people coming and going and daydreaming about what it would be like to casually read a book that you stopped working altogether! Perhaps the library wasn't the ideal spot for your after all.

You pack up your computer, your notebooks, your calendar, and your phone, and you head back out the door, back to the solitary room with the bed, the desk, the computer, and the window, and you settle down once again, into your work day.

11. Choose the best replacement.
    a. No change
    b. exceptionally
    c. except
    d. expect

12. Choose the best replacement.
    a. No change
    b. birds chirping outside
    c. birds, outside
    d. birds, chirping outside

13. Choose the best replacement.
    a. No change
    b. ever-so,
    c. ever so
    d. everso

14. Choose the best replacement word or phrase.
    a. No change
    b. the day no longer seems so sad.
    c. the day brightens your mood.
    d. you smile for the rest of the day.

15. Choose the best replacement.
    a. No change
    b. set up, and
    c. set up; and
    d. set up. And

Questions 16–20 are based on the following passage:

Why must I always be last? I'm the last one in line at the cafeteria, I'm the last one to my locker, and I'm always last (16) one to class. For once, I'd really like to be first. I would even be happy being right smack dab in the middle!

Having also been the last of all the siblings, my older sisters always teased me. (17) I was the last to be potty-trained, the last to get a big girl's bed, and the last to start school. When we entered the teenage years, I was the last to get my driver's license and the last to dye my hair—I went with purple. My older sisters did spoil me, though. When they got their driver's licenses, they would take me out for ice cream or to the movies, and I even got to tag along from time to time when they went out on dates. I guess they thought I was cute or something. Each time I would complain about being last, my mother would say, "Don't ever worry about being last. That means you're the best!" I never quite knew what that meant. My father's famous line was, 'Last, but not least'! (18) and I'm still trying to figure *that* one out too.

Of course, I'm not being completely honest with you and (19) I've never been the last in my class. In fact, I always seem to score in the top tenth percentile which isn't too bad at all. I have come in first, second, and even third in different swimming events, and when it comes to dinner, I am usually the *first* to finish my plate. And now that I think about it, maybe being last doesn't really matter anyway. After all (20), being the youngest in my family means that I get away with a lot too. I don't have to do a lot of the chores my older sisters have to do. Yes, come to think of it, maybe being last isn't so bad after all.

16. Choose the best replacement.
    a. No change
    b. the last
    c. late
    d. the last,

17. Choose the best replacement.
    a. No change
    b. my older sisters would always tease me.
    c. I was always teased by my older sisters.
    d. my older sisters, always, teased me.

18. Choose the best replacement.
    a. No change
    b. 'Last but not least!'
    c. "Last, but not least"!
    d. "Last but not least!"

19. Choose the best replacement.
    a. No change
    b. ;
    c. even though
    d. but

20. Choose the best replacement.
    a. No change
    b. Therefore,
    c. However,
    d. On the contrary,

Questions 21–25 are based on the following passage:

Café Adriatico was the best place in town to go for authentic Italian cuisine. The entire restaurant consisted of the kitchen in the far back of the restaurant, one small bathroom, and a quaint dining area that comfortably held ten tables. There was always Italian classical music playing in the background, and the atmosphere always put me in mind of being in my own grandparents (21) home.

Angela, the sole owner and head chef, was a unique character. She loved her guests and treated them like her extended family. She would saunter (22) around from table to table, talking with the diners and occasionally pulling up a chair to gab a little longer. One famous story she used to tell was of her younger years in Italy, and how once she won a national beauty contest. (23) She wouldn't tell the story in an arrogant way. She would just matter-of-factly tell us what it was like when she was honored with this award. She was reminiscing about the good old days, I guess, and who doesn't like to do that?

People didn't expect to be fed right away at Café Adriatico. Everything was cooked from scratch and made from the freshest and highest-quality of ingredients. Going to Café Adriatico was the equivalent of going out for the evening. Boy, was Angela something else! It wasn't the least bit uncommon to see her walking around the restaurant with a cigarette hanging out of her mouth and her tiny poodle walking along at her ankles. This <u>doesn't</u> (24) seem to bother any of the locals though. Angela was too <u>well-liked</u> (25). People went there for the comradery, the excellent cuisine, and the reminder of what it was like to be served an authentic Italian meal in the comfort of one's home. If you ever visit Hamilton, Ontario, Canada, be sure to visit Café Adriatico. And if Angela is still sauntering from table to table, please tell her I say hello!

21. Choose the best replacement.
    a. No change
    b. grandparent's
    c. grandparents'
    d. grandparents's

22. Choose the best replacement.
    a. No change
    b. sauntered
    c. was sauntering
    d. had sauntered

23. Choose the best replacement.
    a. No change
    b. National Beauty Contest
    c. National beauty contest
    d. national Beauty contest

24. Choose the best replacement.
    a. No change
    b. hadn't
    c. didn't
    d. never

25. Choose the best replacement.
    a. No change
    b. well liked
    c. much liked
    d. liked

Questions 26–30 are based on the following passage:

The school bell was about to ring at Juniper High, but Matt wasn't ready. He wasn't ready because he hadn't studied for today's science test in Mr. <u>Jones</u> (26) class. He planned on studying. He brought his textbook, notebook, and study guide home, but something unexpected happened last night that changed his plans. About an hour after he arrived home from school, Matt's dad walked in the door with a brand-new synthesizer, and that was the end of studying before it even began.

Ever since Matt was five years old, he was obsessed with music—any kind of music. He would listen to his parents' kitchen radio for hours, <u>play</u> (27) their records and singing away to all the classics. When he was a teenager, he started collecting CDs, and within no time at all he had an impressive collection of artists. <u>Therefore</u> (28) what he really wanted was his very own synthesizer so he could create his own music—and that day had finally come! Matt forgot all about his science test and had begun (29) to experiment with his new toy immediately. He played for hours and tested every button on his synthesizer. Before he knew it, the clock struck midnight, and he fell into a deep sleep with music playing in his head.

The bell rang. Matt slowly walked into school, wondering how he was going to pass this test. It was the last test of the semester, and it was the one that was supposed to bring up his average. If he failed this test, he failed (30) the semester. He went to his locker, grabbed what he needed, and headed off to class. Within minutes, Matt knew this was his lucky day. The classroom door opened, and a substitute teacher walked in—the test would be delayed by a day! Now all Matt had to do was go home and study, even though that synthesizer would be tempting him all night.

26. Choose the best replacement.
    a. No change
    b. Jones's
    c. Jones'
    d. Jone's

27. Choose the best replacement.
    a. No change
    b. playing
    c. played
    d. would be playing

28. Choose the best replacement.
    a. No change
    b. However
    c. However,
    d. Nonetheless,

29. Choose the best replacement word or phrase.
    a. No change
    b. was beginning
    c. begins
    d. began

30. Choose the best replacement phrase.
    a. No change
    b. had failed
    c. was going to fail
    d. would fail

Questions 31–35 are based on the following passage:

Everybody has their own version of the perfect peanutbutter (31) and banana sandwich. Some will tell the type of bread you must use and whether the bread should be toasted. Some will tell you that you must cut off the crust, while others will tell you to keep it intact. Crunchy or smooth is always a heated debate, along with just how much peanut butter a sandwich needs.

I'm no different; I consider myself an expert when it comes to peanut butter and banana sandwiches. My grandmother started making me and my brothers peanut butter and banana sandwiches (32) before we could speak. I'd watch my grandmother in the kitchen closely. She would start by setting out the slices of bread along the counter, and then carefully add (33) one thin layer of butter to each slice. She would then start back at the top and add one fine layer of smooth peanut butter. Next was the final and best step. (34) Instead of slicing the bananas, my grandmother would mash them! She would take a bowl, place the bananas inside, and mash them up with the back of a fork! Then she would spread the slices of bread with the mashed banana. In the end (35), grandma would cut the sandwich in two halves diagonally—the only way to go.

Those peanut butter and banana sandwiches with a glass of milk will live on in my memory forever. You can have your sliced bananas. You can have your toasted sandwiches. You can have your crunchy peanut butter. It is 40 years since I used to eat my grandma's sandwiches, but it seems like yesterday. When I got older and had children of my own, I started making peanut butter and banana sandwiches for them just like my grandmother did for me. I've seen different styles and recipes, but I'll stick to my grandma's peanut butter and mashed banana sandwiches. They were—and still are—simply the best.

31. Choose the best replacement.
    a. No change
    b. Peanut-butter
    c. Peanut butter
    d. Peanutbutter,

32. Choose the best replacement sentence.
    a. No change
    b. making peanut butter and banana sandwiches my brothers and me
    c. making me peanut butter and my brothers banana sandwiches
    d. making I and my brothers peanut butter and banana sandwiches

33. Choose the best replacement.
    a. No change
    b. then she would carefully add
    c. then carefully adding
    d. then carefully added

34. Choose the best replacement word or phrase.
    a. No change
    b. The next step was the best
    c. The final step was the best step
    d. Next, the final step was the best

35. Choose the best replacement word or phrase.
    a. No change
    b. Suddenly
    c. Immediately
    d. Finally

Questions 36–40 are based on the following passage:

Writing academic papers is a skill that develops over time. From the earliest grades in school, students learn to spell correctly. They learn how to write coherent sentences with proper grammar and punctuation, and they learn the importance of beginning, middle, and end. In higher education, these skills become more nuanced.

When writing an academic paper, spelling, punctuation, and content development are key. (36) What will your paper be about? How will you introduce the subject? What information will you include in the body? How do you transition from one point to another? How will you wrap it all up so that the paper seamlessly flows from beginning to end?

Transitions are an extremely important part of writing. Without proper transitions, a paper might seem choppy and disconnected. (37) The points the author is trying to make might be less easily understood, and the paper itself risks losing credibility. Transitional phrases have to be carefully planned out and strategically placed. For example, if the paper is giving a chronological order of events or step-by-step instructions, transitional phrases that indicate what order to follow will likely be used. There are transitional words and phrases that emphasize a point, demonstrating (38) an opposing point of view, present a condition, introduce examples in support of an argument, and so much more. There are transitional phrases authors use to show a consequence or an effect following an event. There are even transitional phrases that are used to indicate a specific time or place. Transitional words act as a bridge connecting (39) ideas and strengthening a paper's coherency. Nonetheless (40), transitional words are a very important part of any academic paper, and the more skilled authors become in knowing when to use the most appropriate transitions, the stronger their papers will be.

36. Choose the best replacement phrase.
    a. No change
    b. students should consider spelling, punctuation, and content development.
    c. spelling, punctuation, and content development, are key.
    d. the paper needs to contain proper spelling, punctuation, and convent development.

37. Choose the best replacement passage.
    a. No change
    b. Transitions are an extremely important part of writing without which, a paper might seem choppy and disconnected.
    c. Transitions are an extremely important part of writing. Additionally, without them, a paper might seem choppy and disconnected.
    d. Transitions are an extremely important part of writing without them, a paper might seem choppy and disconnected.

38. Choose the best replacement sentence.
    a. No change
    b. to demonstrate
    c. demonstrate
    d. for demonstrating

39. Choose the best replacement.
   a. No change
   b. bridge of connection
   c. bridge connecting,
   d. bridge to connections

40. Choose the best replacement word or phrase.
   a. No change
   b. Overall,
   c. Additionally,
   d. Next,

Questions 41–45 are based on the following passage:

It seems as though many people are still confused about global warming and climate change, and a lot of <u>there</u> (41) confusion stems from misinformation. There are many skeptics who think that both of these terms are synonymous, but they are not. The terms <u>"global warming" and "climate change"</u> (42) refer to two completely different phenomena. The former refers to the rising of average global temperatures, which has been trending for a long time. "Climate change," <u>on the other hand,</u> (43) refers to changes in climate around the world, like precipitation changes, and the prevalence of droughts, heat waves, and other extreme climate phenomena.

The confusion is <u>twofold.</u> (44) Firstly, since skeptics don't see steadily warmer seasons, they don't believe in global warming. <u>Additionally,</u> (45) because they believe both terms are synonymous, they actually think that scientists purposely changed the term "global warming" to "climate change," assuming they knew they were wrong about the Earth getting warmer. Nothing could be further from the truth. When people hear the term "global warming," they tend to focus on the connotation of the word "warming." Since they still feel they are experiencing the four seasons of winter, spring, summer, and fall, and since each season still goes through approximately the same phases—cold, wet, hot, and cool—many tend to believe there is no truth to the theory behind global warming. And because they hold fast to these false beliefs, they are further skeptical of the scientific evidence behind both global warming and climate change.

There are many challenges that face scientists today, and some of those challenges are in striving to educate the public on issues that greatly affect us all. The more informed the general public becomes, the better chances we have in slowing down the effects of both climate change and global warming. We need to become more conscious about what steps we can take to leave less of a carbon footprint and hopefully leave this world better than we found it—or at the very least, prevent it from becoming worse.

41. Choose the best replacement word or phrase.
   a. No change
   b. these
   c. they're
   d. their

42. Choose the best replacement.
   a No change
   b. "global warming and climate change"
   c. "global warming, and climate change"
   d. 'global warming' and 'climate change'

43. Choose the best replacement word or phrase
   a. No change
   b. however,
   c. on the whole,
   d. likewise,

44. Correct the punctuation error.
   a. No change
   b. twofold:
   c. twofold;
   d. twofold,

45. Choose the best replacement word or phrase.
   a. No change
   b. On the other hand,
   c. Secondly,
   d. Lastly,

Questions 46–50 are based on the following passage.

Don't drive <u>to quickly</u> (46) down neighborhood streets in the state of New York or you might run into speed bumps. In Arizona, though, you may run into speed *humps*.

Depending on what part of the country <u>we</u> (47) live in, you come across different nuances with our language. If you happen to live in one of the northern states, you might even find your speech patterns are similar to those of our Canadian neighbors to the north. <u>It's</u> (48) really fascinating. We are all speaking the same language and living in approximately the same area, and yet we speak so differently. We have different accents, different terms for things, and a whole slew of different expressions. In Canada, you'll hear "I'm putting on my running shoes," while in Western New York, you'll hear "I'm putting on my sneakers." Canadians like to drink "pop" and Western New Yorkers prefer "soda."

The English language is truly dynamic, and it takes on a life of its own depending on the region in which it is being spoken. One could say that language in general is alive and must be flexible enough to change as our societies change. For example, the introduction of the Internet brought with it a new vocabulary, and with every new invention, new words must be created to describe and explain the invention. So, if language is always changing, should we be that fussy about grammar and spelling? Yes, we should. Although our vocabulary may expand to allow for new inventions to be defined and explained, some things need to remain the same. <u>Although frustrating at times, we must insist upon grammar and spelling.</u> (49) If we all of a sudden decided to do away with punctuation, for example, we would have a very difficult time making sense of anything we read. What a mess that would make! How could we possibly differentiate between "there," "their," and "they're," or "buy" and "by"? So, although languages are meant to grow and change along with the people <u>who</u> (50) use them, some things must remain the same.

46. Choose the best replacement.
    a. No change
    b. two quickly
    c. too quickly
    d. to quickly,

47. Choose the best replacement word or phrase.
    a. No change
    b. you
    c. people
    d. some

48. Choose the best replacement.
    a. No change
    b. Its
    c. Its'
    d. It's'

49. Choose the best replacement word or phrase.
    a. No change
    b. Although frustrating at times, grammar and spelling are essential
    c. Although frustrating at times, we must always use proper grammar and spelling
    d. Although frustrating at times, we need grammar and spelling

50. Choose the best replacement.
    a. No change
    b. whom
    c. that
    d. which

Questions 51–55 are based on the following passage.

The morning sun shone down upon the land and the sky was a deep blue. The snowcapped mountains in the distance seemed a thousand miles away, and the valleys down below were covered in the most beautiful yellow flowers. It was springtime (51) once again in the Arctic, and (52) just like countless times before, signs of life were beginning to reappear. The polar bear, foxes, seal, and hare's (53) would soon appear, along with countless newborns to carry on the lifeline.

Throughout the winter months, it seems as though no life exists at all in the Arctic. The winter winds are chillingly cold, and the water is frozen solid. The only sounds for months on end are the distant cries of a lone wolf (54) or the howling wind ravaging through the land. Everything is white and crisp, but some startling changes are beginning to show. The sea ice is melting at an alarming rate, and for the people who call the Arctic their home, this is devastating news. Melting ice means it will be more and more difficult to hunt. Hunters use the frozen fjord as their highway to hunt for fresh food. As the frozen land disappears, so too is their way of life. (55) The people of the Arctic have lived off of the land and have an immense respect for nature, an attitude that has passed been down through the generations. They are worried that the rest of the world has become detached, separated from the Earth. It is because of this detachment that the Earth now suffers. In years past, in springtime there would still be enough frozen land to allow hunters to travel to the best hunting grounds, but now there is so much water that the hunters cannot travel the necessary distance. Their entire villages rely on their skills, but without enough frozen land, hunters are returning home with very little food. The one message the people of the Arctic would like to pass on to the rest of the world is very simple—no matter what you are doing in this moment, take the time to get reconnected to nature. There is still time.

51. Choose the best replacement.
    a. No change
    b. spring time
    c. Springtime
    d. spring-time

52. Choose the best replacement.
    a. No change
    b. arctic and
    c. arctic, and
    d. arctic, and,

53. Choose the best replacement.
    a. No change
    b. rabbits
    c. hare
    d. hair

54. Choose the best replacement.
    a. No change
    b. loan wolf
    c. Lone Wolf
    d. lone-wolf

55. Choose the best replacement sentence.
    a. No change
    b. As the frozen land disappears, so is their way of life.
    c. As the frozen land disappears, so too, is their way of life.
    d. As the frozen land disappears, so too does their way of life.

Questions 56–60 are based on the following passage.

There was nothing more precious to Mom than sitting in the backyard behind the garage on the makeshift deck, in front of the enormous pine tree. Every morning, Mom and Dad would wake up, make their coffee, and head outside. They'd sit with their coffees in their chairs (56) for a good hour or so, sipping, reading, talking, and watching the countless signs of life in their beloved pine tree. Mom was a wildlife rehabilitator, so she loved watching the squirrels scurry up and down the tree, collecting nuts and making their nests. She had cared for some of those same squirrels since they were newborns, and she had released them right there, underneath the great old pine. There were blue jays that would grace Mom and Dad with their presence now and then, and Mom swore that the blue jays were mates for life, just like she and Dad (57) were. When the red cardinals would come in, Mom and Dad taken (58) out their binoculars to get a closer look.

There were even lazy weekends in the <u>summertime</u> (59) when Mom and Dad would get their blanket, lay it out underneath that great old pine, and take a long afternoon nap. Under the pine is where we would gather with friends and family. We would eat, drink, laugh, and spend all afternoon right there in that corner of our yard. Now that I think of it, that was probably the most precious place on our entire property. I'm not sure if we knew it at the time.

We moved away from that old home many years ago. In fact, we moved thousands of miles away to the South, but every so often, I sit back and think about those days. I wonder how the old pine is doing, and I yearn for the day when I can go back to visit. Time waits for <u>no one</u> (60), and that is certainly true no matter who you are. But our memories are what keep us young, and this is one of those memories I will treasure my entire life.

56. Choose the best replacement.
    a. No change
    b. They'd sit in their chairs with their coffees
    c. They'd sit with their chairs and their coffees
    d. They'd sit, coffees and chairs

57. Choose the best replacement.
    a. No change
    b. her and Dad
    c. she and him
    d. hers and Dad's

58. Choose the best replacement.
    a. No change
    b. would take
    c. had taken
    d. will take

59. Choose the best replacement.
    a. No change
    b. summer-time
    c. Summer Time
    d. Summer-Time

60. Choose the best replacement.
    a. No change
    b. no-one
    c. Noone
    d. No-One

Answer Explanations

1. A: Choice *A* is the best answer. To write a direct quotation, double quotation marks are placed at the beginning and ending of the quoted words. All punctuation specifically related to the quoted phrase or sentence is kept inside the quotation marks. Single quotation marks, as those used in Choices *B* and *C*, are not used for dialogue in American English.

2. B: The best answer is Choice *B*. The reciprocal pronoun "one another" generally refers to more than two people.

3. C: The best answer is Choice *C*: "Lulu brought this kitten home only six weeks ago, and she was so fond of her." There are two independent clauses in one sentence, so separating the two clauses with a comma is the best option. The other sentences have incorrect comma placement. Additionally, the information that Lulu was only five years old is not necessary in this particular sentence; it is out of context.

4. C: Choice *C* is the best answer. We would use the preposition "through," when talking about staying with someone "through the night."

5. C: Choice *C* is the best answer here: "Lulu continued to play so that she wouldn't scare the kitten away." Usually the words "in order" are used with the word "to." Choice *A* is incorrect because it uses the word "that" together with "to." In Choices *B* and *D*, there is no need for a comma after "so that."

6. D: The best answer is Choice *D*: "without a set list of chores, and without ensuring the chores are completed, our household would be utterly chaotic." In this sentence, there is an interrupting phrase that must be set apart by two commas on either side of it.

7. A: Choice *A* is the best answer, no change, because "13-year-old" is correct. When an age is describing a noun and precedes the noun, hyphenation is required.

8. B. Choice *B* is the best answer, "approximately once a week." Choice *A* is not the best answer because the adverb "approximately" interrupts the phrase "once a week," producing the awkward, if grammatically correct phrase, "once, approximately, a week." Choice *C* is not correct, because "bi-weekly" means twice a week or every two weeks. Choice *D* is not the best answer because it changes the meaning to a firm "once a week."

9. B: The best answer is Choice *B*: "Both my husband and I drive our daughter." "I" is a subject pronoun, and in this sentence, "I" is part of a compound subject, "my husband and I." A helpful way to answer this question is to take out the phrase "my husband" and read the sentence with each choice. Choice *A*, "myself drive our daughter" is incorrect, and so is Choice *D*, "myself, drive our daughter." Choice *C*, "me drive our daughter" is also incorrect. "I drive our daughter" is the best answer.

10. B: The best answer is Choice *B*: "who do you think has the most chores?" Since the sentence is comparing three people's chores, it is necessary to use a superlative adjective. If there were only two people, it would be correct to use the comparative adjective "more."

11. C: Choice *C* is the best answer. The words "accept" and "except" are often confused. There is no need to use the adverb "exceptionally" here, so Choice *B* is incorrect. Choice *C* is incorrect because "expect" is a verb with a different meaning than "except."

12. B: The best answer is Choice *B.* The word "birds" doesn't necessarily evoke a sound, but any number of actions birds make will create sound. It is best to use a qualifier here, to indicate which bird sound the person hears.

13. C: The best answer is Choice *C.* "Ever so" is a common phrase in English. The other options are spelled incorrectly.

14. D: Choice *D* is the best answer: "you smile for the rest of the day." As the sentence is currently written, there is a dangling modifier. The sentence implies that the day, and not the person, is feeling optimistic.

15. B: The best choice is Choice *B.* The sentence is composed of two independent clauses; therefore, it requires a comma before the conjunction.

16. B: The best answer is Choice *B:* "and I'm always the last one to class." Choice *C* changes the meaning, and Choice *D* uses an unnecessary comma. The original phrase is awkward; it is not parallel with the previous instance of the phrase "the last one." Additionally, English speakers would never omit the article "the" from the phrase "the last one."

17. C: The best answer is Choice *C:* "I was always teased by my older sisters." The sentence in the passage demonstrates a common mistake with dangling participles. The sentence starts with the participial phrase "Having also been the last of all the siblings." That phrase can only modify the speaker, the youngest sibling. But in the original passage, the participial phrase incorrectly modifies "my older sisters." Choice *C* is the only answer that corrects this mistake.

18. D: Choice *D* is the best answer: "Last but not least!" In American English, double quotation marks are used at the beginning and ending of a quotation, and all punctuation relating specifically to the quotation must remain inside the quotation marks. There is also no need for a comma in a simple sentence containing one simple thought, making Choice *C* incorrect.

19. B: The best answer is Choice *B,* the semicolon. The way ideas are connected is important. In this case, the second independent clause gives an example of the first. The author says she has been dishonest; the next clause shows the instance of her dishonesty. The word "and" is not grammatically wrong, but it suggests the two facts are on the same level. Choices *C* and *D* suggest different relationships between the two facts, and these relationships do not make the most sense.

20. A: Choice *A* is the best answer. The phrase "after all" means "despite earlier problems or doubts," which is the perfect transition here because the speaker is changing their mind about being the last sibling. "Therefore" denotes a conclusion, and "however" and "on the contrary" denote a counter, and the speaker is not concluding or countering here.

21. C: The best answer is Choice *C,* which is grandparents'. To show possession of a plural noun that ends in an "s," the apostrophe is placed directly after the final "s."

22. A: The original choice is the best answer here. Sauntering is not something Angela did just one time; she did it often. When referring to repeated past actions, the word "would" is used with the present tense verb, in this case "saunter."

23. A: The best answer is Choice *A.* There is no need to use any uppercase letters in the name because the specific name of a contest is not being used. The words in a proper noun would be capitalized.

24. C: Choice *C* is the best answer: "didn't." The entire story is written in the past tense, and the verbs must reflect this tense in order to keep the passage coherent.

25. B: The best answer is Choice *B*: "well liked." The phrase "well liked" when it appears before the verb: "The well-liked hostess sauntered around the room." In this sentence, "well liked" appears after the verb, so no hyphen is needed.

26. B: Choice *B* is the best answer: "Jones's class." The teacher's name is Mr. Jones, and since the name "Jones" ends in an "s," in order to show possession, it is necessary to place the apostrophe directly after the "s" in the name "Jones," followed by the possessive "s."

27. B: The best answer is Choice *B*: "He would listen to his parents' kitchen radio for hours, playing their records and singing away to all the classics." In order to keep the verb tense consistent and maintain parallel structure, the verb form "playing" should match the verb form "singing."

28. C: The best answer is Choice *C*: "However, what he really wanted . . ." In the choices provided, "however" as a transitional word followed by a comma is the most appropriate choice because it provides a contrast inherent in the context: that he loved music but wanted to create his own.

29. D: Choice *D* is the best answer. The two actions occur one after the other, in a sequence; first Matt forgot, then he began to play. In the original sentence, the verb "had begun" suggests Matt somehow started playing before he forgot about his test.

30. D: The best answer is Choice *D*. This sentence is set up in the past conditional, starting with the condition "if." Therefore, the use of the conditional "would fail" is necessary.

31. C: The best answer is Choice *C*. "Peanut butter" is spelled as two separate words without a hyphen.

32. A: The best answer is Choice *A*. The original sentence is correct. There is a joke in which a boy says to a genie, "Make me a peanut butter sandwich." The genie interprets the sentence incorrectly, and the boy is turned into a sandwich. However, there is nothing grammatically wrong with the way the boy expresses his wish, and there is nothing wrong with the original sentence in Question 32. The word "sandwiches" is the direct object of the verb "make"; sandwiches are what the grandmother makes. The words "me and my brothers" are indirect objects of the verb "make"; the grandmother makes sandwiches for them. Choice *D* is incorrect, because the pronoun "I" is a subject, and the sentence requires the indirect object "me."

33. B: The best answer is Choice *B*. To refer to repeated past actions as in this case, the word "would" is used, and in an independent clause like this one, the pronoun "she" needs to be repeated.

34. B: Choice *B* is the best answer because it is the most coherent. The step listed in the passage is actually *not* the final step, so the word "final" is out of place. All the other choices are confusing because they include the word "final."

35. D: The best answer is Choice *D*. This truly is the final step in the process, when the grandmother would cut the sandwiches. "Finally" is a terminal transitional word. The original phrase "in the end" is more appropriate for an action that happened one time. The author is describing a process the grandmother did repeatedly, so "finally" is a better choice than "in the end."

36. B: Choice *B* is the best answer: "students should consider spelling, punctuation, and content development." This is an example of a dangling modifier. The writing of the paper is not done by the spelling, punctuation, or content development. The writing is done by the students, so the clause has to use "students" as the subject.

37. A: The best answer is Choice *A*. Choice *B* is incorrect because there is a comma after "without which," and this comma interrupts the sentence. Choice *C* is incorrect because of the superfluous word "additionally." Additionally is not needed here because the second sentence acts as an explanation to the first. Choice *D* is incorrect because it is a run-on sentence; there should be a semicolon or period after "writing" and before "without them."

38. C: The best answer is Choice *C*. Each verb in the relative clause should show parallelism with the use of the third-person singular present tense. The clause is about words that demonstrate, words that emphasize, words that present, and so on. Therefore "demonstrate" should be parallel with the verbs "emphasize," "present," and "introduce." Choices *A*, *B*, and *D* are not parallel.

39. A: The best answer is Choice *A*. The gerund form of the verb "to connect" is "connecting." Using this form of the verb creates a parallel structure, because the sentence also uses the gerund form of "to strengthen," which is "strengthening."

40. B: The best answer is Choice *B*. None of the other transitional words make sense. Choice *C*, "additionally," is not a good answer because the sentence is not adding a new point. Likewise, Choice *D*, "next," is not appropriate because the sentence is not about something that comes next in a sequence. Because this is the final summation of the argument, the transitional word "overall" is the best answer.

41. D: Choice *D* is the best answer. "Their" is the possessive form of the pronoun "they" and is the only coherent choice. Choices *A* and *D* are homophones; "there" and "they're" sound like "their," but they have different meanings.

42. A: The best answer is Choice *A*. No change is needed here, because the two terms being introduced are separate and distinct, and therefore need to be separated by their own sets of double quotation marks. Choice *D* is incorrect because it uses single quotation marks.

43. B: The best answer is Choice *B*. "On the other hand" refers to the opposite point of view from the previous one discussed, or the opposite in meaning. Climate change is not the opposite of global warming, although there are differences. "However" is the best transitional word to use here.

44. B: Choice *B* is the best answer. In this case, the colon is being used to introduce several sentences that are examples of the twofold confusion.

45. C: The best answer is *C*. "Secondly" is a transitional term used to strengthen the argument, and it is parallel with the term "firstly," used to introduce the previous idea.

46. C: The best answer is *C*. The adverb "too," which means "more than is desirable," is the only coherent choice in this case. Choice *B*, which is the number "two," and Choice *D*, the preposition "to," do not make sense in this passage.

47. B: Choice *B* is the best answer, because it matches the author's use of pronouns in the rest of the paragraph. In order to write coherently and not confuse the readers, it is important to be consistent with pronouns. In this case, the author is using second-person pronoun "you" in the rest of the sentence and the rest of the paragraph: "*you* come across different nuances."

48. A: The best answer is *A*. No change is required here since the appropriate word is being used. "It's" is a contraction of "it is." Choice *B*, "its," is the possessive and does not make sense here. The punctuation in Choices *C* and *D* is not correct.

49. B: The best answer is *B*. What is "frustrating" is the grammar and spelling, so the terms "grammar" and "spelling" must immediately follow the beginning of the sentence to avoid dangling or misplaced modifiers.

50. A: The best answer is *A*. The word "people" is being used as a subject, and therefore its corresponding pronoun would be "who." The pronoun "whom," in Choice *B*, is meant to be used as an object. "Whom" would be appropriate in a phrase like "the people of whom it is said." Choices *C* and *D*, "that" and "which," are relative pronouns that can be subjects, but they are usually not used to refer to people.

51. A: The best answer is *A*. No change is needed here, because "springtime" is one compound word. It is not two words, as in Choice *B*. It does not require a hyphen, as in Choice *C*. It is not capitalized, as in Choice *D*.

52. A: Choice *A* is the best answer. No change is needed here, because "Arctic," when used as a noun, is capitalized.

53. C: The best answer is *C*. A hare is a long-eared animal similar to a rabbit. The word "hare" can be plural; it is also correct to use "hares" as the plural. Choice *A* is not the best answer because it is in the possessive form. Choice *B* is not the best answer because rabbits are slightly different than hares; for example, most rabbits live underground, while hares nest above ground.

54. A: The best answer is *A*. No change is needed. Choice *B* is incorrect, because the word "loan" refers to something borrowed. Choice *C* is incorrect, because there is no reason to capitalize the phrase "lone wolf." No hyphen is needed, so Choice *D* is incorrect.

55. D: Choice *D* is the best answer. The verb "so does" has to follow the phrase that comes before it—that the land is disappearing. Let's put it in a sentence: "their way of life does disappear" or "their way of life is disappear"? With the latter we would have to change the verb to "disappearing."

56. B: The best answer is *B*. The original passage is grammatically correct, and most people would understand its meaning. However, the original passage gives the awkward image of the coffee cups sitting in the chairs, when it is the parents who sit in the chairs. Choice *C* removes the pertinent information that the parents are in the chairs. Choice *D* is a very different construction, which does not make sense here. Choice *D* would only make sense in a sentence like "They'd sit, coffees and binoculars in hand…"

57. A: Choice *A* is the best answer. There is no change needed since "she and Dad" are part of a complex subject. Choice *C* is not the best answer, because "him" is an object, not a subject like "she." ("She and he" would be grammatically correct, but it would not make sense, because it would not be clear who "he" refers to.)

58. B: The best answer is *B*. When referring to repeated past actions, the word "would" is often used. "Would" is the past form of the modal verb "will."

59. A: The best answer is *A*. There is no change needed, because "summertime" is a compound word that does not require capitalization or hyphenation.

60. A: Choice *A* is the best answer. There is no change needed, because "no one" is spelled as two separate words with no hyphenation.

Essay Prompt

Read the two articles below, then write a 500-word essay where you synthesize the two essays and give your opinion on the subject. Choose a specific side on the issue and use specific examples and evidence from each passage to support your own opinion. Be sure to use the authors' last names or the title of the passage when referring to each specific text.

Passage A

(from "Free Speech in War Time" by James Parker Hall, written in 1921, published in Columbia Law Review, Vol. 21 No. 6)

In approaching this problem of interpretation, we may first put out of consideration certain obvious limitations upon the generality of all guaranties of free speech. An occasional unthinking malcontent may urge that the only meaning not fraught with danger to liberty is the literal one that no utterance may be forbidden, no matter what its intent or result; but in fact it is nowhere seriously argued by anyone whose opinion is entitled to respect that direct and intentional incitations to crime may not be forbidden by the state. If a state may properly forbid murder or robbery or treason, it may also punish those who induce or counsel the commission of such crimes. Any other view makes a mockery of the state's power to declare and punish offences. And what the state may do to prevent the incitement of serious crimes which are universally condemned, it may also do to prevent the incitement of lesser crimes, or of those in regard to the bad tendency of which public opinion is divided. That is, if the state may punish John for burning straw in an alley, it may also constitutionally punish Frank for inciting John to do it, though Frank did so by speech or writing. And if, in 1857, the United States could punish John for helping a fugitive slave to escape, it could also punish Frank for inducing John to do this, even though a large section of public opinion might applaud John and condemn the Fugitive Slave Law.

Passage B

(from "Freedom of Speech in War Time" by Zechariah Chafee, Jr. written in 1919, published in Harvard Law Review Vol. 32 No. 8)

The true boundary line of the First Amendment can be fixed only when Congress and the courts realize that the principle on which speech is classified as lawful or unlawful involves the balancing against each other of two very important social interests, in public safety and in the search for truth. Every reasonable attempt should be made to maintain both interests unimpaired, and the great interest in free speech should be sacrificed only when the interest in public safety is really imperiled, and not, as most men believe, when it is barely conceivable that it may be slightly affected. In war time, therefore, speech should be unrestricted by the censorship or by punishment, unless it is clearly liable to cause direct and dangerous interference with the conduct of the war.

Thus our problem of locating the boundary line of free speech is solved. It is fixed close to the point where words will give rise to unlawful acts. We cannot define the right of free speech with the precision of the Rule against Perpetuities or the Rule in Shelley's Case, because it involves national policies which are much more flexible than private property, but we can establish a workable principle of classification in this method of balancing and this broad test of certain danger. There is a similar balancing in the determination of what is "due process of law." And we can with certitude declare that the First Amendment forbids the punishment of words merely for their injurious tendencies. The history of the Amendment and the political function of free speech corroborate each other and make this conclusion plain.

Mathematics

Numbers and Operations on Numbers

Using Properties of Operations with Real Numbers

Rational and Irrational Numbers

Rational numbers can be **whole** or **negative** numbers, **fractions**, or **repeating decimals** because these numbers can all be written as a fraction. Examples of rational numbers include $\frac{1}{2}$, $\frac{5}{4}$, and 8. Whole numbers can be written as fractions, where 25 and 17 can be written as $\frac{25}{1}$ and $\frac{17}{1}$. One way of interpreting these fractions is to say that they are **ratios**, or comparisons of two quantities. The fractions given may represent 25 students to 1 classroom, or 17 desks to 1 computer lab. Repeating decimals can also be written as fractions of integers, such as 0.3333 and 0.6666667. These repeating decimals can be written as the fractions $\frac{1}{3}$ and $\frac{2}{3}$.

Fractions can be described as having a part-to-whole relationship. The $\frac{1}{3}$ may represent 1 piece of pizza out of the whole cut into 3 pieces. The fraction $\frac{2}{3}$ may represent 2 pieces of the same whole pizza. One operation to perform on rational numbers, or fractions, can be addition. Adding the fractions $\frac{1}{3}$ and $\frac{2}{3}$ can be as simple as adding the numerators, 1 and 2. Because the denominator on both fractions is 3, that means the parts on the top have the same meaning or are the same size piece of pizza. When adding these fractions, the result is $\frac{3}{3}$, or 1. Both of these numbers are rational and represent a whole, or in this problem, a whole pizza.

Other than fractions, rational numbers also include whole numbers and negative integers. When whole numbers are added, other than zero, the result is always greater than the addends. For example, the equation $4 + 18 = 22$ shows 4 increased by 18, with a result of 22. When subtracting rational numbers, sometimes the result is a negative number.

For example, the equation $5 - 12 = -7$ shows that taking 12 away from 5 results in a negative answer because 5 is smaller than 12. The difference is -7 because the starting number is smaller than the number taken away. For multiplication and division, similar results are found. Multiplying rational numbers may look like the following equation: $5 \times 7 = 35$, where both numbers are positive and whole, and the result is a larger number than the factors. The number 5 is counted 7 times, which results in a total of 35. Sometimes, the equation looks like $-4 \times 3 = -12$, so the result is negative because a positive number times a negative number gives a negative answer. The rule is that any time a negative number and a positive number are multiplied or divided, the result is negative.

Operations with Rational Numbers

The four basic operations include addition, subtraction, multiplication, and division. The result of addition is a **sum**, the result of subtraction is a **difference**, the result of multiplication is a **product**, and the result of division is a **quotient**. Each type of operation can be used when working with rational numbers; however, the basic operations need to be understood first while using simpler numbers before working with fractions and decimals.

Performing these operations should first be learned using whole numbers. Addition needs to be done column by column. To add two whole numbers, add the ones column first, then the tens columns, then the hundreds, etc. If the sum of any column is greater than 9, a one must be carried over to the next column.

For example, the following is the result of $482 + 924$:

$$
\begin{array}{r}
1 \\
482 \\
+924 \\
\hline
1406
\end{array}
$$

Notice that the sum of the ten's column was 10, so a one was carried over to the hundred's column. Subtraction is also performed column by column. Subtraction is performed in the one's column first, then the tens, etc. If the number on top is smaller than the number below, a one must be borrowed from the column to the left. For example, the following is the result of $5,424 - 756$:

$$
\begin{array}{r}
4\ 13\ 11\ 14 \\
\cancel{5}\ \cancel{4}\ \cancel{2}\ \cancel{4} \\
-\ 7\ 5\ 6 \\
\hline
4\ 6\ 6\ 8
\end{array}
$$

Notice that a one is borrowed from the tens, hundreds, and thousands place. After subtraction, the answer can be checked through addition. A check of this problem would be to show that $756 + 4,668 = 5,424$.

Multiplication of two whole numbers is performed by writing one on top of the other. The number on top is known as the **multiplicand,** and the number below is the **multiplier.** Perform the multiplication by multiplying the multiplicand by each digit of the multiplier. Make sure to place the ones value of each result under the multiplying digit in the multiplier. Each value to the right is then a 0. The product is found by adding each product. For example, the following is the process of multiplying 46 times 37 where 46 is the multiplicand and 37 is the multiplier:

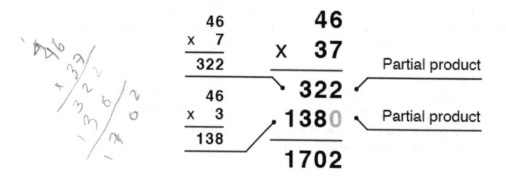

Finally, division can be performed using long division. When dividing a number by another number, the first number is known as the **dividend,** and the second is the **divisor.** For example, with $a \div b = c$, a is the dividend, b is the divisor, and c is the quotient. For long division, place the dividend within the division symbol and the divisor on the outside. For example, with $8,764 \div 4$, refer to the first problem in the diagram below. First, there are 2 4's in the first digit, 8. This number 2 gets written above the 8. Then, multiply 4 times 2 to get 8, and that product goes below the 8. Subtract to get 8, and then carry down the 7. Continue the same steps. $7 \div 4 = 1$ R3, so 1 is written above the 7. Multiply 4 times 1 to get 4, and write it below the 7. Subtract to get 3, and carry the 6 down next to the 3. Resulting steps give a 9 and a 1. The final subtraction results in a 0, which means that 8,764 is divisible by 4. There are no remaining numbers.

The second example shows that $4,536 \div 216 = 21$. The steps are a little different because 216 cannot be contained in 4 or 5, so the first step is placing a 2 above the 3 because there are 2 216's in 453. Finally, the third example shows that $546 \div 31 = 17$ R19. The 19 is a remainder. Notice that the final subtraction does not result in a 0, which means that 546 is not divisible by 31. The remainder can also be written as a fraction over the divisor to say that:

$$546 \div 31 = 17\frac{19}{31}$$

```
    2191              21              17 r 19
  4│8764        216│4536        31│546
    8                432              31
    07               216              236
    4                216              217
    36                 0               19
    36
       04
        4
        0
```

If a division problem relates to a real-world application, and a remainder does exist, it can have meaning. For example, consider the third example above, $546 \div 31 = 17R19$. Let's say that we had \$546 to spend on calculators that cost \$31 each, and we wanted to know how many we could buy. The division problem would answer this question. The result states that 17 calculators could be purchased, with \$19 left over. Notice that the remainder will never be greater than or equal to the divisor.

Once the operations are understood with whole numbers, they can be used with integers. There are many rules surrounding operations with negative numbers. First, consider addition with integers. The sum of two numbers can first be shown using a number line. For example, to add $-5 + (-6)$, plot the point -5 on the number line. Then, because a negative number is being added, move 6 units to the left. This process results in landing on -11 on the number line, which is the sum of -5 and -6. If adding a positive number, move to the right. Visualizing this process using a number line is useful for understanding; however, it is not efficient. A quicker process is to learn the rules. When adding two numbers with the same sign, add the absolute values of both numbers, and use the common sign of both numbers as the sign of the sum. For example, to add $-5 + (-6)$, add their absolute values $5 + 6 = 11$. Then, introduce a negative symbol because both addends are negative. The result is -11. To add two integers with unlike signs, subtract the lesser absolute value from the greater absolute value, and apply the sign of the number with the greater absolute value to the result. For example, the sum $-7 + 4$ can be computed by finding the difference $7 - 4 = 3$ and then applying a negative because the value with the larger absolute value is negative. The result is -3. Similarly, the sum $-4 + 7$ can be found by computing the same difference but leaving it as a positive result because the addend with the larger absolute value is positive. Also, recall that any number plus 0 equals that number. This is known as the **Addition Property of 0**.

Subtracting two integers can be computed by changing to addition to avoid confusion. The rule is to add the first number to the opposite of the second number. The opposite of a number is the number on the other side of 0 on the number line, which is the same number of units away from 0. For example, -2 and 2 are opposites. Consider $4 - 8$. Change this to adding the opposite as follows: $4 + (-8)$. Then, follow the rules of addition of integers to obtain -4. Secondly, consider $-8 - (-2)$. Change this problem to adding the opposite as $-8 + 2$, which equals -6. Notice that subtracting a negative number is really adding a positive number.

Multiplication and division of integers are actually less confusing than addition and subtraction because the rules are simpler to understand. If two factors in a multiplication problem have the same sign, the result is positive. If one factor is positive and one factor is negative, the result, known as the **product**, is negative. For example, $(-9)(-3) = 27$ and $9(-3) = -27$. Also, a number times 0 always results in 0. If a problem consists of more than a single multiplication, the result is negative if it contains an odd number of negative factors, and the result is positive if it contains an even number of negative factors. For example, $(-1)(-1)(-1)(-1) = 1$ and $(-1)(-1)(-1)(-1)(-1) = -1$. These two examples of multiplication also bring up another concept. Both are examples of repeated multiplication, which can be written in a more compact notation using exponents. The first example can be written as $(-1)^4 = 1$, and the second example can be written as $(-1)^5 = -1$. Both are exponential expressions, -1 is the base in both instances, and 4 and 5 are the respective exponents. Note that a negative number raised to an odd power is always negative, and a negative number raised to an even power is always positive. Also, $(-1)^4$ is not the same as -1^4. In the first expression, the negative is included in the parentheses, but it is not in the second expression. The second expression is found by evaluating 1^4 first to get 1 and then by applying the negative sign to obtain -1.

A similar theory applies within division. First, consider some vocabulary. When dividing 14 by 2, it can be written in the following ways: $14 \div 2 = 7$ or $\frac{14}{2} = 7$. 14 is the **dividend,** 2 is the **divisor,** and 7 is the **quotient**. If two numbers in a division problem have the same sign, the quotient is positive. If two numbers in a division problem have different signs, the quotient is negative. For example:

$$14 \div (-2) = -7, \text{ and } -14 \div (-2) = 7$$

To check division, multiply the quotient times the divisor to obtain the dividend. Also, remember that 0 divided by any number is equal to 0. However, any number divided by 0 is undefined. It just does not make sense to divide a number by 0 parts.

If more than one operation is to be completed in a problem, follow the **Order of Operations**. The mnemonic device, **PEMDAS**, for the order of operations states the order in which addition, subtraction, multiplication, and division needs to be done. It also includes when to evaluate operations within

106

grouping symbols and when to incorporate exponents. PEMDAS, which some remember by thinking "please excuse my dear Aunt Sally," refers to *parentheses, exponents, multiplication, division, addition,* and *subtraction.* First, within an expression, complete any operation that is within parentheses, or any other grouping symbol like brackets, braces, or absolute value symbols. Note that this does not refer to the case when parentheses are used to represent multiplication like (2)(5). An operation is not within parentheses like it is in (2 × 5). Then, any exponents must be computed. Next, multiplication and division are performed from left to right.

Finally, addition and subtraction are performed from left to right. The following is an example in which the operations within the parentheses need to be performed first, so the order of operations must be applied to the exponent, subtraction, addition, and multiplication within the grouping symbol:

$$9-3(3^2-3+4\cdot3)$$

$$9-3(3^2-3+4\cdot3)$$ Work within the parentheses first

$$=9-3(9-3+12)$$

$$=9-3(18)$$

$$=9-54$$

$$=-45$$

Operations can be performed on rational numbers in decimal form. Recall that to write a fraction as an equivalent decimal expression, divide the numerator by the denominator. For example:

$$\frac{1}{8}=1\div8=0.125$$

With the case of decimals, it is important to keep track of place value. To add decimals, make sure the decimal places are in alignment so that the numbers are lined up with their decimal points and add vertically. If the numbers do not line up because there are extra or missing place values in one of the numbers, then zeros may be used as placeholders. For example, 0.123 + 0.23 becomes:

$$
\begin{array}{r}
0.123 \\
+\ 0.230 \\
\hline
0.353
\end{array}
$$

Subtraction is done the same way. Multiplication and division are more complicated. To multiply two decimals, place one on top of the other as in a regular multiplication process and do not worry about lining up the decimal points. Then, multiply as with whole numbers, ignoring the decimals. Finally, in the solution, insert the decimal point as many places to the left as there are total decimal values in the original problem. Here is an example of a decimal multiplication:

$$
\begin{array}{rl}
0.52 & \textit{2 decimal places} \\
\times \quad 0.2 & \textit{1 decimal place} \\
\hline
0.104 & \textit{3 decimal places}
\end{array}
$$

The answer to 52 times 2 is 104, and because there are three decimal values in the problem, the decimal point is positioned three units to the left in the answer.

The decimal point plays an integral role throughout the whole problem when dividing with decimals. First, set up the problem in a long division format. If the divisor is not an integer, the decimal must be moved to the right as many units as needed to make it an integer. The decimal in the dividend must be moved to the right the same number of places to maintain equality. Then, division is completed normally. Below is an example of long division with decimals. The problem is $12.72 \div 0.06$:

Long division with decimals

$$
\begin{array}{r}
212 \\
6\,\overline{)\,1272} \\
\underline{12} \downarrow \\
07 \\
\underline{6} \\
12
\end{array}
$$

Because the decimal point is moved two units to the right in the divisor of 0.06 to turn it into the integer 6, it is also moved two units to the right in the dividend of 12.72 to make it 1,272. The result is 212, and remember that a division problem can always be checked by multiplying the answer times the divisor to see if the result is equal to the dividend.

Sometimes it is helpful to round answers that are in decimal form. First, find the place to which the rounding needs to be done. Then, look at the digit to the right of it. If that digit is 4 or less, the number in the place value to its left stays the same, and everything to its right becomes a 0. This process is known as **rounding down.** If that digit is 5 or higher, round up by increasing the place value to its left by 1, and every number to its right becomes a 0. If those 0's are in decimals, they can be dropped.

For example, 0.145 rounded to the nearest hundredth place would be rounded up to 0.15, and 0.145 rounded to the nearest tenth place would be rounded down to 0.1.

Another operation that can be performed on rational numbers is the **square root**. Dealing with real numbers only, the **positive square root** of a number is equal to one of the two repeated positive factors of that number. For example, $\sqrt{49} = \sqrt{7 \times 7} = 7$. A **perfect square** is a number that has a whole number as its square root. Examples of perfect squares are 1, 4, 9, 16, 25, etc. If a number is not a perfect square, an approximation can be used with a calculator. For example, $\sqrt{67} = 8.185$, rounded to the nearest thousandth place. The square root of a fraction involving perfect squares involves breaking up the problem into the square root of the numerator separate from the square root of the denominator.

For example:

$$\sqrt{\frac{16}{25}} = \frac{\sqrt{16}}{\sqrt{25}} = \frac{4}{5}$$

If the fraction does not contain perfect squares, a calculator can be used. Therefore, $\sqrt{\frac{2}{5}} = 0.632$, rounded to the nearest thousandth place. A common application of square roots involves the **Pythagorean theorem**. Given a right triangle, the sum of the squares of the two legs equals the square of the hypotenuse.

For example, consider the following right triangle:

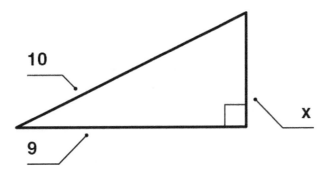

The missing side, x, can be found using the Pythagorean theorem.

$$9^2 + x^2 = 10^2$$

$$81 + x^2 = 100$$

$$x^2 = 19$$

To solve for x, take the square root of both sides. Therefore, $x = \sqrt{19} = 4.36$, which has been rounded to two decimal places.

In addition to the square root, the cube root is another operation. If a number is a **perfect cube**, the cube root of that number is equal to one of the three repeated factors. For example:

$$\sqrt[3]{27} = \sqrt[3]{3 \times 3 \times 3} = 3$$

Also, unlike square roots, a negative number has a *cube root*. The result is a negative number. For example:

$$\sqrt[3]{-27} = \sqrt[3]{(-3)(-3)(-3)} = -3$$

Similar to square roots, if the number is not a perfect cube, a calculator can be used to find an approximation. Therefore, $\sqrt[3]{\frac{2}{3}} = 0.873$, rounded to the nearest thousandth place.

Higher-order roots also exist. The number relating to the root is known as the **index.** Given the following root, $\sqrt[3]{64}$, 3 is the index, and 64 is the **radicand.** The entire expression is known as the **radical.** Higher-order roots exist when the index is larger than 3. They can be broken up into two groups: even and odd roots. **Even roots**, when the index is an even number, follow the properties of square roots. A negative number does not have an even root, and an even root is found by finding the single factor that is repeated the same number of times as the index in the radicand. For example, the fifth root of 32 is equal to 2 because:

$$\sqrt[5]{32} = \sqrt[5]{2 \times 2 \times 2 \times 2 \times 2} = 2$$

Odd roots, when the index is an odd number, follow the properties of cube roots. A negative number has an odd root. Similarly, an odd root is found by finding the single factor that is repeated that many times to obtain the radicand. For example, the 4th root of 81 is equal to 3 because $3^4 = 81$. This radical is written as:

$$\sqrt[4]{81} = 4$$

When performing operations in rational numbers, sometimes it might be helpful to round the numbers in the original problem to get a rough check of what the answer should be. For example, if you walked into a grocery store and had a $20 bill, your approach might be to round each item to the nearest dollar and add up all the items to make sure that you will have enough money when you check out. This process involves obtaining an estimation of what the exact total would be. In other situations, it might be helpful to round to the nearest $10 amount or $100 amount. **Front-end rounding** might be helpful as well in many situations. In this type of rounding, each number is rounded to the highest possible place value. Therefore, all digits except the first digit become 0. Consider a situation in which you are at the furniture store and want to estimate your total on three pieces of furniture that cost $434.99, $678.99, and $129.99. Front-end rounding would round these three amounts to $400, $700, and $100. Therefore, the estimate of your total would be $400 + $700 + $100 = $1,200, compared to the exact total of $1,243.97. In this situation, the estimate is not that far off the exact answer.

Rounding is useful in both approximating an answer when an exact answer is not needed and for comparison when an exact answer is needed. For instance, if you had a complicated set of operations to complete and your estimate was $1,000, if you obtained an exact answer of $100,000, something is off. You might want to check your work to see if a mistake was made because an estimate should not be that different from an exact answer. Estimates can also be helpful with square roots. If a square root of a number is not known, the closest perfect square can be found for an approximation. For example, $\sqrt{50}$ is not equal to a whole number, but 50 is close to 49, which is a perfect square, and $\sqrt{49} = 7$. Therefore, $\sqrt{50}$ is a little bit larger than 7. The actual approximation, rounded to the nearest thousandth, is 7.071.

Various Strategies and Algorithms Used to Perform Operations on Rational Numbers
As mentioned, **rational numbers** are any numbers that can be written as a fraction of integers. Operations to be performed on rational numbers include adding, subtracting, multiplying, and dividing. Essentially, this refers to performing these operations on fractions. Adding and subtracting fractions must be completed by first finding the least common denominator. For example, the problem $\frac{3}{5} + \frac{6}{7}$ requires that the common multiple be found between 5 and 7. The smallest number that divides evenly by 5 and 7 is 35. For the denominators to become 35, they must be multiplied by 7 and 5 respectively. The fraction $\frac{3}{5}$ can be multiplied by 7 on the top and bottom to yield the fraction $\frac{21}{35}$. The fraction $\frac{6}{7}$ can be multiplied by 5 to yield the fraction $\frac{30}{35}$. Now that the fractions have the same denominator, the numerators can be added. The answer to the addition problem becomes:

$$\frac{3}{5} + \frac{6}{7} = \frac{21}{35} + \frac{30}{35} = \frac{41}{35}$$

The same technique can be used for subtraction of rational numbers. The operations multiplication and division may seem easier to perform because finding common denominators is unnecessary. If the problems reads $\frac{1}{3} \times \frac{4}{5}$, then the numerators and denominators are multiplied by each other and the answer is found to be $\frac{4}{15}$. For division, the problem must be changed to multiplication before performing operations. The following words can be used to remember to leave, change, and flip before multiplying. If the problems reads $\frac{3}{7} \div \frac{3}{4}$, then the first fraction is *left* alone, the operation is *changed* to multiplication, and then the last fraction is *flipped*. The problem becomes:

$$\frac{3}{7} \times \frac{4}{3} = \frac{12}{21}$$

Other rational numbers include negative numbers, where two may be added. When two negative numbers are added, the result is a negative number with an even greater magnitude. When a negative number is added to a positive number, the result depends on the value of each addend. For example, $-4 + 8 = 4$ because the positive number is larger than the negative number. For multiplying two negative numbers, the result is positive. For example, $-4 \times -3 = 12$, where the negatives cancel out and yield a positive answer.

Rewriting Expressions Involving Radicals and Rational Exponents

The number 8 is rational because it can be expressed as a fraction also, $\frac{8}{1} = 8$. **Rational exponents** represent one way to show how roots are used to express multiplication of any number by itself. For example, 3^2 has a base of 3 and rational exponent of 2, or $\frac{2}{1}$. It can be rewritten as the square root of 3 raised to the first power, or $\sqrt[2]{3^1}$. Any number with a rational exponent can be written this way. The **numerator,** or number on top of the fraction, becomes the root and the **denominator,** or bottom number of the fraction, becomes the whole number exponent. Another example is $4^{\frac{3}{2}}$. It can be rewritten as the square root of four to the third power, or $\sqrt[2]{4^3}$. This can be simplified by performing the operations 4 to the third power, $4^3 = 4 \times 4 \times 4 = 64$, and then taking the square root of 64, $\sqrt[2]{64}$, which yields an answer of 8. Another way of stating the answer would be 4 to power $\frac{3}{2}$ is eight, or 4 times itself $\frac{3}{2}$ times is eight.

The nth root of a is given as $\sqrt[n]{a}$, which is called a **radical.** Typical values for n are 2 and 3, which represent the square and cube roots. In this form, n represents an integer greater than or equal to 2, and a is a real number. If n is even, a must be nonnegative, and if n is odd, a can be any real number. This radical can be written in exponential form as $a^{\frac{1}{n}}$. Therefore, $\sqrt[4]{15}$ is the same as $15^{\frac{1}{4}}$ and $\sqrt[3]{-5}$ is the same as $(-5)^{\frac{1}{3}}$.

In a similar fashion, the nth root of a can be raised to a power m, which is written as $\left(\sqrt[n]{a}\right)^m$. This expression is the same as $\sqrt[n]{a^m}$. For example:

$$\sqrt[2]{4^3} = \sqrt[2]{64} = 8 = \left(\sqrt[2]{4}\right)^3 = 2^3$$

Because $\sqrt[n]{a} = a^{\frac{1}{n}}$, both sides can be raised to an exponent of m, resulting in:

$$\left(\sqrt[n]{a}\right)^m = \sqrt[n]{a^m} = a^{\frac{m}{n}}$$

This rule allows:

$$\sqrt[2]{4^3} = \left(\sqrt[2]{4}\right)^3 = 4^{\frac{3}{2}} = (2^2)^{\frac{3}{2}} = 2^{\frac{6}{2}} = 2^3 = 8$$

Negative exponents can also be incorporated into these rules. Any time an exponent is negative, the base expression must be flipped to the other side of the fraction bar and rewritten with a positive exponent. For instance, $2^{-3} = \frac{1}{2^3} = \frac{1}{8}$. Therefore, two more relationships between radical and exponential expressions are:

$$a^{-\frac{1}{n}} = \frac{1}{\sqrt[n]{a}} \text{ and } a^{-\frac{m}{n}} = \frac{1}{\sqrt[n]{a^m}} = \frac{1}{\left(\sqrt[n]{a}\right)^m}$$

Thus:

$$8^{-3} = \frac{1}{\sqrt[3]{8}} = \frac{1}{2}$$

All of these relationships are very useful when simplifying complicated radical and exponential expressions. If an expression contains both forms, use one of these rules to change the expression to contain either all radicals or all exponential expressions. This process makes the entire expression much easier to work with, especially if the expressions are contained within equations.

Consider the following example: $\sqrt{x} \times \sqrt[4]{x}$. It is written in radical form; however, it can be simplified into one radical by using exponential expressions first. The expression can be written as $x^{\frac{1}{2}} \times x^{\frac{1}{4}}$. It can be combined into one base by adding the exponents as:

$$x^{\frac{1}{2}+\frac{1}{4}} = x^{\frac{3}{4}}$$

Writing this back in radical form, the result is $\sqrt[4]{x^3}$.

112

Solving Problems Using Scientific Notation

Scientific notation is a system used to represent numbers that are very large or very small. Sometimes, numbers are way too big or small to be written out with multiple zeros behind them or in decimal form, so scientific notation is used as a way to express these numbers in a simpler way.

Scientific notation takes the decimal notation and turns it into scientific notation, like the table below:

Decimal Notation	Scientific Notation
5	5×10^0
500	5×10^2
10,000,000	1×10^7
8,000,000,000	8×10^9
-55,000	-5.5×10^4
.00001	10^{-5}

In scientific notation, the decimal is placed after the first digit and all the remaining numbers are dropped. For example, 5 becomes "5.0×10^0." This equation is raised to the zero power because there are no zeros behind the number "5." Always put the decimal after the first number. Let's say we have the number 125,000. We would write this using scientific notation as follows: 1.25×10^5, because to move the decimal from behind "1" to behind "125,000" takes five counts, so we put the exponent "5" behind the "10." As you can see in the table above, the number ".00001" is too cumbersome to be written out each time for an equation, so we would want to say that it is "10^{-5}." If we count from the place behind the decimal point to the number "1," we see that we go backwards 5 places. Thus, the "-5" in the scientific notation form represents 5 places to the right of the decimal.

Reasoning Quantitatively and Using Units to Solve Problems

Converting Within and Between Standard and Metric Systems
When working with dimensions, sometimes the given units don't match the formula, and conversions must be made. **The metric system** has base units of meter for length, kilogram for mass, and liter for liquid volume. This system expands to three places above the base unit and three places below. These places correspond with prefixes with a base of 10.

The following table shows the conversions:

kilo-	hecto-	deka-	base	deci-	centi-	milli-
1,000 times the base	100 times the base	10 times the base		1/10 times the base	1/100 times the base	1/1000 times the base

To convert between units within the metric system, values with a base ten can be multiplied. The decimal can also be moved in the direction of the new unit by the same number of zeros on the number. For example, 3 meters is equivalent to .003 kilometers. The decimal moved three places (the same number of zeros for kilo-) to the left (the same direction from base to kilo-). Three meters is also equivalent to 3,000 millimeters. The decimal is moved three places to the right because the prefix milli- is three places to the right of the base unit.

The English Standard system used in the United States has a base unit of foot for length, pound for weight, and gallon for liquid volume. These conversions aren't as easy as the metric system because they aren't a base ten model. The following table shows the conversions within this system.

Length	Weight	Capacity
1 foot (ft) = 12 inches (in) 1 yard (yd) = 3 feet 1 mile (mi) = 5280 feet 1 mile = 1760 yards	1 pound (lb) = 16 ounces (oz) 1 ton = 2000 pounds	1 tablespoon (tbsp) = 3 teaspoons (tsp) 1 cup (c) = 16 tablespoons 1 cup = 8 fluid ounces (oz) 1 pint (pt) = 2 cups 1 quart (qt) = 2 pints 1 gallon (gal) = 4 quarts

When converting within the English Standard system, most calculations include a conversion to the base unit and then another to the desired unit. For example, take the following problem: 3 *quarts* = ___ *cups*. There is no straight conversion from quarts to cups, so the first conversion is from quarts to pints. There are 2 pints in 1 quart, so there are 6 pints in 3 quarts.

This conversion can be solved as a proportion:

$$\frac{3\ qt}{x} = \frac{1\ qt}{2\ pints}$$

It can also be observed as a ratio 2:1, expanded to 6:3. Then the 6 pints must be converted to cups. The ratio of pints to cups is 1:2, so the expanded ratio is 6:12. For 6 pints, the measurement is 12 cups. This problem can also be set up as one set of fractions to cancel out units. It begins with the given information and cancels out matching units on top and bottom to yield the answer. Consider the following expression:

$$\frac{3\ quarts}{1} \times \frac{2\ pints}{1\ quart} \times \frac{2\ cups}{1\ pint}$$

It's set up so that units on the top and bottom cancel each other out:

$$\frac{3\ \cancel{quarts}}{1} \times \frac{2\ \cancel{pints}}{1\ \cancel{quart}} \times \frac{2\ cups}{1\ \cancel{pint}}$$

The numbers can be calculated as $3 \times 2 \times 2$ on the top and 1 on the bottom. It still yields an answer of 12 cups.

This process of setting up fractions and canceling out matching units can be used to convert between standard and metric systems. A few common equivalent conversions are 2.54 cm = 1 inch, 3.28 feet = 1 meter, and 2.205 pounds = 1 kilogram. Writing these as fractions allows them to be used in conversions. For the fill-in-the-blank problem 5 meters = ___ feet, an expression using conversions starts with the expression $\frac{5\ meters}{1} \times \frac{3.28\ feet}{1\ meter}$, where the units of meters will cancel each other out, and the final unit is feet. Calculating the numbers yields 16.4 feet. This problem only required two fractions. Others may require longer expressions, but the underlying rule stays the same. When there's a unit on the top of the fraction that's the same as the unit on the bottom, then they cancel each other out. Using this logic and the conversions given above, many units can be converted between and within the different systems.

The conversion between Fahrenheit and Celsius is found in a formula:

$$°C = (°F - 32) \times \frac{5}{9}$$

For example, to convert 78°F to Celsius, the given temperature would be entered into the formula:

$$°C = (78 - 32) \times \frac{5}{9}$$

Solving the equation, the temperature comes out to be 25.56°C. To convert in the other direction, the formula becomes:

$$°F = °C \times \frac{9}{5} + 32$$

Remember the order of operations when calculating these conversions.

Applying Estimation Strategies and Rounding Rules to Real-World Problems
Sometimes it is helpful to find an estimated answer to a problem rather than working out an exact answer. An **estimation** might be much quicker to find, and given the scenario, an estimation might be all that is required. For example, if Aria goes grocery shopping and has only a $100 bill to cover all of her purchases, it might be appropriate for her to estimate the total of the items she is purchasing to determine if she has enough money to cover them. Also, an estimation can help determine if an answer makes sense. For instance, if an answer in the 100s is expected, but the result is a fraction less than 1, something is probably wrong in the calculation.

The first type of estimation involves rounding. As mentioned, **rounding** consists of expressing a number in terms of the nearest decimal place like the tenth, hundredth, or thousandth place, or in terms of the nearest whole number unit like tens, hundreds, or thousands place. When rounding to a specific place value, look at the digit to the right of the place. If it is 5 or higher, round the number to its left up to the next value, and if it is 4 or lower, keep that number at the same value. For instance, 1,654.2674 rounded to the nearest thousand is 2,000, and the same number rounded to the nearest thousandth is 1,654.267. Rounding can be used in the scenario when grocery totals need to be estimated. Items can be rounded to the nearest dollar. For example, a can of corn that costs $0.79 can be rounded to $1.00, and then all other items can be rounded in a similar manner and added together.

When working with larger numbers, it might make more sense to round to higher place values. For example, when estimating the total value of a dealership's car inventory, it would make sense to round the car values to the nearest thousands place. The price of a car that is on sale for $15,654 can be estimated at $16,000. All other cars on the lot could be rounded in the same manner, and then their sum can be found. Depending on the situation, it might make sense to calculate an over-estimate. For example, to make sure Aria has enough money at the grocery store, rounding up every time for each item would ensure that she will have enough money when it comes time to pay. A $.40 item rounded up to $1.00 would ensure that there is a dollar to cover that item. Traditional rounding rules would round $0.40 to $0, which does not make sense in this particular real-world setting. Aria might not have a dollar available at checkout to pay for that item if she uses traditional rounding. It is up to the customer to decide the best approach when estimating.

Estimating is also very helpful when working with measurements. Bryan is updating his kitchen and wants to retile the floor. Again, an over-measurement might be useful. Also, rounding to nearest half-unit might be helpful. For instance, one side of the kitchen might have an exact measurement of 14.32 feet, and the most useful measurement needed to buy tile could be estimating this quantity to be 14.5 feet. If the kitchen was rectangular and the other side measured 10.9 feet, Bryan might round the other side to 11 feet. Therefore, Bryan would find the total tile necessary according to the following area calculation: $14.5 \times 11 = 159.5$ square feet. To make sure he purchases enough tile, Bryan would probably want to purchase at least 160 square feet of tile. This is a scenario in which an estimation might be more useful than an exact calculation. Having more tile than necessary is better than having an exact amount, in case any tiles are broken or otherwise unusable.

Finally, estimation is helpful when exact answers are necessary. Consider a situation in which Sabina has many operations to perform on numbers with decimals, and she is allowed a calculator to find the result. Even though an exact result can be obtained with a calculator, there is always a possibility that Sabina could make an error while inputting the data. For example, she could miss a decimal place, or misuse a parenthesis, causing a problem with the actual order of operations. In this case, a quick estimation at the beginning would be helpful to make sure the final answer is given with the correct number of units. Sabina has to find the exact total of 10 cars listed for sale at the dealership. Each price has two decimal places included to account for both dollars and cents. If one car is listed at $21, 234.43 but Sabina incorrectly inputs into the calculator the price of $2,123.443, this error would throw off the final sum by almost $20,000. A quick estimation at the beginning, by rounding each price to the nearest thousands place and finding the sum of the prices, would give Sabina an amount to compare the exact amount to. This comparison would let Sabina see if an error was made in her exact calculation.

Choosing a Level of Accuracy Appropriate to Limitations on Measurement

Accuracy is defined to be the closeness of a given measurement to the actual measurement of a certain object. The accuracy of a measurement depends on both the tools used to obtain the measurement and the units that are provided. For example, if you have a scale at home that does not utilize any decimal values, it actually rounds your weight to the nearest whole number. Therefore, it might say that you weigh 125 pounds, when your true weight is 125.4 pounds. The desired accuracy depends on the situation. For instance, you might not care that your weight is rounded to the nearest whole number. However, if you are using that scale to weigh your cat and it states he weighs 8 pounds, but he actually weighs 8.4 pounds, a much larger disparity percentage-wise exists between the measurements. The difference between the two weights has to do with precision. **Precision** is defined to be the level of exactness of measurements, or how closely they are to one another. For example, if you step on the scale numerous times and you get weights of 125, 127, and 123 pounds, the scale is not that precise. It is important to note that measurements can be precise, but not accurate. Imagine using a scale that is improperly calibrated. You might weigh yourself five times and get 140 pounds every time. Therefore, the measurements are precise. However, if your true weight is 125 pounds, the measurements you took were not very accurate.

A way to discuss accuracy in more detail is to use both absolute and relative accuracy. **Absolute accuracy** is the absolute value of the difference between an exact measurement and the approximate measurement. For instance, in the examples above, the absolute accuracy was 0.4 because both the actual and measured amounts differed by 0.4 pound. The **relative accuracy** takes into consideration the size of the object and is equal to the absolute accuracy divided by the exact value, turned into a percent. Therefore, the relative accuracy of your measurement was:

$$\frac{0.4}{125.4} = 0.32\%$$

The relative accuracy of your cat's measurement was:

$$\frac{0.4}{8.4} = 4.77\%$$

Solving Multi-Step Real-World and Mathematical Problems Involving Rational Numbers and Proportional Relationships

Solving Real-World Problems Involving Ratios and Rates of Change

Recall that a ratio is the comparison of two different quantities. Comparing 2 apples to 3 oranges results in the ratio 2:3, which can be expressed as the fraction $\frac{2}{3}$. Note that order is important when discussing ratios. The number mentioned first is the **numerator**, and the number mentioned second is the **denominator**. The ratio 2:3 does not mean the same quantity as the ratio 3:2. Also, it is important to make sure than when discussing ratios that have units attached to them, the two quantities use the same units. For example, to think of 8 feet to 4 yards, it would make sense to convert 4 yards to feet by multiplying by 3. Therefore, the ratio would be 8 feet to 12 feet, which can be expressed as the fraction $\frac{8}{12}$. Also, note that it is proper to refer to ratios in lowest terms. Therefore, the ratio of 8 feet to 4 yards is equivalent to the fraction $\frac{2}{3}$. Many real-world problems involve ratios. Often, problems with ratios involve proportions, as when two ratios are set equal to find the missing amount. However, some problems involve deciphering single ratios. For example, consider an amusement park that sold 345 tickets last Saturday. If 145 tickets were sold to adults and the rest of the tickets were sold to children, what would the ratio of the number of adult tickets to children's tickets be? A common mistake would be to say the ratio is 145:345. However, 345 is the total number of tickets sold. There were $345 - 145 = 200$ tickets sold to children. The correct ratio of adult to children's tickets is 145:200. As a fraction, this expression is written as $\frac{145}{200}$, which can be reduced to $\frac{29}{40}$.

While a ratio compares two measurements using the same units, rates compare two measurements with different units. Examples of rates would be $200 of work for 8 hours, or 500 miles per 20 gallons. Because the units are different, it is important to always include the units when discussing rates. Rates can be easily seen because if they are expressed in words, the two quantities are usually split up using one of the following words: *for, per, on, from, in.* Just as with ratios, it is important to write rates in lowest terms. A common rate that can be found in many real-life situations is cost per unit. This quantity describes how much one item or one unit costs. This rate allows the best buy to be determined, given a couple of different sizes of an item with different costs. For example, if 2 quarts of soup was sold for $3.50 and 3 quarts was sold for $4.60, to determine the best buy, the cost per quart should be found.

$$\frac{\$3.50}{2} = \$1.75 \; per \; quart$$

And

$$\frac{\$4.60}{3} = \$1.53 \; per \; quart$$

Therefore, the better deal would be the 3-quart option.

Rate of change problems involve calculating a quantity per some unit of measurement. Usually the unit of measurement is time. For example, meters per second is a common rate of change. To calculate this measurement, find the amount traveled in meters and divide by total time traveled. The calculation is an average of the speed over the entire time interval. Another common rate of change used in the real world is miles per hour. Consider the following problem that involves calculating an average rate of change in temperature. Last Saturday, the temperature at 1:00 a.m. was 34 degrees Fahrenheit, and at noon, the temperature had increased to 75 degrees Fahrenheit. What was the average rate of change over that time interval? The average rate of change is calculated by finding the change in temperature and dividing by the total hours elapsed. Therefore, the rate of change was equal to:

$$\frac{75 - 34}{12 - 1} = \frac{41}{11} \, degrees \, per \, hour$$

This quantity rounded to two decimal places is equal to 3.72 degrees per hour.

A common rate of change that appears in algebra is the slope calculation. Given a linear equation in one variable, $y = mx + b$, the slope, m, is equal to:

$$\frac{rise}{run} \text{ or } \frac{change \ in \ y}{change \ in \ x}$$

In other words, **slope** is equivalent to the ratio of the vertical and horizontal changes between any two points on a line. The vertical change is known as the **rise**, and the horizontal change is known as the **run**. Given any two points on a line (x_1, y_1) and (x_2, y_2), slope can be calculated with the formula:

$$m = \frac{y_2 - y_1}{x_2 - x_1} = \frac{\Delta y}{\Delta x}$$

Common real-world applications of slope include determining how steep a staircase should be, calculating how steep a road is, and determining how to build a wheelchair ramp.

Many times, problems involving rates and ratios involve proportions. A **proportion** states that two ratios (or rates) are equal. The property of cross products can be used to determine if a proportion is true, meaning both ratios are equivalent. If $\frac{a}{b} = \frac{c}{d}$, then to clear the fractions, multiply both sides times the least common denominator, bd. This results in $ac = cd$, which is equal to the result of multiplying along both diagonals. For example, $\frac{4}{40} = \frac{1}{10}$ grants the cross product $4 \times 10 = 40 \times 1$. $40 = 40$ shows that this proportion is true. Cross products are used when proportions are involved in real-world problems. Consider the following: If 3 pounds of fertilizer will cover 75 square feet of grass, how many pounds are needed for 375 square feet? To solve this problem, a proportion can be set up using two ratios. Let x equal the unknown quantity, pounds needed for 375 feet. Then, the equation found by setting the two given ratios equal to one another is $\frac{3}{75} = \frac{x}{375}$. Cross multiplication gives $3 \times 375 = 75x$. Therefore, $1,125 = 75x$. Divide both sides by 75 to get $x = 15$. Therefore, 15 gallons of fertilizer is needed to cover 75 square feet of grass.

Another application of proportions involves similar triangles. If two triangles have the same measurement as two triangles in another triangle, the triangles are said to be **similar**. If two are the same, the third pair of angles are equal as well because the sum of all angles in a triangle is equal to 180 degrees. Each pair of equivalent angles are known as **corresponding angles**. **Corresponding sides** face the corresponding angles, and it is true that corresponding sides are in proportion.

For example, consider the following set of similar triangles:

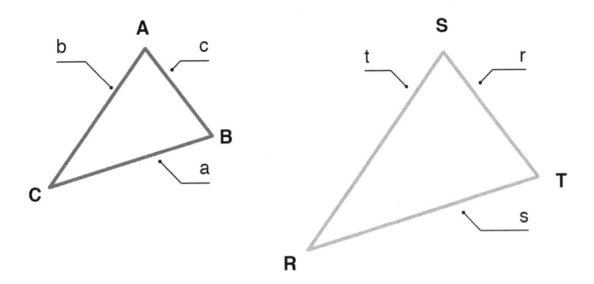

Angles A and R have the same measurement, angles C and T have the same measurement, and angles B and S have the same measurement.

Therefore, the following proportion can be set up from the sides:

$$\frac{c}{t} = \frac{a}{r} = \frac{b}{s}$$

This proportion can be helpful in finding missing lengths in pairs of similar triangles. For example, if the following triangles are similar, a proportion can be used to find the missing side lengths, *a* and *b*.

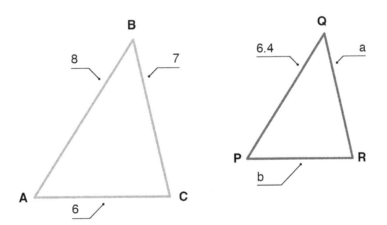

The proportions $\frac{8}{6.4} = \frac{6}{b}$ and $\frac{8}{6.4} = \frac{7}{a}$ can both be cross multiplied and solved to obtain *a* = 5.6 and *b* = 4.8.

A real-life situation that uses similar triangles involves measuring shadows to find heights of unknown objects. Consider the following problem: A building casts a shadow that is 120 feet long, and at the same time, another building that is 80 feet high casts a shadow that is 60 feet long. How tall is the first building?

Each building, together with the sun rays and shadows casted on the ground, forms a triangle. They are similar because each building forms a right angle with the ground, and the sun rays form equivalent angles. Therefore, these two pairs of angles are both equal. Because all angles in a triangle add up to 180 degrees, the third angles are equal as well. Both shadows form corresponding sides of the triangle, the buildings form corresponding sides, and the sun rays form corresponding sides. Therefore, the triangles are similar, and the following proportion can be used to find the missing building length:

$$\frac{120}{x} = \frac{60}{80}$$

Cross-multiply to obtain the cross products, $9600 = 60x$. Then, divide both sides by 60 to obtain $x = 160$. This solution means that the other building is 160 feet high.

Solving Real-World Problems Involving Proportions

Fractions appear in everyday situations, and in many scenarios, they appear in the real-world as ratios and in proportions. A **ratio** is formed when two different quantities are compared. For example, in a group of 50 people, if there are 33 females and 17 males, the ratio of females to males is 33 to 17. This expression can be written in the fraction form, $\frac{33}{17}$, or by using the ratio symbol, 33:17. The order of the number matters when forming ratios. In the same setting, the ratio of males to females is 17 to 33, which is equivalent to $\frac{17}{33}$ or 17:33. A **proportion** is an equation involving two ratios. The equation $\frac{a}{b} = \frac{c}{d}$, or $a:b = c:d$ is a proportion, for real numbers a, b, c, and d. Usually, in one ratio, one of the quantities is unknown, and cross-multiplication is used to solve for the unknown. Consider $\frac{1}{4} = \frac{x}{5}$. To solve for x, cross-multiply to obtain $5 = 4x$. Divide each side by 4 to obtain the solution $x = \frac{5}{4}$. It is also true that percentages are ratios in which the second term is 100. For example:

$$65\% \text{ is } 65:100 \text{ or } \frac{65}{100}$$

Therefore, when working with percentages, one is also working with ratios.

Real-world problems frequently involve proportions. For example, consider the following problem: If 2 out of 50 pizzas are usually delivered late from a local Italian restaurant, how many would be late out of 235 orders? The following proportion would be solved with x as the unknown quantity of late pizzas: $\frac{2}{50} = \frac{x}{235}$. Cross multiplying results in $470 = 50x$. Divide both sides by 50 to obtain $x = \frac{470}{50}$, which in lowest terms is equal to $\frac{47}{5}$. In decimal form, this improper fraction is equal to 9.4. Because it does not make sense to answer this question with decimals (portions of pizzas do not get delivered) the answer must be rounded. Traditional rounding rules would say that 9 pizzas would be expected to be delivered late. However, to be safe, rounding up to 10 pizzas out of 235 would probably make more sense.

Solving Single- and Multistep Problems Involving Percentages

Percentages are defined to be parts per one hundred. To convert a decimal to a percentage, move the decimal point two units to the right and place the percent sign after the number. Percentages appear in many scenarios in the real world. It is important to make sure the statement containing the percentage is translated to a correct mathematical expression. Be aware that it is extremely common to make a mistake when working with percentages within word problems.

An example of a word problem containing a percentage is the following: 35% of people speed when driving to work. In a group of 5,600 commuters, how many would be expected to speed on the way to their place of employment? The answer to this problem is found by finding 35% of 5,600. First, change the percentage to the decimal 0.35. Then compute the product: $0.35 \times 5,600 = 1,960$. Therefore, it would be expected that 1,960 of those commuters would speed on their way to work based on the data given. In this situation, the word "of" signals to use multiplication to find the answer. Another way percentage is used is in the following problem: *Teachers work 8 months out of the year. What percent of the year do they work?* To answer this problem, find what percent *of* 12 the number 8 *is*, because there are 12 months in a year. Therefore, divide 8 by 12, and convert that number to a percentage:

$$\frac{8}{12} = \frac{2}{3} = 0.66\overline{6}$$

The percentage rounded to the nearest tenth place tells us that teachers work 66.7% of the year. Percentages also appear in real-world application problems involving finding missing quantities like in the following question: 60% of what number is 75? To find the missing quantity, an equation can be used. Let x be equal to the missing quantity. Therefore, $0.60x = 75$. Divide each side by 0.60 to obtain 125. Therefore, 60% of 125 is equal to 75.

Sales tax is an important application relating to percentages because tax rates are usually given as percentages. For example, a city might have an 8% sales tax rate. Therefore, when an item is purchased with that tax rate, the real cost to the customer is 1.08 times the price in the store. For example, a $25 pair of jeans costs the customer $25 \times 1.08 = \$27$. Sales tax rates can also be determined if they are unknown when an item is purchased. If a customer visits a store and purchases an item for $21.44, but the price in the store was $19, they can find the tax rate by first subtracting $21.44 - $19 to obtain $2.44, the sales tax amount. The sales tax is a percentage of the in-store price. Therefore, the tax rate is $\frac{2.44}{19} = 0.128$, which has been rounded to the nearest thousandths place. In this scenario, the actual sales tax rate given as a percentage is 12.8%.

Measurement/Geometry

Transformations in the Plane

Two-dimensional figures can undergo various types of transformations in the plane. They can be shifted horizontally and vertically, reflected, compressed, or stretched.

A **shift**, also known as a slide or a translation, moves the shape in one direction. Here is a picture of a shift:

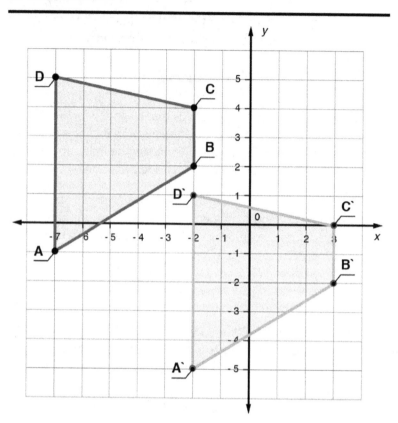

A Translation

Notice that the size of the original shape has not changed at all. If a shift exists within a shape drawn on the **Cartesian coordinate system**, the shift can be represented by adding or subtracting values onto the x, y, or both the x and y coordinates of the points. However, all vertices will move the same number of units because the shape and size of the shape do not change.

A figure can also be **reflected** or flipped, and this transformation involves reflecting over a given line, known as the **line of reflection**. For instance, consider the following picture:

A Reflection Over the Y-Axis

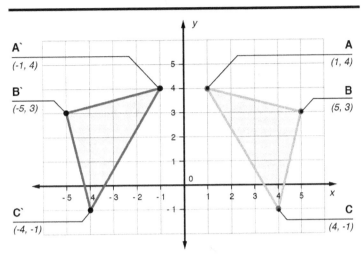

The original shapes have been reflected over the line of reflection and notice again that the shapes remain the same size. If a reflection occurs within the Cartesian coordinate system, the coordinates change. For instance, if a shape gets reflected over the *y*-axis, as above, the *x*-coordinate stays the same, but the *y*-coordinates are made negative. For example, the triangle above starts in the first and second quadrants, but it is reflected over *y*-axis to the third and fourth quadrants. Point A has the initial coordinates of (1, 4), but in the reflection, the point A′ becomes (-1, 4).

Similarly, if the shape is reflected over the *x*-axis, the *y*-coordinate stays the same, but the *x*-coordinates are made negative. For instance, in the graphic below, the point C at (3, 5) becomes C′ at (3, -5).

A Reflection Over the X-Axis

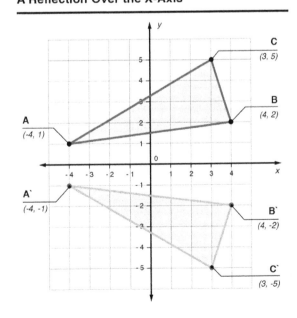

123

Thirdly, a **compression** or a **stretch** of a figure involves changing the size of the original figure, and together they can be classified as a **dilation**. A compression shrinks the size of the figure. We can think about this as a multiplication process by multiplying times a value between 0 and 1. A stretch of a figure results in a figure larger than the original shape. If we consider multiplication, the factor would be greater than 1. Here is a picture of a dilation that is comprised of a stretch in which the original square doubled in size:

A Dilation with a Scale Factor of 2

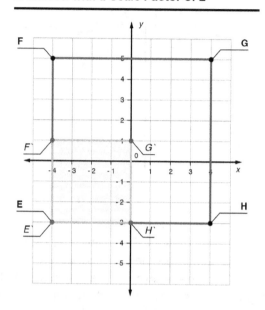

If a shape within the Cartesian coordinate system gets stretched, its coordinates get multiplied by a number greater than 1, and if a shape gets compressed, its coordinates get multiplied by a number between 0 and 1.

A figure can undergo any combination of transformations. For instance, it can be shifted, reflected, and stretched at the same time.

Properties of Polygons and Circles

Shapes are defined by their angles and number of sides. A shape with one continuous side, where all points on that side are equidistant from a center point is called a **circle**. A shape made with three straight line segments is a **triangle**. A shape with four sides is called a **quadrilateral,** but more specifically a **square, rectangle, parallelogram**, or **trapezoid,** depending on the interior angles. These shapes are two-dimensional and only made of straight lines and angles.

Solids can be formed by combining these shapes and forming three-dimensional figures. These figures have another dimension because they add one more direction. Examples of solids may be prisms or spheres.

There are four figures below that can be described based on their sides and dimensions. *Figure 1* is a **cone** because it has three dimensions, where the bottom is a circle and the top is formed by the sides combining to one point. *Figure 2* is a **triangle** because it has two dimensions, made up of three line segments. *Figure 3* is a **cylinder** made up of two base circles and a rectangle to connect them in three dimensions. *Figure 4* is an **oval** because it is one continuous line in two dimensions, not equidistant from the center.

Shapes and Solids

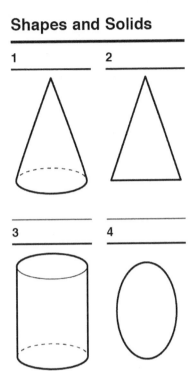

Figure 5 below is made up of squares in three dimensions, combined to make a **cube**. *Figure 6* is a **rectangle** because it has four sides that intersect at right angles. More specifically, it can be described as a **square** because the four sides have equal measures.

Figure 7 is a **pyramid** because the bottom shape is a square and the sides are all triangles. These triangles intersect at a point above the square. **Figure 8** is a **circle** because it is made up of one continuous line where the points are all equidistant from one center point.

Shapes and Solids

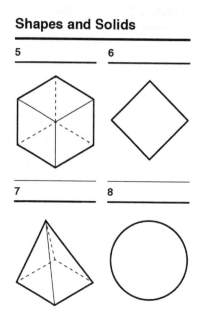

Perimeter and Area

Perimeter and area are two commonly used geometric quantities that describe objects. **Perimeter** is the distance around an object. The perimeter of an object can be found by adding the lengths of all sides. Perimeter may be used in problems dealing with lengths around objects such as fences or borders. It may also be used in finding missing lengths, or working backwards. If the perimeter is given, but a length is missing, subtraction can be used to find the missing length. Given a square with side length s, the formula for perimeter is $P = 4s$. Given a rectangle with length l and width w, the formula for perimeter is $P = 2l + 2w$. The perimeter of a triangle is found by adding the three side lengths, and the perimeter of a trapezoid is found by adding the four side lengths. The units for perimeter are always the original units of length, such as meters, inches, miles, etc. When discussing a circle, the distance around the object is referred to as its **circumference,** not perimeter. The formula for circumference of a circle is $C = 2\pi r$, where r represents the radius of the circle. This formula can also be written as $C = d\pi$, where d represents the diameter of the circle.

Area is the two-dimensional space covered by an object. These problems may include the area of a rectangle, a yard, or a wall to be painted. Finding the area may be a simple formula, or it may require multiple formulas to be used together.

The units for area are square units, such as square meters, square inches, and square miles. Given a square with side length s, the formula for its area is $A = s^2$. Some other common shapes are shown below:

Shape	Formula	Graphic
Rectangle	$Area = length \times width$	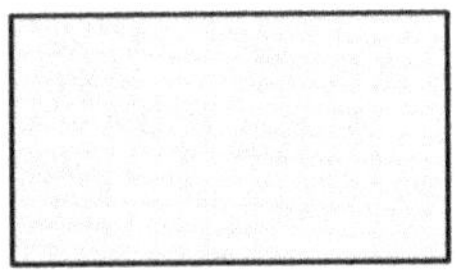
Triangle	$Area = \dfrac{1}{2} \times base \times height$	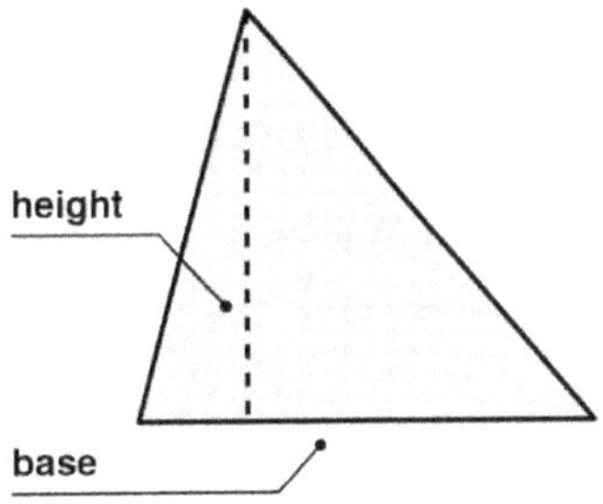
Circle	$Area = \pi \times radius^2$	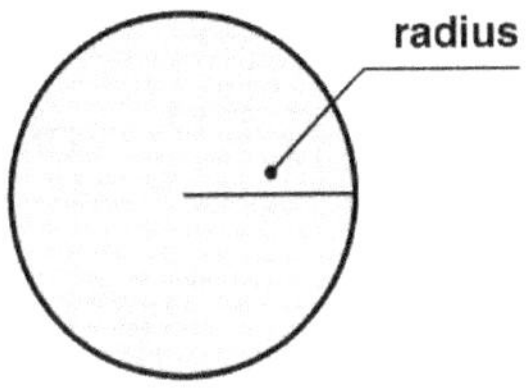

The following formula, not as widely used as those shown in the table but still very important, is the area of a trapezoid:

Area of a Trapezoid

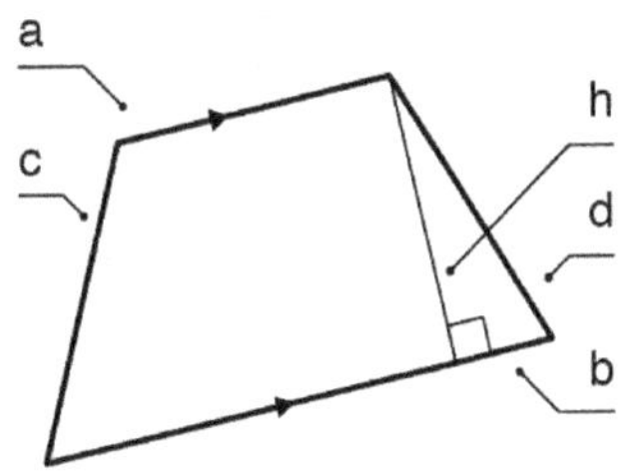

$$A = \frac{1}{2}(a+b)h$$

To find the area of the shapes above, use the given dimensions of the shape in the formula. Complex shapes might require more than one formula. To find the area of the figure below, break the figure into two shapes. The rectangle has dimensions 6 cm by 7 cm. The triangle has dimensions 6 cm by 6 cm. Plug the dimensions into the rectangle formula: $A = 6 \times 7$. Multiplication yields an area of 42 cm². The triangle area can be found using the formula $A = \frac{1}{2} \times 4 \times 6$. Multiplication yields an area of 12 cm². Add the areas of the two shapes to find the total area of the figure, which is 54 cm².

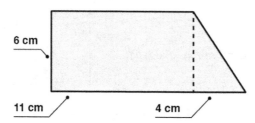

Instead of combining areas, some problems may require subtracting them, or finding the difference.

To find the area of the shaded region in the figure below, determine the area of the whole figure. Then the area of the circle can be subtracted from the whole.

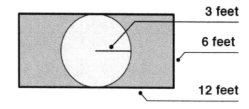

The following formula shows the area of the outside rectangle: $A = 12 \times 6 = 72\ ft^2$. The area of the inside circle can be found by the following formula: $A = \pi(3)^2 = 9\pi = 28.3\ ft^2$. As the shaded area is outside the circle, the area for the circle can be subtracted from the area of the rectangle to yield an area of $43.7\ ft^2$.

While some geometric figures may be given as pictures, others may be described in words. If a rectangular playing field with dimensions 95 meters long by 50 meters wide is measured for perimeter, the distance around the field must be found. The perimeter includes two lengths and two widths to measure the entire outside of the field. This quantity can be calculated using the following equation: $P = 2(95) + 2(50) = 290\ m$. The distance around the field is 290 meters.

With any geometric calculations, it's important to determine what dimensions are given and what quantities the problem is asking for. If a connection can be made between them, the answer can be found.

The Pythagorean Theorem

The **Pythagorean Theorem** is an important relationship between the three sides of a right triangle. It states that the square of hypotenuse is equal to the sum of the squares of the other two sides. When using the Pythagorean Theorem, the **hypotenuse** is labeled as side *c*, the *opposite* is labeled as side *a*, and the *adjacent* side is side *b*.

The theorem can be seen in the following diagram:

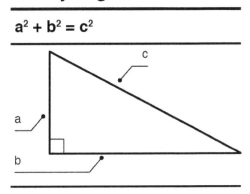

The Pythagorean Theorum

$$a^2 + b^2 = c^2$$

Both the trigonometric ratios and Pythagorean Theorem can be used in problems that involve finding either a missing side or missing angle of a right triangle. Look to see what sides and angles are given and select the correct relationship that will assist in finding the missing value. These relationships can also be used to solve application problems involving right triangles. Often, it is helpful to draw a figure to represent the problem to see what is missing.

Using Congruence and Similarity Criteria for Triangles

Two figures are **congruent** if they have the same shape and same size, meaning same angle measurements and equal side lengths. Two figures are **similar** if they have the same angle measurement but not side lengths. Basically, angles are congruent in similar triangles and their side lengths are constant multiples of each other. Proving two shapes are similar involves showing that all angles are the same; proving two shapes are congruent involves showing that all angles are the same *and* that all sides are the same. If two pairs of angles are congruent in two triangles, then those triangles are similar because their third angle has to be equal due to the fact that all three angles add up to 180 degrees.

There are five main theorems that are used to show triangles are congruent. Each theorem involves showing different combinations of sides and angles are the same in two triangles, which proves the triangles are congruent. The **side-side-side (SSS) theorem** states that if all sides are equal in two triangles, the triangles are congruent. The **side-angle-side (SAS) theorem** states that if two pairs of sides are equal and the included angles are congruent in two triangles, then the triangles are congruent. Similarly, the **angle-side-angle (ASA) theorem** states that if two pairs of angles are congruent and the included side lengths are equal in two triangles, the triangles are similar. The **angle-angle-side (AAS) theorem** states that two triangles are congruent if they have two pairs of congruent angles and a pair of corresponding equal side lengths that are not included. Finally, the **hypotenuse-leg (HL) theorem** states that if two right triangles have equal hypotenuses and an equal pair of shorter sides, the triangles are congruent. An important item to note is that **angle-angle-angle (AAA)** is not enough information to have congruence because if three angles are equal in two triangles, the triangles can only be described as similar.

Using Volume Formulas to Solve Problems

Perimeter and area are two-dimensional descriptions; volume is three-dimensional. **Volume** describes the amount of space that an object occupies, but it's different from area because it has three dimensions instead of two. The units for volume are **cubic units**, such as cubic meters, cubic inches, and cubic miles. Volume can be found by using formulas for common objects such as cylinders and boxes. The following chart shows a diagram and formula for the volume of two objects:

Shape	Formula	Diagram
Rectangular Prism (box)	$V = length \times width \times height$	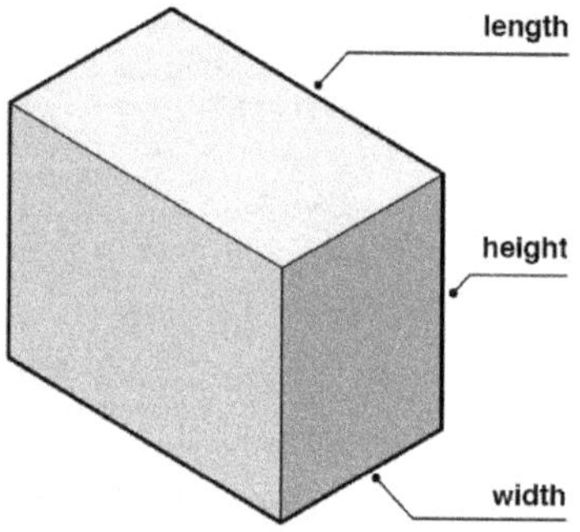
Cylinder	$V = \pi \times radius^2 \times height$	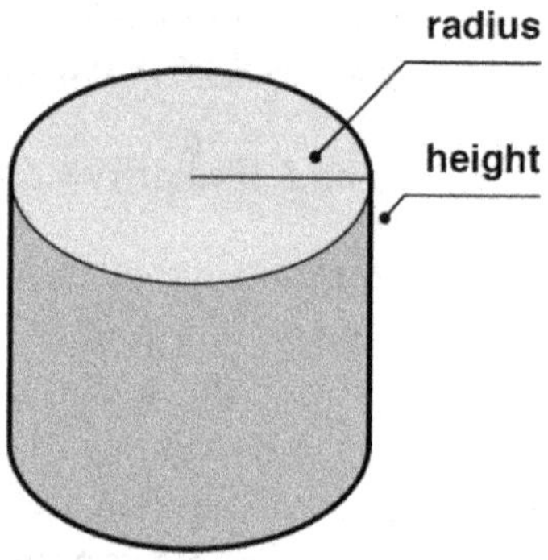

Volume formulas of these two objects are derived by finding the area of the bottom two-dimensional shape, such as the circle or rectangle, and then multiplying times the height of the three-dimensional shape.

Other volume formulas include the volume of a cube with side length:

$$V = s^3$$

the volume of a sphere with radius:

$$V = \frac{4}{3}\pi r^3$$

and the volume of a cone with radius r and height

$$V = \frac{1}{3}\pi r^2 h$$

If a soda can has a height of 5 inches and a radius on the top of 1.5 inches, the volume can be found using one of the given formulas. A soda can is a cylinder. Knowing the given dimensions, the formula can be completed as follows: $V = \pi \,(radius)^2 \times height = \pi \,(1.5)^2 \times 5 = 35.325\ inches^3$. Notice that the units for volume are inches cubed because it refers to the number of cubic inches required to fill the can.

Volume Formulas

Right rectangular prisms are those prisms in which all sides are rectangles and all angles are right, or equal to 90 degrees. The volume for these objects can be found by multiplying the length by the width by the height. The formula is $V = lwh$. For the following prism, the volume formula is:

$$V = 6\frac{1}{2} \times 3 \times 9$$

When dealing with fractional edge lengths, it is helpful to convert the length to an improper fraction. The length 6 ½ cm becomes $\frac{13}{2}$ cm. Then the formula becomes:

$$V = \frac{13}{2} \times 3 \times 9$$

$$\frac{13}{2} \times \frac{3}{1} \times \frac{9}{1} = \frac{351}{2}$$

This value for volume is better understood when turned into a mixed number, which would be 175 ½ cm^3.

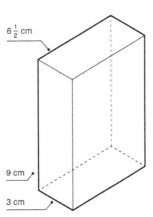

When dimensions for length are given with fractional parts, it can be helpful to turn the mixed number into an improper fraction, then multiply to find the volume, then convert back to a mixed number. When finding surface area, the conversion to improper fractions can also be helpful. The surface area can be found for the same prism above by breaking down the figure into basic shapes. These shapes are rectangles, made up of the two bases, two sides, and the front and back. The formula for the surface area uses the area for each of these shapes for the terms in the following equation:

$$SA = 6\frac{1}{2} \times 3 + 6\frac{1}{2} \times 3 + 3 \times 9 + 3 \times 9 + 6\frac{1}{2} \times 9 + 6\frac{1}{2} \times 9$$

Because there are so many terms in a surface area formula and because this formula contains a fraction, it can be simplified by combining groups that are the same. Each set of numbers is used twice, to represent areas for the opposite sides of the prism.

The formula can be simplified to:

$$SA = 2\left(6\frac{1}{2} \times 3\right) + 2(3 \times 9) + 2\left(6\frac{1}{2} \times 9\right)$$

$$2\left(\frac{13}{2} \times 3\right) + 2(27) + 2\left(\frac{13}{2} \times 9\right)$$

$$2\left(\frac{39}{2}\right) + 54 + 2\left(\frac{117}{2}\right)$$

$$39 + 54 + 117 = 210 \text{ cm}^2$$

Determining How Changes to Dimensions Change Area and Volume

When the dimensions of an object change, the area and volume are also subject to change. For example, the following rectangle has an area of 98 square centimeters ($A = 7 \times 14 = 98\text{cm}^2$). If the length is increased by 2, to be 16 cm, then the area becomes $A = 7 \times 16 = 112\text{cm}^2$. The area increased by 14 cm, or twice the width because there were two more widths of 7 cm.

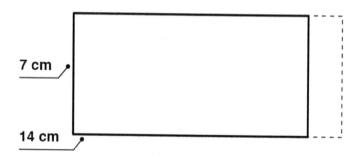

For the volume of an object, there are three dimensions. The given prism has a volume of $V = 4 \times 12 \times 3 = 144\text{m}^2$. If the height is increased by 3, the volume becomes $V = 4 \times 12 \times 6 = 288\text{m}^2$. The increase of 3 for the height, or doubling of the height, resulted in a volume that was doubled. From the original, if the width was doubled, the volume would be $V = 8 \times 12 \times 3 = 288\text{m}^2$. When the width doubled, the volume was doubled also. The same increase in volume would result if the length was doubled.

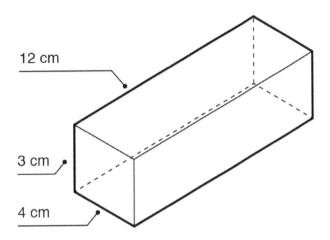

Applying Concepts of Density Based on Area and Volume

Applying Geometric Concepts to Modeling Situations

The **density** of a given type of matter is the ratio of its mass to area or volume. Density deals with the relationship between an object's **mass** and **amount of space** that object takes up. An example of a real-life situation involving density is determining the BTUs of a heating and cooling system for a house. Population density of a neighborhood, city, or country is another real-life example.

The **net of a figure** is the resulting two-dimensional shapes when a three-dimensional shape is broken down. For example, the net of a cone is shown below. The base of the cone is a circle, shown at the bottom. The rest of the cone is a quarter of a circle. The bottom is the circumference of the circle, while the top comes to a point. If the cone is cut down the side and laid out flat, these would be the resulting shapes:

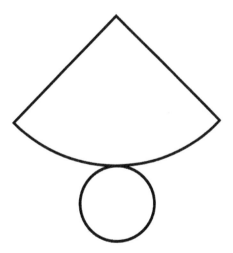

A net for a pyramid is shown in the figure below. The base of the pyramid is a square. The four shapes coming off the pyramid are triangles. When built up together, folding the triangles to the top results in a pyramid:

The Net of a Pyramid

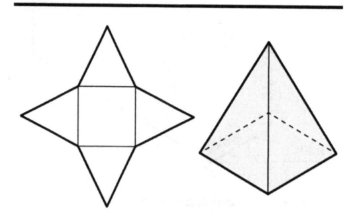

Another example of a net of a figure, in this case a cylinder, is shown below. When the cylinder is broken down, the bases are circles and the side is a rectangle wrapped around the circles. The circumference of the circle turns into the length of the rectangle:

The circumference of the circle is equal to the length of the rectangle

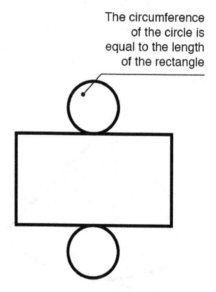

Nets can be used in calculating different values for given shapes. One useful way to calculate surface area is to find the net of the object, then find the area of each of the shapes and add them together. Nets are also useful in composing shapes and decomposing objects so as to view how objects connect and can be used in conjunction with each other.

Using Nets to Determine the Surface Area of Three-Dimensional Figures

Surface area of three-dimensional figures is the total area of each of the faces of the figures. Nets are used to lay out each face of an object. On the following page, the dimensions are labeled for each of the faces of the triangular prism. The area for each of the two triangles can be determined by the formula:

$$A = \frac{1}{2}bh$$

$$\frac{1}{2} \times 8 \times 9 = 36cm^2$$

The rectangle areas can be described by the equation:

$$A = lw = 8 \times 5 + 9 \times 5 + 10 \times 5$$

$$40 + 45 + 50 = 135cm^2$$

The area for the triangles can be multiplied by two, then added to the rectangle areas to yield a total surface area of $207cm^2$.

The following figure shows a triangular prism. The bases are triangles and the sides are rectangles. The second figure shows the net for this triangular prism:

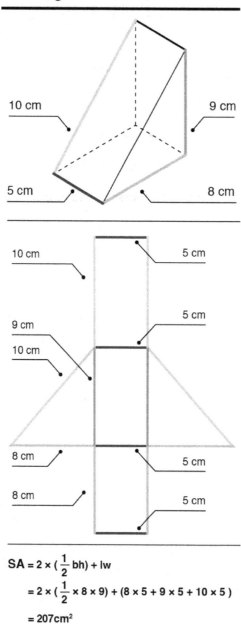

A Triangular Prism and Its Net

10 cm

9 cm

5 cm

8 cm

10 cm

5 cm

5 cm

9 cm

10 cm

8 cm

5 cm

8 cm

5 cm

$$SA = 2 \times (\frac{1}{2} bh) + lw$$

$$= 2 \times (\frac{1}{2} \times 8 \times 9) + (8 \times 5 + 9 \times 5 + 10 \times 5)$$

$$= 207 cm^2$$

Other figures that have rectangles or triangles in their nets include pyramids, rectangular prisms, and cylinders. When the shapes of these three-dimensional objects are found, and areas are calculated, the sum will result in the surface area.

The following picture shows the net for a rectangular prism. The dimensions for each of the shapes that make up the prism are shown to the right. As a formula, the surface area is the sum of each rectangle added together.

The following equation shows the formula:

$$SA = 5 \times 10 + 5 \times 6 + 6 \times 10 + 5 \times 6 + 5 \times 10 + 6 \times 10$$

$$50 + 30 + 60 + 30 + 50 + 60 = 280\text{m}^2$$

A Rectangular Prism and its Net

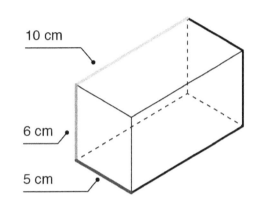

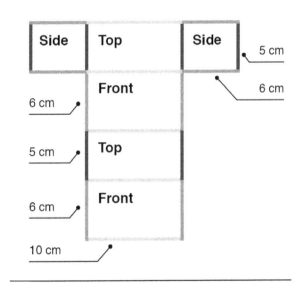

SA = 5×10 + 5×6 + 6×10 + 5×6 + 5×10 + 6×10

= 50 + 30 + 60 + 30 + 50 + 60

= 280cm²

Solving Problems Involving Angles

Geometric figures can be identified by matching the definition with the object. For example, a **line segment** is made up of two endpoints and the line drawn between them. A **ray** is made up of one endpoint and one extending side that goes on forever. A **line** has no endpoints and two sides that extend on forever. These three geometric figures are shown below. What happens at A and B determines the name of each figure.

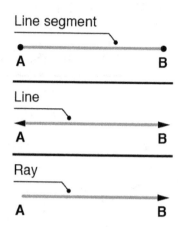

Parallel and perpendicular lines are made up of two lines, like the second figure above. They are distinguished from each other by how the two lines interact. **Parallel** lines run alongside one another, but they never intersect. **Perpendicular** lines intersect at a 90-degree, or a right, angle. An example of these two sets of lines is shown below. Also shown in the figure are non-examples of these two types of lines. Because the first set of lines, in the top left corner, will eventually intersect if they continue, they are not parallel. In the second set, the lines run in the same direction and will never intersect, making them parallel. The third set, in the bottom left corner, intersect at an angle that is not right, or not 90 degrees.

The fourth set is perpendicular because it intersects at exactly a right angle.

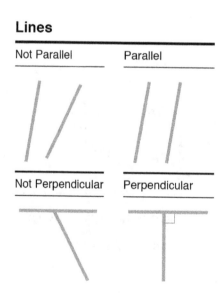

Classifying Angles Based on Their Measure

When two rays are joined together at their endpoints, an **angle** is formed. Angles can be described based on their measure. An angle whose measure is 90 degrees is described as a right angle, just as with perpendicular lines. Ninety degrees is a standard, to which other angles are compared. If an angle is less than ninety degrees, it is an **acute angle**. If it is greater than ninety degrees, it is an **obtuse angle**. If an angle is equal to twice a right angle, or 180 degrees, it is a **straight angle**.

Examples of these types of angles are shown below:

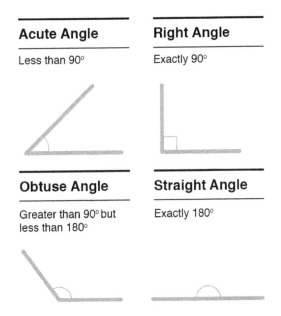

Acute Angle

Less than 90°

Right Angle

Exactly 90°

Obtuse Angle

Greater than 90° but
less than 180°

Straight Angle

Exactly 180°

A **straight angle** is equal to 180 degrees, or a straight line. If the line continues through the **vertex,** or point where the rays meet, and does not change direction, then the angle is straight. This is shown in Figure 1 below. The second figure shows an obtuse angle. Its measure is greater than ninety degrees, but less than that of a straight angle. An estimate for its measure may be 175 degrees. Figure 3 shows an acute angle because it is just less than that of a right angle. Its measure may be estimated to be 80 degrees.

The last image, Figure 4, shows another acute angle. This measure is much smaller, at approximately 35 degrees, but it is still classified as acute because it is between zero and 90 degrees.

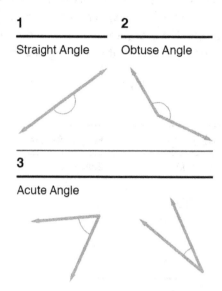

Solving for Missing Values in Triangles, Circles, and Other Figures

Other geometric quantities can include angles inside a triangle. The sum of the measures of three angles in any triangle is 180 degrees. Therefore, if only two angles are known inside a triangle, the third can be found by subtracting the sum of the two known quantities from 180. Two angles whose sum is equal to 90 degrees are known as **complementary angles.** For example, angles measuring 72 and 18 degrees are complementary, and each angle is a complement of the other. Finally, two angles whose sum is equal to 180 degrees are known as **supplementary angles.** To find the supplement of an angle, subtract the given angle from 180 degrees. For example, the supplement of an angle that is 50 degrees is $180 - 50 = 130$ degrees.

These terms involving angles can be seen in many types of word problems. For example, consider the following problem: The measure of an angle is 60 degrees less than two times the measure of its complement. What is the angle's measure? To solve this, let x be the unknown angle. Therefore, its complement is $90 - x$. The problem gives that $x = 2(90 - x) - 60$. To solve for x, distribute the 2, and collect like terms. This process results in $x = 120 - 2x$. Then, use the addition property to add $2x$ to both sides to obtain $3x = 120$. Finally, use the multiplication properties of equality to divide both sides by 3 to get $x = 40$. Therefore, the angle measures 40 degrees. Also, its complement measures 50 degrees.

Solving for missing values in shapes requires knowledge of the shape and its characteristics. For example, a triangle has three sides and three angles that add up to 180 degrees. If two angle measurements are given, the third can be calculated. For the triangle below, the one given angle has a measure of 55 degrees. The missing angle is x. The third angle is labeled with a square, which indicates a measure of 90 degrees. Because all angles must sum to 180 degrees, the following equation can be used to find the missing x-value: $55 + 90 + x = 180$. Adding the two given angles and subtracting the total from 180, the missing angle is found to be 35 degrees.

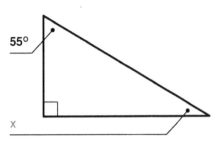

A similar problem can be solved with circles. If the radius is given, but the circumference is unknown, it can be calculated based on the formula $C = 2\pi r$. This example can be used in the figure below. The radius can be substituted for r in the formula. Then the circumference can be found as $C = 2\pi \times 8 = 16\pi = 50.24\ cm$.

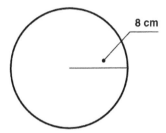

Other figures that may have missing values could be the length of a square, given the area, or the perimeter of a rectangle, given the length and width. All of the missing values can be found by first identifying all the characteristics that are known about the shape, then looking for ways to connect the missing value to the given information.

Data Analysis/Probability/Statistics

Summarizing Data Presented Verbally, Tabularly, and Graphically

Tables, charts, and graphs can be used to convey information about different variables. They are all used to organize, categorize, and compare data, and they all come in different shapes and sizes. Each type has its own way of showing information, whether it is in a column, shape, or picture. To answer a question relating to a table, chart, or graph, some steps should be followed. First, the problem should be read thoroughly to determine what is being asked to determine what quantity is unknown. Then, the title of the table, chart, or graph should be read. The title should clarify what actual data is being summarized in the table. Next, look at the key and both the horizontal and vertical axis labels, if they are given. These items will provide information about how the data is organized. Finally, look to see if there is any more labeling inside the table. Taking the time to get a good idea of what the table is summarizing will be helpful as it is used to interpret information.

Tables are a good way of showing a lot of information in a small space. The information in a table is organized in columns and rows. For example, a table may be used to show the number of votes each candidate received in an election. By interpreting the table, one may observe which candidate won the election and which candidates came in second and third. In using a bar chart to display monthly rainfall amounts in different countries, rainfall can be compared between countries at different times of the year. Graphs are also a useful way to show change in variables over time, as in a line graph, or percentages of a whole, as in a pie graph.

The table below relates the number of items to the total cost. The table shows that 1 item costs $5. By looking at the table further, 5 items cost $25, 10 items cost $50, and 50 items cost $250. This cost can be extended for any number of items. Since 1 item costs $5, then 2 items would cost $10. Though this information isn't in the table, the given price can be used to calculate unknown information.

Number of Items	1	5	10	50
Cost ($)	5	25	50	250

A **bar graph** is a graph that summarizes data using bars of different heights. It is useful when comparing two or more items or when seeing how a quantity changes over time. It has both a horizontal and vertical axis. Interpreting bar graphs includes recognizing what each bar represents and connecting that to the two variables.

The bar graph below shows the scores for six people on three different games. The color of the bar shows which game each person played, and the height of the bar indicates their score for that game. William scored 25 on game 3, and Abigail scored 38 on game 3. By comparing the bars, it's obvious that Williams scored lower than Abigail.

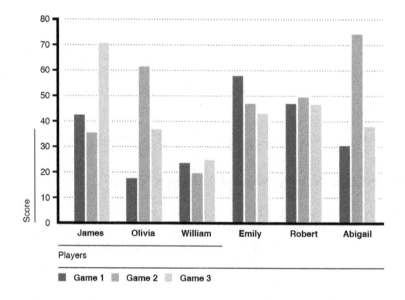

A **line graph** is a way to compare two variables. Each variable is plotted along an axis, and the graph contains both a horizontal and a vertical axis. On a line graph, the line indicates a continuous change. The change can be seen in how the line rises or falls, known as its slope, or rate of change. Often, in line graphs, the horizontal axis represents a variable of time. Readers can quickly see if an amount has grown or decreased over time. The bottom of the graph, or the x-axis, shows the units for time, such as days, hours, months, etc. If there are multiple lines, a comparison can be made between what the two lines represent. For example, the following line graph displays the change in temperature over five days. The top line represents the high, and the bottom line represents the low for each day. Looking at the top line alone, the high decreases for a day, then increases on Wednesday. Then it decreases on Thursday and increases again on Friday. The low temperatures have a similar trend, shown in bottom line. The range in temperatures each day can also be calculated by finding the difference between the top line and bottom line on a particular day. On Wednesday, the range was 14 degrees, from 62 to 76° F.

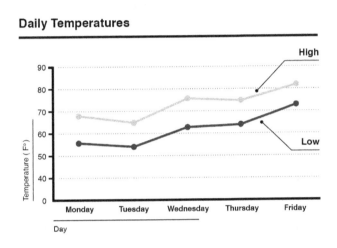

Daily Temperatures

Pie charts are used to show percentages of a whole, as each category is given a piece of the pie, and together all the pieces make up a whole. They are a circular representation of data that are used to highlight numerical proportions. It is true that the arc length of each pie slice is proportional to the amount it individually represents. When a pie chart is shown, a reader can quickly make comparisons by comparing the sizes of the pieces of the pie. They can be useful for comparison between different categories. The following pie chart is a simple example of three different categories shown in comparison to each other.

Light gray represents cats, dark gray represents dogs, and the gray between those two represents other pets. As the pie is cut into three equal pieces, each value represents just more than 33 percent, or $\frac{1}{3}$ of the whole. Values 1 and 2 may be combined to represent $\frac{2}{3}$ of the whole. In an example where the total pie represents 75,000 animals, then cats would be equal to $\frac{1}{3}$ of the total, or 25,000. Dogs would equal 25,000 and other pets would hold equal 25,000.

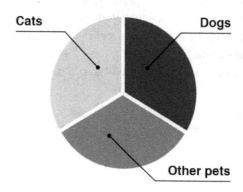

The fact that a circle is 360 degrees is used to create a pie chart. Because each piece of the pie is a percentage of a whole, that percentage is multiplied times 360 to get the number of degrees each piece represents. In the example above, each piece is 1/3 of the whole, so each piece is equivalent to 120 degrees. Together, all three pieces add up to 360 degrees.

Stacked bar graphs are also used fairly frequently when comparing multiple variables at one time. They combine some elements of both pie charts and bar graphs, using the organization of bar graphs and the proportionality aspect of pie charts. The following is an example of a stacked bar graph that represents the number of students in a band playing drums, flute, trombone, and clarinet. Each bar graph is broken up further into girls and boys.

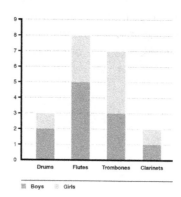

To determine how many boys play trombone, refer to the darker portion of the trombone bar, which indicates 3 boys.

As mentioned, a **scatterplot** is another way to represent paired data. It uses Cartesian coordinates, like a line graph, meaning it has both a horizontal and vertical axis. Each data point is represented as a dot on the graph. The dots are never connected with a line. For example, the following is a scatterplot showing people's age versus height.

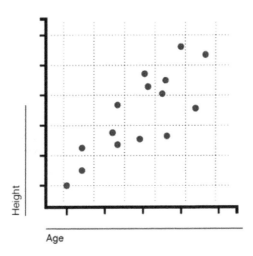

A scatterplot, also known as a **scattergram,** can be used to predict another value and to see if a correlation exists between a set of data. If the data resembles a straight line, the data is **associated** or correlated. The following is an example of a scatterplot in which the data does not seem to have an association:

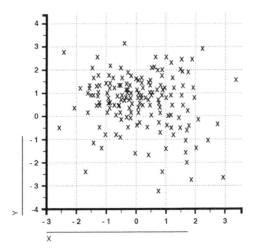

Sets of numbers and other similarly organized data can also be represented graphically. Venn diagrams are a common way to do so. A **Venn diagram** represents each set of data as a circle. The circles overlap, showing that each set of data is overlapping. A Venn diagram is also known as a *logic diagram* because it visualizes all possible logical combinations between two sets. Common elements of two sets are represented by the area of overlap.

The following is an example of a Venn diagram of two sets A and B:

Parts of the Venn Diagram

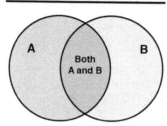

Another name for the area of overlap is the **intersection.** The intersection of A and B, written $A \cap B$, contains all elements that are in both sets A and B. The **union** of A and B, $A \cup B$, contains all elements that are in either set A or set B. Finally, the **complement** of $A \cup B$ is equal to all elements that are not in either set A or set B. These elements are placed outside of the circles.

The following is an example of a Venn diagram in which 22 students were surveyed asking about their siblings. Ten students only had a brother, 7 students only had a sister, and 5 had both a brother and a sister. This number 5 is the intersection and is placed where the circles overlap. Two students did not have a brother or a sister. Therefore, two is the complement and is placed outside of the circles.

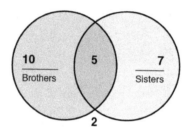

Venn diagrams can have more than two sets of data. The more circles, the more logical combinations are represented by the overlapping. The following is a Venn diagram that represents sock colors worn by a class of students. There were 30 students surveyed. The innermost region represents those students that had green, pink, and blue on their socks (perhaps in a striped pattern). Therefore, 2 students had all three colors. In this example, all students had at least one color on their socks, so no one exists in the complement.

30 students

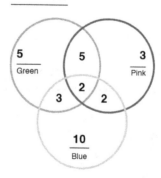

146

Venn diagrams are typically not drawn to scale, but if they are and their area is proportional to the amount of data it represents, it is known as an **area-proportional Venn diagram**.

Recognizing Possible Associations and Trends in the Data
Independent and dependent are two types of variables that describe how they relate to each other. The **independent variable** is the variable controlled by the experimenter. It stands alone and isn't changed by other parts of the experiment. This variable is normally represented by x and is found on the horizontal, or x-axis, of a graph. The **dependent variable** changes in response to the independent variable. It reacts to, or depends on, the independent variable. This variable is normally represented by y and is found on the vertical, or y-axis of the graph.

The relationship between two variables, x and y, can be seen on a scatterplot (as seen on the following page).

The following scatterplot shows the relationship between weight and height. The graph shows the weight as x and the height as y. The first dot on the left represents a person who is 45 kg and approximately 150 cm tall. The other dots correspond in the same way. As the dots move to the right and weight increases, height also increases. A line could be drawn through the middle of the dots to move from bottom left to top right. This line would indicate a **positive correlation** between the variables. If the variables had a **negative correlation**, then the dots would move from the top left to the bottom right.

Height and Weight

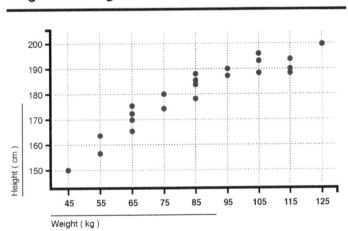

A **scatterplot** is useful in determining the relationship between two variables, but it's not required. Consider an example where a student scores a different grade on his math test for each week of the month. The independent variable would be the weeks of the month. The dependent variable would be the grades, because they change depending on the week. If the grades trended up as the weeks passed, then the relationship between grades and time would be positive. If the grades decreased as the time passed, then the relationship would be negative. (As the number of weeks went up, the grades went down.)

The relationship between two variables can further be described as strong or weak. The relationship between age and height shows a strong positive correlation because children grow taller as they grow up. In adulthood, the relationship between age and height becomes weak, and the dots will spread out. People stop growing in adulthood, and their final heights vary depending on factors like genetics and health. The closer the dots on the graph to the trend line, the stronger the relationship. As they spread apart, the relationship becomes weaker. If they are too spread out to determine a trend (and thus, correlation) up or down, then the variables are said to have no correlation.

Variables are values that change, so determining the relationship between them requires an evaluation of who changes them. If the variable changes because of a result in the experiment, then it's **dependent**. If the variable changes before the experiment, or is changed by the person controlling the experiment, then it's the **independent variable**. As they interact, one is manipulated by the other. The manipulator is the independent, and the manipulated is the dependent. Once the independent and dependent variable are determined, they can be evaluated to have a positive, negative, or no correlation.

Identifying the Line of Best Fit

Data rarely fits into a straight line. Usually, we must be satisfied with approximations and predictions. Typically, when considering linear data with some sort of trend or association, the scatter plot for the data set appears to "fit" a straight line, and this line is known as the **line of best fit**. For instance, consider the following set of data which shows test scores for 10 students in a classroom on a final exam based on the number of hours each student studied: {(8,89), (7,88), (7,78), (7,77), (6,76), (5,76), (4,72), (5,65), (4,64), (3,61)}. Within each ordered pair, the first coordinate is the number of hours studied and the second coordinate is the test score. Plotting these in a scatter plot yields the following:

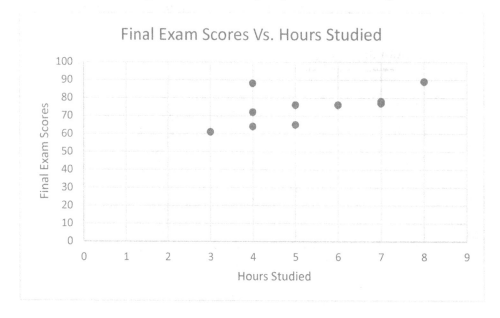

Note that the data does not follow a straight line exactly; however, a straight line can be drawn through the points as shown in the following plot:

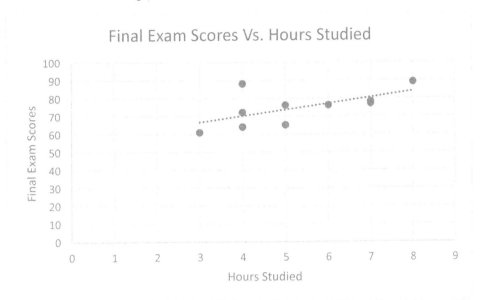

The line that was inserted is the line of best fit and notice that it runs through the middle of the points. It can be found using technology, such as Excel® or a graphing calculator. In our example, Excel® was used for the scatter plot, and that function allowed for us to add a **linear trend line**, which is another name for line of best fit. A **linear regression line** is also another term that is used in place of line of best fit. Once you obtain the line, it can be used to predict other points by looking at ordered pairs that the line runs through. For instance, the plot can be used to predict that a student who studied for 5.5 hours should receive close to a 75 on the final exam.

A good line of fit minimizes the distance from each point to the line, which are known as the **residuals**. A line of fit best fits a data set if those distances are small. For instance, here is a line of fit that has very small residuals because the data points are all relatively close to the line:

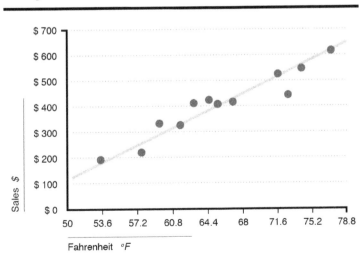

Here is one that has larger residuals:

Backyard Insects

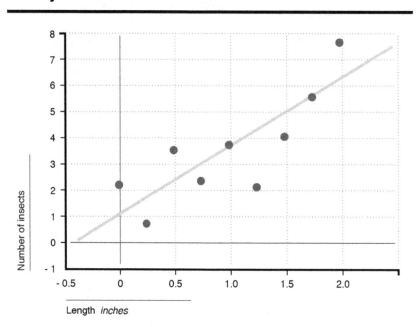

A data set has to follow a linear pattern in order to have a line of best fit. Here is another example of a data set and the corresponding line of best fit:

Example of a Linear Regression or Line of Best Fit

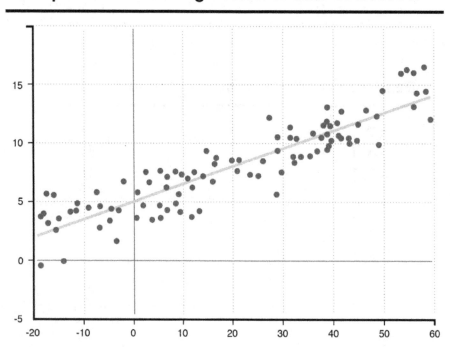

Finding the Probabilities of Single and Compound Events

A **probability experiment** is an action that causes specific results, such as counts or measurements. The result of such an experiment is known as an **outcome**, and the set of all potential outcomes is known as the **sample space**. An **event** consists of one or more of those outcomes. For example, consider the probability experiment of tossing a coin and rolling a six-sided die. The coin has two possible outcomes— a heads or a tails—and the die has six possible outcomes—rolling each number 1–6. Therefore, the sample space has twelve possible outcomes: a heads or a tails paired with each roll of the die.

A **simple event** is an event that consists of a single outcome. For instance, selecting a queen of hearts from a standard fifty-two-card deck is a simple event; however, selecting a queen is not a simple event because there are four possibilities.

Classical, or **theoretical, probability** is when each outcome in a sample space has the same chance to occur. The probability for an event is equal to the number of outcomes in that event divided by the total number of outcomes in the sample space. For example, consider rolling a six-sided die. The probability of rolling a 2 is $\frac{1}{6}$, and the probability of rolling an even number is $\frac{3}{6}$, or $\frac{1}{2}$, because there are three even numbers on the die. This type of probability is based on what should happen in theory but not what actually happens in real life.

Empirical probability is based on actual experiments or observations. For example, if a die is rolled eight times, and a 1 is rolled two times, the empirical probability of rolling a 1 is $\frac{2}{8} = \frac{1}{4}$, which is higher than the theoretical probability. The Law of Large Numbers states that as an experiment is completed repeatedly, the empirical probability of an event should get closer to the theoretical probability of an event.

Probabilities range from 0 to 1. The closer something is to 0, the less likely it will occur. The closer it is to 1, the more likely it is to occur.

The **addition rule** is necessary to find the probability of event A or event B occurring or both occurring at the same time. If events A and B are **mutually exclusive** or **disjoint,** which means they cannot occur at the same time, $P(A \text{ or } B) = P(A) + P(B)$. If events A and B are not mutually exclusive, $P(A \text{ or } B) = P(A) + P(B) - P(A \text{ and } B)$ where $P(A \text{ and } B)$ represents the probability of event A and B both occurring at the same time. An example of two events that are mutually exclusive are rolling a 6 on a die and rolling an odd number on a die. The probability of rolling a 6 or rolling an odd number is:

$$\frac{1}{6} + \frac{3}{6} = \frac{4}{6} = \frac{2}{3}$$

Rolling a 6 and rolling an even number are not mutually exclusive because there is some overlap. The probability of rolling a 6 or rolling an even number is:

$$\frac{1}{6} + \frac{3}{6} - \frac{1}{6} = \frac{3}{6} = \frac{1}{2}$$

Conditional Probability
The **multiplication rule** is necessary when finding the probability that an event A occurs in a first trial and event B occurs in a second trial, which is written as $P(A \text{ and } B)$. This rule differs if the events are independent or dependent. Two events A and B are **independent** if the occurrence of one event does not affect the probability that the other will occur. If A and B are not independent, they are **dependent,** and the outcome of the first event somehow affects the outcome of the second. If events A and B are

independent, $P(A \text{ and } B) = P(A)P(B)$, and if events A and B are dependent, $P(A \text{ and } B) = P(A)P(B|A)$, where $P(B|A)$ represents the probability event B occurs given that event A has already occurred.

$P(B|A)$ represents **conditional probability**, or the probability of event B occurring given that event A has already occurred. $P(B|A)$ can be found by dividing the probability of events A and B both occurring by the probability of event A occurring using the formula $P(B|A) = \frac{P(A \text{ and } B)}{P(A)}$ and represents the total number of outcomes remaining for B to occur after A occurs. This formula is derived from the multiplication rule with dependent events by dividing both sides by $P(A)$. Note that $P(B|A)$ and $P(A|B)$ are not the same. The first quantity shows that event B has occurred after event A, and the second quantity shows that event A has occurred after event B. To incorrectly interchange these ideas is known as **confusion of the inverse.**

Consider the case of drawing two cards from a deck of fifty-two cards. The probability of pulling two queens would vary based on whether the initial card was placed back in the deck for the second pull. If the card is placed back in, the probability of pulling two queens is $\frac{4}{52} \times \frac{4}{52} = 0.00592$. If the card is not placed back in, the probability of pulling two queens is $\frac{4}{52} \times \frac{3}{51} = 0.00452$. When the card is not placed back in, both the numerator and denominator of the second probability decrease by 1. This is due to the fact that, theoretically, there is one less queen in the deck, and there is one less total card in the deck as well.

Conditional probability is used frequently when probabilities are calculated from tables. A **two-way frequency table** displays categorical data with two variables, and it highlights relationships that exist between those two variables. Such tables are used frequently to summarize survey results and are also known as **contingency tables**. Each cell shows a count pertaining to that individual variable paring, known as a **joint frequency**, and the totals of each row and column also are in the table.

Consider the following two-way frequency table:

Distribution of the Residents of a Particular Village

	70 or older	69 or younger	Totals
Women	20	40	60
Men	5	35	40
Total	25	75	100

The table shows the breakdown of ages and sexes of 100 people in a particular village. The total number of people in the data is shown in the bottom right corner. Each total is shown at the end of each row or column, as well. For instance, there were 25 people aged 70 or older and 60 women in the data. The 20 in the first cell shows that out of 100 total villagers, 20 were women aged 70 or older. The 5 in the cell below shows that out of 100 total villagers, 5 were men aged 70 or older.

A two-way table can also show relative frequencies. If instead of the count, the percentage of people in each category was placed into the cells, the two-way table would show relative frequencies. If each frequency is calculated over the entire total of 100, the first cell would be 20% or 0.2. However, the relative frequencies can also be calculated over row or column totals. If row totals were used, the first cell would be:

$$\frac{20}{60} = 0.333 = 33.3\%.$$

If column totals were used, the first cell would be:

$$\frac{20}{25} = 0.8 = 80\%$$

Such tables can be used to calculate conditional probabilities, which are probabilities that an event occurs, given another event. Consider a randomly-selected villager. The probability of selecting a male 70 years old or older is $\frac{5}{100} = 0.05$ because there are 5 males over the age of 70 and 100 total villagers.

Approximating the Probability of a Chance Event

Probability describes how likely it is that an event will occur. Probabilities are always a number from 0 to 1. If an event has a high likelihood of occurrence, it will have a probability close to 1. If there is only a small chance that an event will occur, the likelihood is close to 0. A fair six-sided die has one of the numbers 1, 2, 3, 4, 5, and 6 on each side. When this die is rolled, there is a one in six chance that it will land on 2. This is because there are six possibilities and only one side has a 2 on it. The probability then is $\frac{1}{6}$ or .167. The probability of rolling an even number from this die is three in six, or ½ or .5. This is because there are three sides on the die with even numbers (2, 4, 6), and there are six possible sides. The probability of rolling a number less than 10 is one because every side of the die has a number less than 6, so this is certain to occur. On the other hand, the probability of rolling a number larger than 20 is zero. There are no numbers greater than 20 on the die, so it is certain that this will not occur, thus the probability is zero.

If a teacher says that the probability of anyone passing her final exam is .2, is it highly likely that anyone will pass? No, the probability of anyone passing her exam is low because .2 is closer to 0 than to 1. If another teacher is proud that the probability of students passing his class is .95, how likely is it that a student will pass? It is highly likely that a student will pass because the probability, .95, is very close to 1.

Probability problems require that the total number of events in the sample space must be known. Different methods can be used to count the number of possible outcomes, depending on whether different arrangements of the same items are counted only once or separately. **Permutations** are arrangements in which different sequences are counted separately. Therefore, order matters in permutations. **Combinations** are arrangements in which different sequences are not counted separately. Therefore, order does not matter in combinations. For example, if 123 is considered different from 321, permutations would be discussed. However, if 123 is considered the same as 321, combinations would be considered.

If the sample space contains n different permutations of n different items and all of them must be selected, there are $n!$ different possibilities. For example, five different books can be rearranged $5! = 120$ times. The probability of one person randomly ordering those five books in the same way as another person is $\frac{1}{120}$. A different calculation is necessary if a number less than n is to be selected or if order does not matter. In general, the notation $P(n,r)$ represents the number of ways to arrange r objects from a set of n if order does matter, and:

$$P(n,r) = \frac{n!}{(n-r)!}$$

Therefore, in order to calculate the number of ways five books can be arranged in three slots if order matters, plug $n = 5$ and $r = 3$ in the formula to obtain:

$$P(5,3) = \frac{5!}{(5-3)!} = \frac{5!}{2!} = 60$$

Secondly, $C(n,r)$ represents the total number of r combinations selected out of n items when order does not matter, and:

$$C(n,r) = \frac{n!}{(n-r)!\ r!}$$

Therefore, the number of ways five books can be arranged in three slots if order does not matter is:

$$C(5,3) = \frac{5!}{(5-3)!\ 3!} = 10$$

The following relationship exists between permutations and combinations:

$$C(n,r) = \frac{P(n,r)}{r!}$$

Developing and Using a Probability Model

A **discrete random variable** is a set of values that is either finite or countably infinite. If there are infinitely many values, being *countable* means that each individual value can be paired with a natural number. For example, the number of coin tosses before getting heads could potentially be infinite, but the total number of tosses is countable. Each toss refers to a number, like the first toss, second toss, etc. A **continuous random variable** has infinitely many values that are not countable. The individual items cannot be enumerated; an example of such a set is any type of measurement. There are infinitely many heights of human beings due to decimals that exist within each inch, centimeter, millimeter, etc. Each type of variable has its own **probability distribution**, which calculates the probability for each potential value of the random variable. Probability distributions exist in tables, formulas, or graphs. The expected value of a random variable represents what the mean value should be in either a large sample size or after many trials. According to the Law of Large Numbers, after many trials, the actual mean and that of the probability distribution should be very close to the expected value. The **expected value** is a weighted average that is calculated as $E(X) = \sum x_i p_i$, where x_i represents the value of each outcome, and p_i represents the probability of each outcome.

The expected value if all of the probabilities are equal is:

$$E(X) = \frac{x_1 + x_2 + \cdots + x_n}{n}$$

Expected value is often called the **mean of the random variable** and is known as a **measure of central tendency** like mean and mode.

A **binomial probability distribution** is a probability distribution that adheres to some important criteria. The distribution must consist of a fixed number of trials where all trials are independent, each trial has an outcome classified as either success or failure, and the probability of a success is the same in each trial. Within any binomial experiment, x is the number of resulting successes, n is the number of trials, P is the probability of success within each trial, and $Q = 1 - P$ is the probability of failure within each trial. The probability of obtaining x successes within n trials is:

$$\binom{n}{x} P^x (1 - P)^{n-x}$$

Where:

$$\binom{n}{x} = \frac{n!}{x! \, (n - x)!}$$

is called the **binomial coefficient**. A binomial probability distribution could be used to find the probability of obtaining exactly two heads on five tosses of a coin. In the formula, $x = 2$, $n = 5$, $P = 0.5$, and $Q = 0.5$.

A **uniform probability distribution** exists when there is constant probability. Each random variable has equal probability, and its graph is a rectangle because the height, representing the probability, is constant.

Finally, a **normal probability distribution** has a graph that is symmetric and bell-shaped; an example using body weight is shown here:

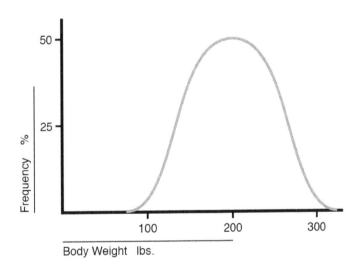

Population percentages can be estimated using normal distributions. For example, the probability that a data point will be less than the mean is 50 percent. The Empirical Rule states that 68 percent of the data falls within 1 standard deviation of the mean, 95 percent falls within 2 standard deviations of the mean, and 99.7 percent falls within 3 standard deviations of the mean. A **standard normal distribution** is a normal distribution with a mean equal to 0 and standard deviation equal to 1. The area under the entire curve of a standard normal distribution is equal to 1.

Using Measures of Central to Draw Inferences About Populations

Information can be interpreted from tables, charts, and graphs through statistics. The three most common calculations for a set of data are the mean, median, and mode. These three are called **measures of central tendency**. Measures of central tendency are helpful in comparing two or more different sets of data. The **mean** refers to the average and is found by adding up all values and dividing the total by the number of values. In other words, the mean is equal to the sum of all values divided by the number of data entries. For example, if you bowled a total of 532 points in 4 bowling games, your mean score was $\frac{532}{4} = 133$ points per game. A common application of mean useful to students is calculating what he or she needs to receive on a final exam to receive a desired grade in a class.

The **median** is found by lining up values from least to greatest and choosing the middle value. If there's an even number of values, then the mean of the two middle amounts must be calculated to find the median. For example, the median of the set of dollar amounts $5, $6, $9, $12, and $13 is $9. The median of the set of dollar amounts $1, $5, $6, $8, $9, $10 is $7, which is the mean of $6 and $8. The **mode** is the value that occurs the most. The mode of the data set {1, 3, 1, 5, 5, 8, 10} actually refers to two numbers: 1 and 5. In this case, the data set is **bimodal** because it has two modes. A data set can have no mode if no amount is repeated. Another useful statistic is range. The **range** for a set of data refers to the difference between the highest and lowest value.

In some cases, some numbers in a list of data might have weights attached to them. In that case, a **weighted mean** can be calculated. A common application of a weighted mean is GPA. In a semester, each class is assigned a number of credit hours, its weight, and at the end of the semester each student receives a grade. To compute GPA, an A is a 4, a B is a 3, a C is a 2, a D is a 1, and an F is a 0. Consider a student that takes a 4-hour English class, a 3-hour math class, and a 4-hour history class and receives all B's. The weighted mean, GPA, is found by multiplying each grade times its weight, number of credit hours, and dividing by the total number of credit hours. Therefore, the student's GPA is:

$$\frac{3 \times 4 + 3 \times 3 + 3 \times 4}{11} = \frac{33}{1} = 3.0$$

The following bar chart shows how many students attend a cycle class on each day of the week. To find the mean attendance for the week, each day's attendance can be added together, $10 + 7 + 6 + 9 + 8 + 14 + 4 = 58$, and the total divided by the number of days, $58 \div 7 = 8.3$. The mean attendance for the week was 8.3 people. The median attendance can be found by putting the attendance numbers in order from least to greatest: 4, 6, 7, 8, 9, 10, 14, and choosing the middle number: 8 people. The mode for attendance is none for this set of data because no numbers repeat. The range is 10, which is found by finding the difference between the lowest number, 4, and the highest number, 14.

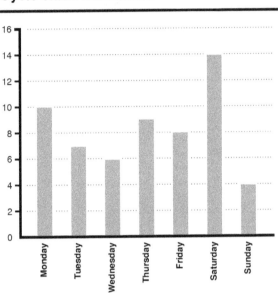

Cycle class attendance

A **histogram** is a bar graph used to group data into "bins" that cover a range on the horizontal, or x-axis. Histograms consist of rectangles whose height is equal to the frequency of a specific category. The horizontal axis represents the specific categories. Because they cover a range of data, these bins have no gaps between bars, unlike the bar graph above. In a histogram showing the heights of adult golden retrievers, the bottom axis would be groups of heights, and the y-axis would be the number of dogs in each range. Evaluating this histogram would show the height of most golden retrievers as falling within a certain range. It also provides information to find the average height and range for how tall golden retrievers may grow. On this histogram on the following page, the horizontal axis represents ranges of the number points scored, and the vertical axis represents the number of students. For example, approximately 33 students scored in the 60 to 70 range.

The following is a histogram that represents exam grades in a given class.

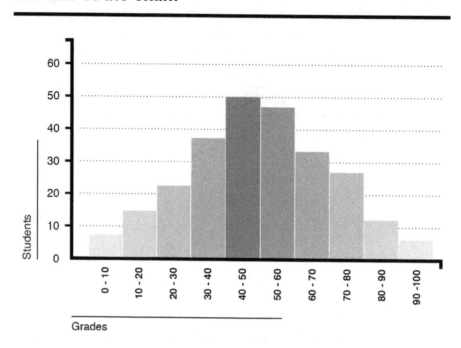

Results of the exam

Measures of central tendency can be discussed using a histogram. If the points scored were shown with individual rectangles, the tallest rectangle would represent the mode. A bimodal set of data would have two peaks of equal height. Histograms can be classified as having data **skewed to the left, skewed to the right**, or **normally distributed**, which is also known as **bell-shaped**.

These three classifications can be seen in the following chart:

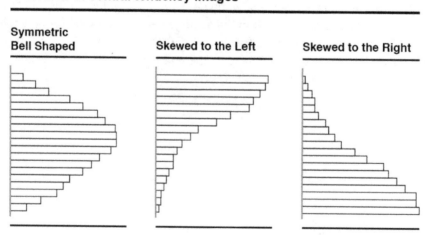

Measures of central tendency images

Symmetric Bell Shaped | Skewed to the Left | Skewed to the Right

When the data is normal, the mean, median, and mode are all very close. They all represent the most typical value in the data set. The mean is typically used as the best measure of central tendency in this case because it does include all data points. However, if the data is skewed, the mean becomes less meaningful. The median is the best measure of central tendency because it is not affected by any outliers, unlike the mean. When the data is skewed, the mean is dragged in the direction of the skew. Therefore, if the data is not normal, it is best to use the median as the measure of central tendency.

The measures of central tendency and the range may also be found by evaluating information on a line graph.

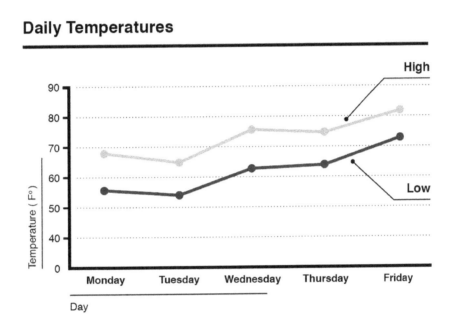

In the line graph above that shows the daily high and low temperatures, the average high temperature can be found by gathering data from each day on the triangle line. The days' highs are 82, 78, 75, 65, and 70. The average is found by adding them together to get 370, then dividing by 5 (because there are 5 temperatures). The average high for the five days is 74. If 74 degrees is found on the graph, then it falls in the middle of the values on the triangle line. The low temperature can be found in the same way.

Solving Problems Involving Measures of Center and Range
As mentioned, a data set can be described by calculating the mean, median, and mode. These values allow the data to be described with a single value that is representative of the data set.

The most common measure of center is the **mean,** also referred to as the **average.**

To calculate the mean:

- Add all data values together
- Divide by the sample size (the number of data points in the set)

The **median** is middle data value, so that half of the data lies below this value and half lies below the data value.

To calculate the median:

- Order the data from least to greatest
- The point in the middle of the set is the median
- If there is an even number of data points, add the two middle points and divide by 2

The **mode** is the data value that occurs most often.

To calculate the mode:

- Order the data from least to greatest
- Find the value that occurs most often

Example: Amelia is a leading scorer on the school's basketball team. The following data set represents the number of points that Amelia has scored in each game this season. Use the mean, median, and mode to describe the data.

16, 12, 26, 14, 13, 28, 14, 12, 15, 25

Solution:

Mean:

$$16 + 12 + 26 + 14 + 28 + 14 + 12 + 15 + 25 = 162$$

$$162 \div 9 = 18$$

Amelia averages 18 points per game.

Median:

12, 12, 14, 14, **15**, 16, 25, 26, 28

Amelia's median score is 15.

Mode:

12, 12, 14, 14, 15, 16, 25, 26, 28

The numbers 12 and 14 each occur twice, so this data set has 2 modes: 12 and 14.

The **range** is the difference between the largest and smallest values in the set. In the example above, the range is $28 - 12 = 16$.

Determining How Changes in Data Affect Measures of Center or Range
An **outlier** is a data point that lies an unusual distance from other points in the data set. Removing an outlier from a data set will change the measures of center. Removing a _large outlier_ (a high number) from a data set will decrease both the mean and the median. Removing a _small outlier_ (a number much lower than most in the data set) from a data set will increase both the mean and the median. For example, given the data set {3, 6, 8, 12, 13, 14, 60}, the data point, 60, is an outlier because it is unusually far from the other points. In this data set, the mean is 16.6. Notice that this mean number is even larger than all other data points in the set except for 60. Removing the outlier, the mean changes to 9.3 and the median

becomes 10. Removing an outlier will also decrease the range. In the data set above, the range is 57 when the outlier is included, but decreases to 11 when the outlier is removed.

Adding an outlier to a data set will affect the centers of measure as well. When a larger outlier is added to a data set, the mean and median increase. When a small outlier is added to a data set, the mean and median decrease. Adding an outlier to a data set will increase the range.

This does not seem to provide an appropriate measure of center when considering this data set. What will happen if that outlier is removed? Removing the extremely large data point, 60, is going to reduce the mean to 9.3. The mean decreased dramatically because 60 was much larger than any of the other data values. What would happen with an extremely low value in a data set like this one, {12, 87, 90, 95, 98, 100}? The mean of the given set is 80. When the outlier, 12, is removed, the mean should increase and should fit more closely to the other data points. Removing 12 and recalculating the mean shows that this is correct. The mean after removing 12 is 94. So, removing a large outlier will decrease the mean while removing a small outlier will increase the mean.

Using Statistics to Gain Information About a Population

Statistics is the branch of mathematics that deals with the collection, organization, and analysis of data. A statistical question is one that can be answered by collecting and analyzing data. When collecting data, expect variability. For example, "How many pets does Yanni own?" is not a statistical question because it can be answered in one way. "How many pets do the people in a certain neighborhood own?" is a statistical question because, to determine this answer, one would need to collect data from each person in the neighborhood, and it is reasonable to expect the answers to vary.

Identify these as statistical or not statistical:

- How old are you?
- What is the average age of the people in your class?
- How tall are the students in Mrs. Jones' sixth grade class?
- Do you like Brussels sprouts?

Questions 1 and 4 are not statistical questions, but questions 2 and 3 are.

Data collection can be done through surveys, experiments, observations, and interviews. A **census** is a type of survey that is done with a whole population. Because it can be difficult to collect data for an entire population, sometimes a **sample** is used. In this case, one would survey only a fraction of the population and make inferences about the data. Sample surveys are not as accurate as a census, but it is an easier and less expensive method of collecting data. An **experiment** is used when a researcher wants to explain how one variable causes change in another variable. For example, if a researcher wanted to know if a particular drug affects weight loss, he or she would choose a **treatment group** that would take the drug, and another group, the **control group**, that would not take the drug. Special care must be taken when choosing these groups to ensure that bias is not a factor. **Bias** occurs when an outside factor influences the outcome of the research. In observational studies, the researcher does not try to influence either variable but simply observes the behavior of the subjects. Interviews are sometimes used to collect data as well. The researcher will ask questions that focus on her area of interest in order to gain insight from the participants. When gathering data through observation or interviews, it is important that the researcher is well trained so that he or she does not influence the results and so that the study is reliable. A study is reliable if it can be repeated under the same conditions and the same results are received each time.

In statistics, a **population** contains all subjects being studied. For example, a population could be every student at a university or all males in the United States. A **sample** consists of a group of subjects from an entire population. A sample would be 100 students at a university or 100,000 males in the United States. **Inferential statistics** is the process of using a sample to generalize information concerning populations. **Hypothesis testing** is the actual process used when evaluating claims made about a population based on a sample.

A **statistic** is a measure obtained from a sample, and a **parameter** is a measure obtained from a population. For example, the mean SAT score of the 100 students at a university would be a statistic, and the mean SAT score of all university students would be a parameter.

The beginning stages of hypothesis testing starts with formulating a **hypothesis,** a statement made concerning a population parameter. The hypothesis may be true, or it may not be true. The experiment will help answer that question. In each setting, there are two different types of hypotheses: the **null hypothesis**, written as H_0, and the **alternative hypothesis**, written as H_1. The null hypothesis represents verbally when there is not a difference between two parameters, and the alternative hypothesis represents verbally when there is a difference between two parameters. Consider the following experiment: A researcher wants to see if a new brand of allergy medication has any effect on drowsiness of the patients who take the medication. He wants to know if the average hours spent sleeping per day increases. The mean for the population under study is 8 hours, so $\mu = 8$. In other words, the population parameter is μ, the mean. The null hypothesis is $\mu = 8$ and the alternative hypothesis is $\mu > 8$. When using a smaller sample of a population, the null hypothesis represents the situation when the mean remains unaffected and the alternative hypothesis represents the situation when the mean increases. The chosen statistical test will apply the data from the sample to actually decide whether the null hypothesis should or should not be rejected.

Algebraic Concepts

Interpreting Parts of an Expression

An **algebraic expression** is a mathematical phrase that may contain numbers, variables, and mathematical operations. An expression represents a single quantity. For example, $3x + 2$ is an algebraic expression.

An **algebraic equation** is a mathematical sentence with two expressions that are equal to each other. That is, an equation must contain an *equals* sign, as in $3x + 2 = 17$. This statement says that the value of the expression on the left side of the equals sign is equivalent to the value of the expression on the right side. In an expression, there are not two sides because there is no equals sign. The equals sign ($=$) is the difference between an expression and an equation.

To distinguish an expression from an equation, just look for the *equals sign*.

Example: Determine whether each of these is an expression or an equation.

    a. $16 + 4x = 9x - 7$ Solution: Equation

    b. $-27x - 42 + 19y$ Solution: Expression

    c. $4 = x + 3$ Solution: Equation

In an algebraic expression, the **variables,** such as x and y, are the unknowns. To add and subtract linear algebra expressions, you must combine like terms. **Like terms** are described as those terms that have the same variable with the same exponent. The other like terms are called **constants** because they have no variable component.

Performing Arithmetic Operations on Polynomials and Rational Expressions

A **rational expression** is a fraction or a ratio in which both the numerator and denominator are polynomials that are not equal to zero. A **polynomial** is a mathematical expression containing the sum and difference of one or more terms that are constants multiplied times variables raised to positive powers. Here are some examples of rational expressions:

$$\frac{2x^2 + 6x}{x}$$

$$\frac{x - 2}{x^2 - 6x + 8}$$

$$\frac{x + 2}{x^3 - 1}$$

Such expressions can be simplified using different forms of division. The first example can be simplified in two ways. First, because the denominator is a monomial, the expression can be split up into two expressions: $\frac{2x^2}{x} + \frac{6x}{x}$, and then simplified using properties of exponents as $2x + 6$. It also can be simplified using factoring and then crossing common factors out of the numerator and denominator. For instance, it can be written as:

$$\frac{2x(x + 3)}{x} = 2(x + 3)$$

$$2x + 6$$

The second expression above can also be simplified using factoring. It can be written as:

$$\frac{x - 2}{(x - 2)(x - 4)} = \frac{1}{x - 4}$$

Finally, the third example can only be simplified using long division, as there are no common factors in the numerator and denominator. First, divide the first term of the denominator by the first term of the numerator, then write that in the quotient. Then, multiply the divisor by that number and write it below the dividend. Subtract, bring down the next term from the dividend, and continue that process with the next first term and first term of the divisor. Continue the process until every term in the divisor is accounted for. Here is the actual long division:

Simplifying Expressions Using Long Division

$$
\begin{array}{r}
x^2 \quad -2x \quad +4 \\
x+2 \overline{\smash{\big)} x^3 \qquad\qquad\qquad -1} \\
\underline{x^3 \;+2x^2} \\
-2x^2 \qquad\quad -1 \\
\underline{-2x^2 \;-4x} \\
4x \quad -1 \\
\underline{4x \;+8} \\
-9
\end{array}
$$

Adding and Subtracting Linear Algebraic Expressions

To add and subtract linear algebra expressions, you must combine like terms. As mentioned, **like terms** have the same variable with the same exponent. In the following example, the x-terms can be added because the variable is the same and the exponent on the variable of one is also the same. These terms add to be $9x$. The other like terms are called **constants** because they have no variable component. These terms will add to be nine.

Example: Add $(3x - 5) + (6x + 14)$

$3x - 5 + 6x + 14$ *Rewrite without parentheses*

$3x + 6x - 5 + 14$ *Commutative property of addition*

$9x + 9$ *Combine like terms*

When subtracting linear expressions, be careful to add the opposite when combining like terms. Do this by distributing -1, which is multiplying each term inside the second parenthesis by negative one. Remember that distributing -1 changes the sign of each term.

Example: Subtract $(17x + 3) - (27x - 8)$

$17x + 3 - 27x + 8$ *Distributive Property*

$17x - 27x + 3 + 8$ *Commutative property of addition*

$-10x + 11$ *Combine like terms*

Example: Simplify by adding or subtracting:

$(6m + 28z - 9) + (14m + 13) - (-4z + 8m + 12)$

$6m + 28z - 9 + 14m + 13 + 4z - 8m - 12$ *Distributive Property*

$6m + 14m - 8m + 28z + 4z - 9 + 13 - 12$ *Commutative Property of Addition*

$12m + 32z - 8$ *Combine like terms*

Writing Expressions in Equivalent Forms to Solve Problems

Two algebraic expressions are equivalent if, even though they look different, they represent the same expression. Therefore, plugging in the same values into the variables in each expression will result in the same result in both expressions. To obtain an equivalent form of an algebraic expression, laws of algebra must be followed. For instance, addition and multiplication are both commutative and associative. Therefore, terms in an algebraic expression can be added in any order and multiplied in any order. For instance, $4x + 2y$ is equivalent to $2y + 4x$ and $y \times 2 + x \times 4$. Also, the distributive law allows a number to be distributed throughout parentheses, as in the following: $a(b + c) = ab + ac$. The two expressions on both sides of the equals sign are equivalent. Also, collecting like terms is important when working with equivalent forms. The simplest version of an expression is always the one easiest to work with, so all like terms (those with the same variables raised to the same powers) must be combined.

Note that an expression is not an equation; therefore, expressions cannot be multiplied by numbers, divided by numbers, or have numbers added to them or subtracted from them and still have equivalent expressions. These processes can only happen in equations when the same step is performed on both sides of the equals sign.

Using the Distributive Property to Generate Equivalent Linear Algebraic Expressions
The Distributive Property: $a(b + c) = ab + ac$

The **distributive property** is a way of taking a factor and multiplying it through a given expression in parentheses. Each term inside the parentheses is multiplied by the outside factor, eliminating the parentheses. The following example shows how to distribute the number 3 to all the terms inside the parentheses.

Example: Use the distributive property to write an equivalent algebraic expression:

$3(2x + 7y + 6)$

$3(2x) + 3(7y) + 3(6)$ *Distributive property*

$6x + 21y + 18$ *Simplify*

Because $a - b$ can be written $a + (-b)$, the distributive property can be applied in the example below:

Example: Use the distributive property to write an equivalent algebraic expression.

$7(5m - 8)$

$7[5m + (-8)]$ *Rewrite subtraction as addition of -8*

$7(5m) + 7(-8)$ *Distributive property*

$35m - 56$ *Simplify*

In the following example, note that the factor of 2 is written to the right of the parentheses but is still distributed as before.

Example: Use the distributive property to write an equivalent algebraic expression:

$$(3m + 4x - 10)2$$

$$(3m)2 + (4x)2 + (-10)2 \qquad \textit{Distributive property}$$

$$6m + 8x - 20 \qquad \textit{Simplify}$$

Example: $\quad -(-2m + 6x)$

In this example, the negative sign in front of the parentheses can be interpreted as $-1(-2m + 6x)$

$$-1(-2m + 6x)$$

$$-1(-2m) + (-1)(6x) \qquad \textit{Distributive property}$$

$$2m - 6x \qquad \textit{Simplify}$$

Factoring a Quadratic Expression to Reveal the Zeros of the Function it Defines

Factorization is the process of breaking up a mathematical quantity, such as a number or polynomial, into a product of two or more factors. For example, a factorization of the number 16 is $16 = 8 \times 2$. If multiplied out, the factorization results in the original number. A **prime factorization** is a specific factorization when the number is factored completely using prime numbers only. For example, the prime factorization of 16 is $16 = 2 \times 2 \times 2 \times 2$. A factor tree can be used to find the prime factorization of any number. Within a factor tree, pairs of factors are found until no other factors can be used, as in the following factor tree of the number 72:

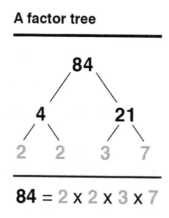

A factor tree

$$84 = 2 \text{ x } 2 \text{ x } 3 \text{ x } 7$$

It first breaks 84 into 21×4, which is not a prime factorization. Then, both 21 and 4 are factored into their primes. The final numbers on each branch consist of the numbers within the prime factorization. Therefore, $84 = 2 \times 2 \times 3 \times 7$. Factorization can be helpful in finding greatest common divisors and least common denominators.

Also, a factorization of an algebraic expression can be found. Throughout the process, a more complicated expression can be decomposed into products of simpler expressions. To factor a polynomial, first determine if there is a greatest common factor. If there is, factor it out. For example, $2x^2 + 8x$ has a

166

greatest common factor of $2x$ and can be written as $2x(x + 4)$. Once the greatest common monomial factor is factored out, if applicable, count the number of terms in the polynomial. If there are two terms, is it a difference of squares, a sum of cubes, or a difference of cubes?

If so, the following rules can be used:

$$a^2 - b^2 = (a + b)(a - b)$$

$$a^3 + b^3 = (a + b)(a^2 - ab + b^2)$$

$$a^3 - b^3 = (a + b)(a^2 + ab + b^2)$$

If there are three terms, and if the trinomial is a perfect square trinomial, it can be factored into the following:

$$a^2 + 2ab + b^2 = (a + b)^2$$

$$a^2 - 2ab + b^2 = (a - b)^2$$

If not, try factoring into a product of two binomials by trial and error into a form of $(x + p)(x + q)$. For example, to factor $x^2 + 6x + 8$, determine what two numbers have a product of 8 and a sum of 6. Those numbers are 4 and 2, so the trinomial factors into $(x + 2)(x + 4)$.

Finally, if there are four terms, try factoring by grouping. First, group terms together that have a common monomial factor. Then, factor out the common monomial factor from the first two terms. Next, look to see if a common factor can be factored out of the second set of two terms that results in a common binomial factor. Finally, factor out the common binomial factor of each expression, for example:

$$xy - x + 5y - 5 = x(y - 1) + 5(y - 1) = (y - 1)(x + 5)$$

After the expression is completely factored, check to see if the factorization is correct by multiplying to try to obtain the original expression. Factorizations are helpful in solving equations that consist of a polynomial set equal to 0. If the product of two algebraic expressions equals 0, then at least one of the factors is equal to 0. Therefore, factor the polynomial within the equation, set each factor equal to 0, and solve. For example, $x^2 + 7x - 18 = 0$ can be solved by factoring into $(x + 9)(x - 2) = 0$. Set each factor equal to 0, and solve to obtain $x = -9$ and $x = 2$.

There are a number of ways to solve a quadratic equation. The first way is through **factoring**. If the equation is in standard form and the polynomial can be factored, set each factor equal to 0 and solve. This can be done because if $ab = 0$, either $a = 0$, $b = 0$, or both are equal to 0. For example:

$$x^2 - 7x + 10 = (x - 5)(x - 2)$$

Therefore, the solutions of $x^2 - 7x + 10 = 0$ are those that satisfy both $x - 5 = 0$ and $x - 2 = 0$, or $x =$ 5 and 2. This is the simplest method to solve quadratic equations; however, not all quadratic polynomials can be factored, so this method does not work for all quadratic equations.

Identifying Zeros of Polynomials
A **polynomial function** is a function containing a polynomial expression, which is an expression containing constants and variables combined using the four mathematical operations. The degree of a polynomial in one variable is the largest exponent seen on any variable in the expression. Typical polynomial functions are **quartic,** with a degree of 4, **cubic,** with a degree of 3, and **quadratic,** with a

degree of 2. Note that the exponents on the variables can only be nonnegative integers. The domain of any polynomial function is all real numbers because any number plugged into a polynomial expression grants a real number output. An example of a quartic polynomial equation is:

$$y = x^4 + 3x^3 - 2x + 1$$

The zeros of a polynomial function are the points where its graph crosses the *y*-axis. In order to find the number of real zeros of a polynomial function, **Descartes' Rule of Sign** can be used. The number of possible positive real zeros is equal to the number of sign changes in the coefficients of the terms in the polynomial. If there is only one sign change, there is only one positive real zero. In the example above, the signs of the coefficients are positive, positive, negative, and positive. Therefore, the sign changes two times and therefore, there are at most two positive real zeros. The number of possible negative real zeros is equal to the number of sign changes in the coefficients when plugging $-x$ into the equation. Again, if there is only one sign change, there is only one negative real zero. The polynomial result when plugging -*x* into the equation is:

$$y^4 - 3x^3 + 2x + 1$$

The sign changes two times, so there are at most two negative real zeros. Another polynomial equation this rule can be applied to is:

$$y = x^3 + 2x - x - 5$$

There is only one sign change in the terms of the polynomial, so there is exactly one real zero. When plugging -*x* into the equation, the polynomial result is:

$$-x^3 - 2x - x - 5$$

There are no sign changes in this polynomial, so there are no possible negative zeros.

Factoring

A polynomial expression can be **factored**. Throughout the process, a more complicated expression can be decomposed into products of simpler expressions. To factor a polynomial, first determine if there is a **greatest common factor**. If there is, factor it out. For example, $2x^2 + 8x$ has a greatest common factor of $2x$ and can be written as $2x(x + 4)$. Once the greatest common monomial factor is factored out, if applicable, count the number of terms in the polynomial. If there are two terms, is it a difference of squares, a sum of cubes, or a difference of cubes? If so, the following rules can be used:

$$a^2 - b^2 = (a + b)(a - b)$$

$$a^3 + b^3 = (a + b)(a^2 - ab + b^2)$$

$$a^3 - b^3 = (a + b)(a^2 + ab + b^2)$$

If there are three terms, and if the trinomial is a perfect square trinomial, it can be factored into the following:

$$a^2 + 2ab + b^2 = (a + b)^2$$

$$a^2 - 2ab + b^2 = (a - b)^2$$

If not, try factoring into a product of two binomials by trial and error into a form of $(x + p)(x + q)$. For example, to factor $x^2 + 6x + 8$, determine what two numbers have a product of 8 and a sum of 6. Those numbers are 4 and 2, so the trinomial factors into $(x + 2)(x + 4)$.

Finally, if there are four terms, try factoring by grouping. First, group terms together that have a common monomial factor. Then, factor out the common monomial factor from the first two terms. Next, look to see if a common factor can be factored out of the second set of two terms that results in a common binomial factor. Finally, factor out the common binomial factor of each expression, for example:

$$xy - x + 5y - 5$$

$$x(y - 1) + 5(y - 1)$$

$$(y - 1)(x + 5)$$

After the polynomial is completely factored, check to see if the factorization is correct by multiplying to try to obtain the original expression. Factorizations are helpful in solving equations that consist of a polynomial set equal to 0. If the product of two algebraic expressions equals 0, then at least one of the factors is equal to 0. Therefore, factor the polynomial within the equation, set each factor equal to 0, and solve. For example, $x^2 + 7x - 18 = 0$ can be solved by factoring into $(x + 9)(x - 2) = 0$. Set each factor equal to 0, and solve to obtain $x = -9$ and $x = 2$.

Once the zeros are identified, they can be used to help sketch a graph of the polynomial.

Solving Linear Equations and Inequalities in One Variable

Solving Equations in One Variable
An **equation in one variable** is a mathematical statement where two algebraic expressions in one variable, usually x, are set equal. To solve the equation, the variable must be isolated on one side of the equals sign. The addition and multiplication principles of equality are used to isolate the variable. The **addition principle of equality** states that the same number can be added to or subtracted from both sides of an equation. Because the same value is being used on both sides of the equals sign, equality is maintained. For example, the equation $2x - 3 = 5x$ is equivalent to both $2x - 3 + 2 = 5x + 2$, and $2x - 3 - 5 = 5x - 5$. This principle can be used to solve the following equation: $x + 5 = 4$. The variable x must be isolated, so to move the 5 from the left side, subtract 5 from both sides of the equals sign. Therefore, $x + 5 - 5 = 4 - 5$. So, the solution is $x = -1$. This process illustrates the idea of an **additive inverse** because subtracting 5 is the same as adding -5. Basically, add the opposite of the number that must be removed to both sides of the equals sign.

The **multiplication principle of equality** states that equality is maintained when a number is either multiplied times both expressions on each side of the equals sign, or when both expressions are divided by the same number. For example, $4x = 5$ is equivalent to both $16x = 20$ and $x = \frac{5}{4}$. Multiplying both sides times 4 and dividing both sides by 4 maintains equality. Solving the equation $6x - 18 = 5$ requires the use of both principles. First, apply the addition principle to add 18 to both sides of the equals sign, which results in $6x = 23$. Then use the multiplication principle to divide both sides by 6, giving the solution $x = \frac{23}{6}$. Using the multiplication principle in the solving process is the same as involving a multiplicative inverse. A **multiplicative inverse** is a value that, when multiplied by a given number, results in 1. Dividing by 6 is the same as multiplying by $\frac{1}{6}$, which is both the reciprocal and multiplicative inverse of 6.

When solving a linear equation in one variable, checking the answer shows if the solution process was performed correctly. Plug the solution into the variable in the original equation. If the result is a false statement, something was done incorrectly during the solution procedure. Checking the example above gives the following: $6 \times \frac{23}{6} - 18 = 23 - 18 = 5$. Therefore, the solution is correct.

Some equations in one variable involve fractions or the use of the distributive property. In either case, the goal is to obtain only one variable term and then use the addition and multiplication principles to isolate that variable. Consider the equation $\frac{2}{3}x = 6$. To solve for x, multiply each side of the equation by the reciprocal of $\frac{2}{3}$, which is $\frac{3}{2}$. This step results in:

$$\frac{3}{2} \times \frac{2}{3}x = \frac{3}{2} \times 6$$

This simplifies into the solution $x = 9$. Now consider the equation:

$$3(x + 2) - 5x = 4x + 1$$

Use the distributive property to clear the parentheses. Therefore, multiply each term inside the parentheses by 3. This step results in:

$$3x + 6 - 5x = 4x + 1$$

Next, collect like terms on the left-hand side. **Like terms** are terms with the same variable or variables raised to the same exponent(s). Only like terms can be combined through addition or subtraction. After collecting like terms, the equation is $-2x + 6 = 4x + 1$. Finally, apply the addition and multiplication principles. Add $2x$ to both sides to obtain $6 = 6x + 1$. Then, subtract 1 from both sides to obtain $5 = 6x$. Finally, divide both sides by 6 to obtain the solution $\frac{5}{6} = x$.

Two other types of solutions can be obtained when solving an equation in one variable. The final result could be that there is either no solution or that the solution set contains all real numbers. Consider the equation $4x = 6x + 5 - 2x$. First, the like terms can be combined on the right to obtain $4x = 4x + 5$. Next, subtract $4x$ from both sides. This step results in the false statement $0 = 5$. There is no value that can be plugged into x that will ever make this equation true. Therefore, there is no solution. The solution procedure contained correct steps, but the result of a false statement means that no value satisfies the equation. The symbolic way to denote that no solution exists is ∅.

Next, consider the equation:

$$5x + 4 + 2x = 9 + 7x - 5$$

Combining the like terms on both sides results in:

$$7x + 4 = 7x + 4$$

The left-hand side is exactly the same as the right-hand side. Using the addition principle to move terms, the result is $0 = 0$, which is always true. Therefore, the original equation is true for any number, and the solution set is all real numbers. The symbolic way to denote such a solution set is $\mathbb{R}$, or in interval notation, $(-\infty, \infty)$.

A **linear equation in** x can be written in the form $ax + b = 0$. A **linear inequality** is very similar, although the equals sign is replaced by an inequality symbol such as $<, >, \leq,$ or $\geq$. In any case, a can never be 0. Some examples of linear inequalities in one variable are $2x + 3 < 0$ and $4x - 2 \leq 0$. Solving an inequality involves finding the set of numbers that when plugged into the variable, make the inequality a true statement. These numbers are known as the **solution set** of the inequality. To solve an inequality, use the same properties that are necessary in solving equations. First, add or subtract variable terms and/or constants to obtain all variable terms on one side of the equals sign and all constant terms on the other side. Then, either multiply both sides times the same number, or divide both sides by the same number, to obtain an inequality that gives the solution set. When multiplying times, or dividing by, a negative number in an inequality, change the direction of the inequality symbol. The solution set can be graphed on a number line. Consider the linear inequality $-2x - 5 > x + 6$. First, add 5 to both sides and subtract $-x$ off of both sides to obtain $-3x > 11$. Then, divide both sides by -3, making sure to change the direction of the inequality symbol. These steps result in the solution $x < -\frac{11}{3}$. Therefore, any number less than $-\frac{11}{3}$ satisfies this inequality.

Solving Quadratic Equations in One Variable

A **quadratic equation** in standard form, $ax^2 + bx + c = 0$, can have either two solutions, one solution, or two complex solutions (no real solutions). This is determined using the determinant $b^2 - 4ac$. If the determinant is positive, there are two real solutions. If the determinant is negative, there are no real solutions. If the determinant is equal to 0, there is one real solution. For example, given the quadratic equation $4x^2 - 2x + 1 = 0$, its determinant is:

$$(-2)^2 - 4(4)(1) = 4 - 16 = -12$$

So, it has two complex solutions, meaning no real solutions.

There are quite a few ways to solve a quadratic equation. The first is by **factoring**. If the equation is in standard form and the polynomial can be factored, set each factor equal to 0, and solve using the Principle of Zero Products. For example:

$$x^2 - 4x + 3 = (x - 3)(x - 1)$$

Therefore, the solutions of $x^2 - 4x + 3 = 0$ are those that satisfy both $x - 3 = 0$ and $x - 1 = 0$, or $x = 3$ and $x = 1$. This is the simplest method to solve quadratic equations; however, not all polynomials inside the quadratic equations can be factored.

Another method is **completing the square**. The polynomial $x^2 + 10x - 9$ cannot be factored, so the next option is to complete the square in the equation $x^2 + 10x - 9 = 0$ to find its solutions. The first step is to add 9 to both sides, moving the constant over to the right side, resulting in $x^2 + 10x = 9$. Then the coefficient of x is divided by 2 and squared. This result is then added to both sides of the equation. In this example, $\left(\frac{10}{2}\right)^2 = 25$ is added to both sides of the equation to obtain:

$$x^2 + 10x + 25 = 9 + 25 = 34$$

The left-hand side can then be factored into $(x + 5)^2 = 34$. Solving for x then involves taking the square root of both sides and subtracting 5. This leads to the two solutions:

$$x = \pm\sqrt{34} - 5$$

171

The third method is the **quadratic formula.** Given a quadratic equation in standard form, $ax^2 + bx + c = 0$, its solutions always can be found using the formula:

$$x = \frac{-b \pm \sqrt{b^2 - 4ac}}{2a}$$

Solving Simple Rational and Radical Equations in One Variable

A linear equation in one variable can be solved using the following steps:

1. Simplify the algebraic expressions on both sides of the equals sign by removing all parentheses, using the distributive property, and then collect all like terms.

2. Collect all variable terms on one side of the equals sign and all constant terms on the other side by adding the same quantity to both sides of the equals sign, or by subtracting the same quantity from both sides of the equals sign.

3. Isolate the variable by either dividing both sides of the equation by the same number, or by multiplying both sides by the same number.

4. Check the answer.

The only difference between solving linear inequalities versus equations is that when multiplying by a negative number or dividing by a negative number, the direction of the inequality symbol must be reversed.

If an equation contains multiple fractions, it might make sense to clear the equation of fractions first by multiplying all terms by the least common denominator. Also, if an equation contains several decimals, it might make sense to clear the decimals as well by multiplying times a factor of 10. If the equation has decimals in the hundredth place, multiply every term in the equation by 100.

Equivalent Expressions Involving Rational Exponents and Radicals
Writing radical expressions into equivalent forms involving rational exponents can help in simplifying complex radical expressions.

The rule that helps this conversion is:

$$\sqrt[n]{x^m} = x^{\frac{m}{n}}$$

If $m = 1$, the rule is simply:

$$\sqrt[n]{x} = x^{\frac{1}{n}}$$

For instance, consider the following expression:

$$\sqrt[4]{x}\sqrt[2]{y}$$

It can be written as one radical expression, but first it needs to be converted to an equivalent expression using rational expressions. The equivalent expression is $x^{\frac{1}{4}}y^{\frac{1}{2}}$. The goal is to have one radical, which means one index n, so a common denominator of the exponents must be found. The common denominator is 4, so an equivalent expression is:

$$x^{\frac{1}{4}}y^{\frac{2}{4}}$$

The exponential rule $a^m b^m = (ab)^m$ can be used to, in a sense, factor out a $\frac{1}{4}$ out of both exponents. This process results in the expression $(xy^2)^{\frac{1}{4}}$, and its equivalent radical form is

$$\sqrt[4]{xy^2}.$$

Converting to rational exponents has allowed the entire expression to be written as one radical.

Another type of problem could involve going in the opposite direction—starting with rational exponents and using an equivalent radical form to simplify the expression. For instance, $32^{\frac{1}{5}}$ might not seem obviously equal to 2. However, putting it in its equivalent radical form $\sqrt[5]{32}$ shows that it is equivalent to the fifth root of 32, which is 2.

Solving Systems of Equations

Creating, Solving, or Interpreting Systems of Linear Inequalities in Two Variables
A system of linear inequalities *in two variables* consists of two inequalities in two variables, x and y. For example, the following is a system of linear inequalities in two variables:

$$\begin{cases} 4x + 2y < 1 \\ 2x - y \leq 0 \end{cases}$$

The brace on the left side shows that the two inequalities are grouped together. A solution of a single inequality in two variables is an ordered pair that satisfies the inequality. For example, (1, 3) is a solution of the linear inequality $y \geq x + 1$ because when plugged in, it results in a true statement. The graph of an inequality in two variables consists of all ordered pairs that make the solution true. Therefore, the entire solution set of a single inequality contains many ordered pairs, and the set can be graphed by using a half plane. A **half plane** consists of the set of all points on one side of a line. If the inequality consists of $>$ or $<$, the line is dashed because no solutions actually exist on the line shown. If the inequality consists of $\geq$ or $\leq$, the line is solid and solutions are on the line shown. To graph a linear inequality, graph the corresponding equation found by replacing the inequality symbol with an equals sign. Then pick a test point that exists on either side of the line. If that point results in a true statement when plugged into the original inequality, shade in the side containing the test point. If it results in a false statement, shade in the opposite side.

Solving a system of linear inequalities must be done graphically. Follow the process as described above for both given inequalities. The solution set to the entire system is the region that is in common to every graph in the system.

For example, here is the solution to the following system:

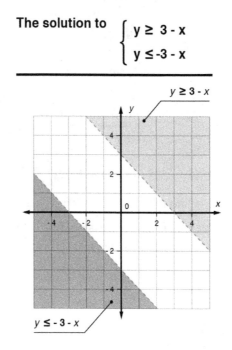

The solution to
$$\begin{cases} y \geq 3 - x \\ y \leq -3 - x \end{cases}$$

Note that there is no region in common, so this system has no solution.

Creating, Solving, or Interpreting Systems of Two Linear Equations in Two Variables
An example of a system of two linear equations in two variables is the following:

$$2x + 5y = 8$$

$$5x + 48y = 9$$

A solution to a system of two linear equations is an ordered pair that satisfies both the equations in the system. A system can have one solution, no solution, or infinitely many solutions. The solution can be found through a **graphing** technique. The solution of a system of equations is actually equal to the point of intersection of both lines. If the lines intersect at one point, there is one solution and the system is said to be consistent. However, if the two lines are parallel, they will never intersect and there is no solution. In this case, the system is said to be inconsistent. Third, if the two lines are actually the same line, there are

174

infinitely many solutions and the solution set is equal to the entire line. The lines are dependent. Here is a summary of the three cases:

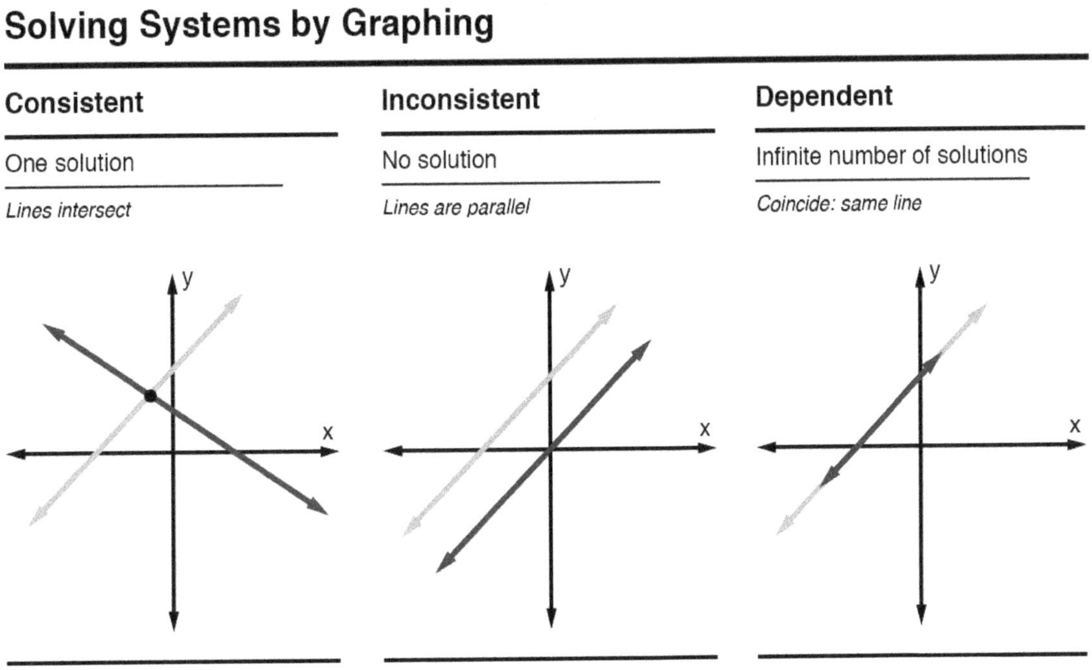

Solving Systems by Graphing

Consistent	Inconsistent	Dependent
One solution	No solution	Infinite number of solutions
Lines intersect	Lines are parallel	Coincide: same line

Consider the following system of equations:

$$y + x = 3$$
$$y - x = 1$$

To find the solution graphically, graph both lines on the same xy-plane. Graph each line using either a table of ordered pairs, the x- and y-intercepts, or slope and the y-intercept. Then, locate the point of intersection.

The graph is shown here:

The System of Equations $\begin{cases} y + x = 3 \\ y - x = 1 \end{cases}$

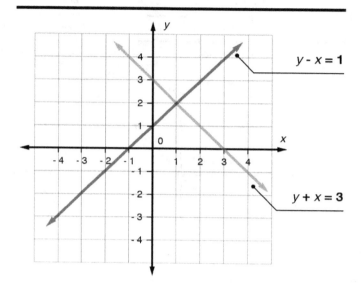

It can be seen that the point of intersection is the ordered pair (1, 2). This solution can be checked by plugging it back into both original equations to make sure it results in true statements. This process results in:

$$2 + 1 = 3$$
$$2 - 1 = 1$$

Both are true equations, so therefore the point of intersection is truly the solution.

The following system has no solution:

$$y = 4x + 1$$
$$y = 4x - 1$$

Both lines have the same slope and different y-intercepts, and therefore they are parallel. This means that they run alongside each other and never intersect.

Finally, the following solution has infinitely many solutions:

$$2x - 7y = 12$$
$$4x - 14y = 24$$

Note that the second equation is equal to the first equation times 2. Therefore, they are the same line. The solution set can be written in set notation as $\{(x, y) | 2x - 7y = 12\}$, which represents the entire line.

Algebraically Solving Systems of Two Linear Equations in Two Variables

There are two algebraic methods to finding solutions. The first is substitution. This process is better suited for systems when one of the equations is already solved for one variable, or when solving for one variable is easy to do. The equation that is already solved for is substituted into the other equation for that variable, and this process results in a linear equation in one variable. This equation can be solved for the given variable, and then that solution can be plugged into one of the original equations, which can then be solved for the other variable. This last step is known as **back-substitution** and the end result is an ordered pair.

A system that is best suited for substitution is the following:

$$y = 4x + 2$$

$$2x + 3y = 9$$

The other method is known as **elimination,** or the **addition method**. This is better suited when the equations are in standard form $Ax + By = C$. The goal in this method is to multiply one or both equations times numbers that result in opposite coefficients. Then, add the equations together to obtain an equation in one variable. Solve for the given variable, then take that value and back-substitute to obtain the other part of the ordered pair solution.

A system that is best suited for elimination is the following:

$$2x + 3y = 8$$

$$4x - 2y = 10$$

Note that in order to check an answer when solving a system of equations, the solution must be checked in both original equations to show that it solves not only one of the equations, but both of them.

If throughout either solution procedure the process results in an untrue statement, there is no solution to the system. Finally, if throughout either solution procedure the process results in the variables dropping out, which gives a statement that is always true, there are infinitely many solutions.

Graphing Functions

Functions can be displayed graphically to analyze behavior and patterns within data. Different functions have certain characteristics that make the corresponding graph distinctive.

Polynomial Functions

A polynomial equation is a polynomial set equal to another polynomial, or in standard form, a polynomial is set equal to 0. A **polynomial function** is a polynomial set equal to y. For instance, $x^2 + 2x - 8$ is a polynomial, $x^2 + 2x - 8 = 0$ is a polynomial equation, and $y = x^2 + 2x - 8$ is the corresponding polynomial function. To solve a polynomial equation, the x-values in which the graph of the corresponding polynomial function crosses the x-axis are sought. These coordinates are known as the *zeros* of the polynomial function because they are the coordinates in which the y-coordinates are 0. One way to find the zeros of a polynomial is to find its factors, then set each individual factor equal to 0, and solve each equation to find the zeros. A **factor** is a linear expression, and to completely factor a polynomial, the polynomial must be rewritten as a product of individual linear factors. The polynomial listed above can be factored as $(x + 4)(x - 2)$. Setting each factor equal to zero results in the zeros $x = -4$ and $x = 2$.

Here is the graph of the zeros of the polynomial:

The Graph of the Zeros of $x^2 + 2x - 8 = 0$

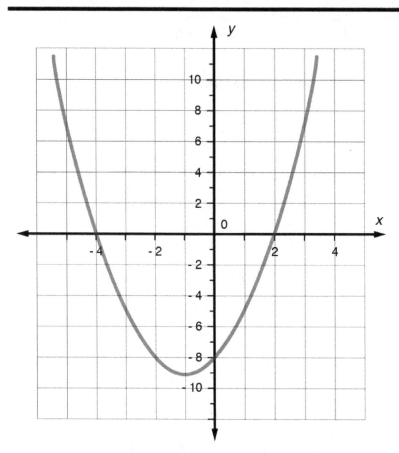

Nonlinear Relationships

A common nonlinear relationship between two variables involves inverse variation where one quantity varies inversely with respect to another. The equation for inverse variation is $y = \frac{k}{x}$, where k is still known as the constant of variation. Here is the graph of the curve $y = \frac{3}{x}$:

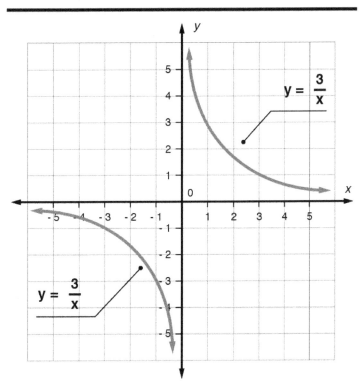

The Graph of $y = \dfrac{3}{x}$

Other common nonlinear functions include the squaring function $f(x) = x^2$, seen here:

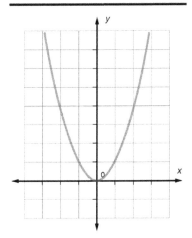

$f(x) = x^2$

Notice that as the independent variable x increases when $x > 0$, the dependent variable y also increases. However, y does not increase at a constant rate.

Root Functions

A **radical expression** is an expression involving a square root, a cube root, or a higher order root such as fourth root, fifth root, etc. The expression underneath the radical is known as the radicand, and the index is the number corresponding to the root. An index of 2 corresponds to a square root. A radical function is a function that involves a radical expression. For example, $\sqrt{x + 1}$ is a radical expression, $x + 1$ is the radicand, and the corresponding function is $y = \sqrt{x + 1}$. The function can also be written in function notation as $f(x) = \sqrt{x + 1}$. If the root is even, meaning a square root, fourth root, etc., the radicand must be positive. Therefore, in order to find the domain of a radical function with an even index, set the radicand greater than or equal to zero and find the set of numbers that satisfies that inequality. The domain of $f(x) = \sqrt{x + 1}$ is all numbers greater than or equal to -1. The range of this function is all nonnegative real numbers because the square root, or any even root, can never output a negative number. The domain of an odd root is all real numbers because the radicand can be negative in an odd root. Take a look at this square root function:

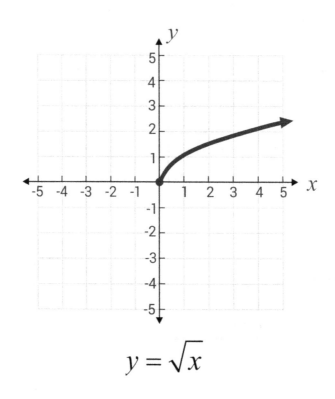

$$y = \sqrt{x}$$

Here's a cube root function:

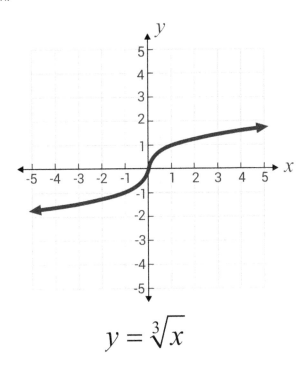

$$y = \sqrt[3]{x}$$

Absolute value functions contain an expression that uses absolute value symbols. The vertex is (0, 0), and the axis of symmetry divides the graph on the y-axis as seen below:

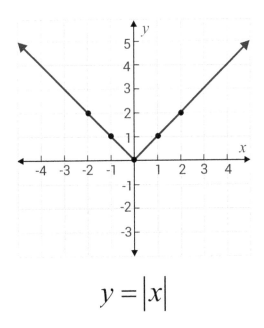

$$y = |x|$$

A **piecewise function** is basically a function that is defined in pieces or sections. The graph of the function behaves differently over different intervals along the *x*-axis, or different intervals of its domain.

Therefore, the function is defined using different mathematical expressions over these intervals. The function is not defined by only one equation. In a piecewise function, the function is actually defined by two or more equations, where each equation is used over a specific interval of the domain. Here is a graph of a piecewise function:

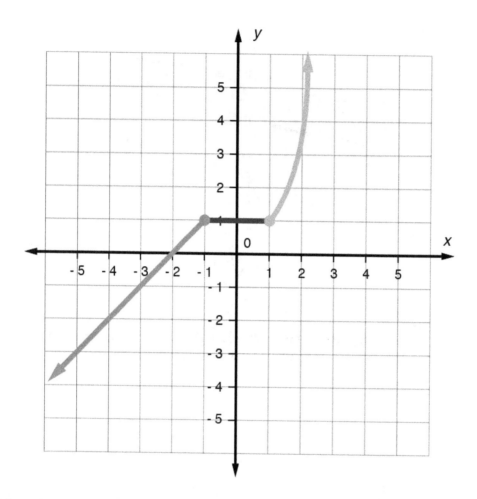

Notice that from $(-\infty, -1]$, the graph is a line with positive slope. From $[-1, 1]$ the graph is a horizontal line. Finally, from $[1, \infty)$ the graph is a nonlinear curve. Both the domain and range of this piecewise defined function is all real numbers, which is expressed as $(-\infty, \infty)$.

Piecewise functions can also have discontinuities, which are jumps in the graph. When drawing a graph, if the pencil must be picked up at any point to continue drawing, the graph has a discontinuity. Here is the graph of a piecewise function with discontinuities at $x = 1$ and $x = 2$:

A Piecewise Function

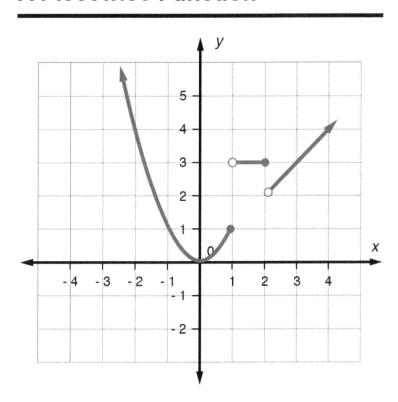

The open circle at a point indicates that the endpoint is not included in that part of the graph, and the closed circle indicates that the endpoint is included. The domain of this function is all real numbers; however, the range is all non-negative real numbers $[0, \infty)$.

Logarithmic Functions
For $x > 0$, $b > 0$, $b \neq 1$, the function $f(x) = \log_b x$ is known as the **logarithmic function** with base b. With $y = \log_b x$, its exponential equivalent is $b^y = x$. In either case, the **exponent** is y and the **base** is b. Therefore, $3 = \log_2 8$ is the same as $2^2 = 8$. So, in order to find the logarithm with base 2 of 8, find the exponent that when 2 is raised to that value results in 8. Similarly, $\log_3 243 = 5$. In order to do this mentally, ask the question, what exponent does 3 need to be raised to that results in 234? The answer is 5. Most logarithms do not have whole number results. In this case, a calculator can be used. A calculator typically has buttons with base 10 and base e, so the change of base formula can be used to calculate these logs. For instance, $\log_3 55 = \frac{\log 55}{\log 3} = 3.64$. Similarly, the natural logarithm with base e could be used to obtain the same result. $\log_3 55 = \frac{\ln 55}{\ln 3} = 3.64$.

The domain of a logarithmic function $f(x) = \log_b x$ is all positive real numbers. This is because the exponent must be a positive number. The range of a logarithmic function $f(x) = \log_b x$ is all real numbers. The graphs of all logarithmic functions of the form $f(x) = \log_b x$ always pass through the point

(1, 0) because anything raised to the power of 0 is 1. Therefore, such a function always has an *x*-intercept at 1. If the base is greater than 1, the graph increases from the left to the right along the *x*-axis. If the base is between 0 and 1, the graph decreases from the left to the right along the *x*-axis. In both situations, the *y*-axis is a vertical asymptote. The graph will never touch the *y*-axis, but it does approach it closely. Here are the graphs of the two cases of logarithmic functions:

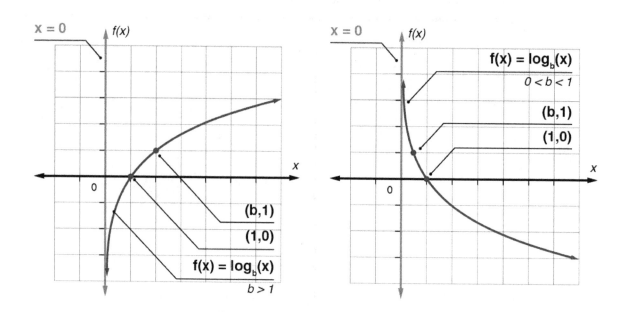

Creating Equations and Inequalities to Represent Relationships and Using Them to Solve Problems

Translating Phrases and Sentences into Expressions, Equations, and Inequalities
When presented with a real-world problem that must be solved, the first step is always to determine what the **unknown quantity** is that must be solved for. Use a variable, such as x or t, to represent that unknown quantity. Sometimes there can be two or more unknown quantities. In this case, either choose an additional variable, or if a relationship exists between the unknown quantities, express the other quantities in terms of the original variable. After choosing the variables, form algebraic expressions and/or equations that represent the verbal statement in the problem.

The following table shows examples of vocabulary used to represent the different operations:

Addition	Sum, plus, total, increase, more than, combined, in all
Subtraction	Difference, less than, subtract, reduce, decrease, fewer, remain
Multiplication	Product, multiply, times, part of, twice, triple
Division	Quotient, divide, split, each, equal parts, per, average, shared

The combination of operations and variables form both mathematical expression and equations. The difference between expressions and equations are that there is no equals sign in an expression, and that expressions are evaluated to find an unknown quantity, while equations are *solved* to find an unknown quantity. Also, inequalities can exist within verbal mathematical statements. Instead of a statement of equality, expressions state quantities are *less than* ($<$), *less than or equal to* ($\leq$), *greater than* ($>$), or *greater than or equal to* ($\geq$). Another type of inequality is when a quantity is said to be not equal to another quantity. The symbol used to represent "not equal to" is $\neq$.

The steps for solving inequalities in one variable are the same steps for solving equations in one variable. The addition and multiplication principles are used. However, to maintain a true statement when using the $<, \leq, >$, and $\geq$ symbols, if a negative number is either multiplied times both sides of an inequality or divided from both sides of an inequality, the sign must be flipped. For instance, consider the following inequality: $3 - 5x \leq 8$. First, 3 is subtracted from each side to obtain $-5x \leq 5$. Then, both sides are divided by -5, while flipping the sign, to obtain $x \geq -1$. Therefore, any real number greater than or equal to -1 satisfies the original inequality.

Interpreting a Linear Function Given Its Context

A linear function of the form $f(x) = mx + b$ has two important quantities: m and b. The quantity m represents the **slope** of the line, and the quantity b represents the **y-intercept** of the line. When the function represents an actual real-life situation, or mathematical model, these two quantities are very meaningful. The slope, m, represents the rate of change, or the amount y increases or decreases given an increase in x. If m is positive, the rate of change is positive, and if m is negative, the rate of change is negative. The y-intercept, b, represents the amount of the quantity y when x is 0. In many applications, if the x-variable is never a negative quantity, the y-intercept represents the initial amount of the quantity y. Often the x-variable represents time, so it makes sense that the x-variable is never negative.

Consider the following example. These two equations represent the cost, C, of t-shirts, x, at two different printing companies:

$$C(x) = 7x$$

$$C(x) = 5x + 25$$

The first equation represents a scenario that shows the cost per t-shirt is $7. In this equation, x varies directly with y. There is no y-intercept, which means that there is no initial cost for using that printing company. The rate of change is 7, which is price per shirt. The second equation represents a scenario that has both an initial cost and a cost per t-shirt. The slope 5 shows that each shirt is $5. The y-intercept 25 shows that there is an initial cost of using that company. Therefore, it makes sense to use the first company at $7 a shirt when only purchasing a small number of t-shirts. However, any large orders would be cheaper by going with the second company because eventually that initial cost will be negligible.

Rearranging Formulas/Equations to Highlight a Quantity of Interest

When solving equations, it is important to note which quantity must be solved for. This quantity can be referred to as the **quantity of interest.** The goal of solving is to isolate the variable in the equation using logical mathematical steps. The **addition property of equality** states that the same real number can be added to both sides of an equation and equality is maintained. Also, the same real number can be subtracted from both sides of an equation to maintain equality. Second, the **multiplication property of equality** states that the same nonzero real number can multiply both sides of an equation, and still, equality is maintained. Because division is the same as multiplying times a reciprocal, an equation can be divided by the same number on both sides as well.

When solving inequalities, the same ideas are used. However, when multiplying by a negative number on both sides of an inequality, the inequality symbol must be flipped in order to maintain the logic. The same is true when dividing both sides of an inequality by a negative number.

Basically, in order to isolate a quantity of interest in either an equation or inequality, the same thing must be done to both sides of the equals sign, or inequality symbol, to keep everything mathematically correct.

Functions and Function Notation

A **relation** is any set of ordered pairs (x, y). The first set of points, known as the x-coordinates, make up the domain of the relation. The second set of points, known as the y-coordinates, make up the range of the relation. A relation in which every member of the domain corresponds to only one member of the range is known as a **function.** A function cannot have a member of the domain corresponding to two members of the range. Functions are most often given in terms of equations instead of ordered pairs. For instance, here is an equation of a line: $y = 2x + 4$. In function notation, this can be written as $f(x) = 2x + 4$. The expression $f(x)$ is read "f of x" and it shows that the inputs, the x-values, get plugged into the function and the output is $y = f(x)$. The set of all inputs are in the domain and the set of all outputs are in the range.

The x-values are known as the **independent variables** of the function and the y-values are known as the **dependent variables** of the function. The y-values depend on the x-values. For instance, if $x = 2$ is plugged into the function shown above, the y-value depends on that input.

$$f(2) = 2 \times 2 + 4 = 8.$$

Therefore, $f(2) = 8$, which is the same as writing the ordered pair (2, 8). To graph a function, graph it in equation form. Therefore, replace $f(x)$ with h and plot ordered pairs.

Due to the definition of a function, the graph of a function cannot have two first components being paired to the same second component. Therefore, all graphs of functions pass the **vertical line test**. If any vertical line intersects a graph in more than one place, the graph is not that of a function. For instance, the graph of a circle is not a function because one can draw a vertical line through a circle and the line would intersect the circle twice. Common functions include lines and polynomials, and they all pass the vertical line test.

Key features of functions and their graphs include the following:

Intervals of Increasing/Decreasing: Intervals where the x-values of a function are increasing or decreasing

Absolute maximum/minimum: The highest and lowest y-value in a function

Relative maximum/minimum: Turning points where the graph changes from increasing to decreasing or decreasing to increasing

Positive/Negative: A function is positive when the function rises above the x-axis, and negative when the function falls below the x-axis.

Symmetry: Symmetry occurs when there is a line (axis of symmetry) that divides the function into half so that both sides are the same.

End Behavior: Descriptive of what happens to the graph as it extends in either direction.

Periodicity: Values in a function repeating in a regular pattern

Evaluating Functions for Inputs in their Domains and Writing a Function
An **equation** occurs when two algebraic expressions are set equal to one another. Functions can be represented in equation form. Given an equation in two variables, x and y, it can be expressed in function form if solved for y. For example, the linear equation $2x + y = 5$ can be solved for y to obtain $y = -2x +$

5, otherwise known as **slope-intercept** form. To place the equation in function form, replace y with $f(x)$, which is read "f of x." Therefore, $f(x) = -2x + 5$. This notation clarifies the input–output relationship of the function. The function f is a function of x, so an x value can be plugged into the function to obtain an output. For example:

$$f(2) = -2 \times 2 + 5 = 1$$

Therefore, an input of 2 corresponds to an output of 1.

A function can be graphed by plotting ordered pairs in the xy-plane in the same way that the equation form is graphed. The graph of a function always passes the **Vertical Line Test**.

Inequalities look like equations, but instead of an equals sign, $<, >, \leq, \geq$, or $\neq$ are used. Here are some examples of inequalities:

$$2x + 7 < y, 3x^2 \geq 5$$

$$x \neq 4$$

Inequalities show relationships between algebraic expressions when the quantities are different. Inequalities can also be expressed in function form if they are solved for y. For instance, the first inequality listed above can be written as:

$$2x + y < f(x)$$

The Domain and Range of a Function

As mentioned, a relation consists of a set of inputs (the x-values), known as the **domain**, and a set of outputs (the y-values), known as the **range**. A function is a relation in which each member of the domain is paired to only one other member of the range. In other words, each input has only one output.

Here is an example of a relation that is not a function:

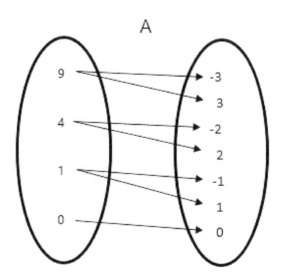

A

Every member of the first set, the domain, is mapped to two members of the second set, the range. Therefore, this relation is not a function.

In addition to a diagram representing sets, a function can be represented by a table of ordered pairs, a graph of ordered pairs (a scatterplot), or a set of ordered pairs as shown in the following:

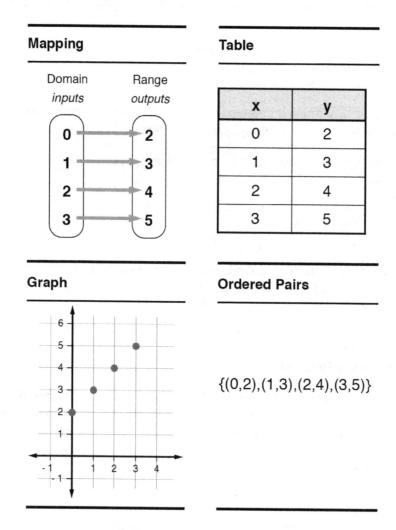

Mapping

Domain
inputs

Range
outputs

Table

x	y
0	2
1	3
2	4
3	5

Graph

Ordered Pairs

$\{(0,2),(1,3),(2,4),(3,5)\}$

Note that this relation is a function because every member of the domain is mapped to exactly one member of the range.

Writing a Function that Describes and Models a Relationship Between Two Quantities

A function can be written that models the relationship between the corresponding terms of numerical patterns. Sets of numerical patterns can be found by starting with a number and following a given rule. If two sets are generated, the corresponding terms in each set can be found to relate to one another by one or more operations. For example, the following table shows two sets of numbers that each follow their own pattern.

The first column shows a pattern of numbers increasing by 1. The second column shows the numbers increasing by 4. Because the numbers are lined up, corresponding numbers are side by side for the two sets. A question to ask is, "How can the number in the first column be turned into the number in the second column?"

1	4
2	8
3	12
4	16
5	20

This answer will lead to the relationship between the two sets. By recognizing the multiples of 4 in the right column and the counting numbers in order in the left column, the relationship of multiplying by four is determined. The first set is multiplied by 4 to get the second set of numbers. To confirm this relationship, each pair of corresponding numbers can be checked. For any two sets of numerical patterns, the corresponding numbers can be lined up to find how each one relates to the other. In some cases, the relationship is simply addition or subtraction, multiplication or division. In other relationships, these operations are used in conjunction with each together. As seen in the following table, the relationship uses multiplication and addition. The following expression shows this relationship: $3x + 2$. The x represents the numbers in the first column.

1	5
2	8
3	11
4	14

A linear function that models a linear relationship between two quantities is of the form $y = mx + b$, or in function form $f(x) = mx + b$. In a linear function, the value of y depends on the value of x, and y increases or decreases at a constant rate as x increases. Therefore, the independent variable is x, and the dependent variable is y. The graph of a linear function is a line, and the constant rate can be seen by looking at the steepness, or **slope**, of the line. If the line increases from left to right, the slope is positive. If the line slopes downward from left to right, the slope is negative. In the function, m represents slope. Each point on the line is an **ordered pair** (x, y), where x represents the x-coordinate of the point and y represents the y-coordinate of the point. The point where $x = 0$ is known as the y-intercept, and it is the place where the line crosses the y-axis. If $x = 0$ is plugged into $f(x) = mx + b$, the result is $f(0) = b$, so therefore, the point $(0, b)$ is the y-intercept of the line. The derivative of a linear function is its slope.

Consider the following situation. A taxicab driver charges a flat fee of $2 per ride and $3 a mile. This statement can be modeled by the function $f(x) = 3x + 2$ where x represents the number of miles and $f(x) = y$ represents the total cost of the ride. The total cost increases at a constant rate of $2 per mile, and that is why this situation is a linear relationship. The slope $m = 3$ is equivalent to this rate of change. The flat fee of $2 is the y-intercept. It is the place where the graph crosses the x-axis, and it represents the cost when $x = 0$, or when no miles have been traveled in the cab. The y-intercept in this situation represents the flat fee.

When given data in ordered pairs, choosing an appropriate function or equation to model the data is important. Besides linear relationships, other common relationships that exist are quadratic and exponential. A helpful way to determine what type of function to use is to find the difference between consecutive dependent variables. Basically, find pairs of ordered pairs where the x-values increase by 1, and take the difference of the y-values. If the first differences are always the same value, then the function is **linear**. If the first differences are not the same, the function could be **quadratic** or **exponential.** If the differences are not the same, find differences of those differences. If consecutive differences are the same, then the function is **quadratic**. If consecutive differences are not the same, try taking ratios of consecutive y-values. If the ratios are the same, the data have an exponential relationship and an exponential function should be used.

For example, the ordered pairs (1, 4), (2, 6), (3 ,8), and (4,10) have a linear relationship because the difference in y-values is 2. The ordered pairs (1, 0), (2, 3), (3, 10), and (4, 21) have a nonlinear relationship. The first differences in y-values are 3, 7, and 11, however, consecutive second differences are both 4. Third, the ordered pairs (1, 10), (2, 30), (3, 90), and (4, 270) have an exponential relationship. Taking ratios of consecutive y-values leads to a common ratio of 4.

The general form of a **quadratic equation** is $y = ax^2 + bx + c$, and its vertex form is:

$$y = a(x - h)^2 + k, \text{ with vertex } (h, k)$$

If the vertex and one other point are known, the vertex form should be used to solve for a. If three points, not the vertex, are known, the general form should be used. The three points create a system of three equations in three unknowns that can be solved for.

The general form of an exponential function is $y = b \times a^x$, where a is the base and b is the y-intercept.

Explaining the Steps in Solving a Simple Equation

One-step problems take only one mathematical step to solve. For example, solving the equation $5x = 45$ is a one-step problem because the one step of dividing both sides of the equation by 5 is the only step necessary to obtain the solution $x = 9$. The multiplication principle of equality is the one step used to isolate the variable. The equation is of the form $ax = b$, where a and b are rational numbers. Similarly, the addition principle of equality could be the one step needed to solve a problem. In this case, the equation would be of the form $x + a = b$ or $x - a = b$, for real numbers a and b.

A multi-step problem involves more than one step to find the solution, or it could consist of solving more than one equation. An equation that involves both the addition principle and the multiplication principle is a two-step problem, and an example of such an equation is $2x - 4 = 5$. Solving involves adding 4 to both sides and then dividing both sides by 2. An example of a two-step problem involving two separate equations is $y = 3x, 2x + y = 4$. The two equations form a system of two equations that must be solved together in two variables. The system can be solved by the substitution method. Since y is already solved for in terms of x, plug $3x$ in for y into the equation $2x + y = 4$, resulting in $2x + 3x = 4$. Therefore, $5x = 4$ and $x = \frac{4}{5}$. Because there are two variables, the solution consists of both a value for x and for y. Substitute $x = \frac{4}{5}$ into either original equation to find y. The easiest choice is $y = 3x$. Therefore,

$$y = 3 \times \frac{4}{5} = \frac{12}{5}$$

The solution can be written as the ordered pair $\left(\frac{4}{5}, \frac{12}{5}\right)$.

190

Real-world problems can be translated into both one-step and multi-step problems. In either case, the word problem must be translated from the verbal form into mathematical expressions and equations that can be solved using algebra. An example of a one-step real-world problem is the following: A cat weighs half as much as a dog living in the same house. If the dog weighs 14.5 pounds, how much does the cat weigh? To solve this problem, an equation can be used. In any word problem, the first step must be defining variables that represent the unknown quantities. For this problem, let x be equal to the unknown weight of the cat. Because two times the weight of the cat equals 14.5 pounds, the equation to be solved is: $2x = 14.5$. Use the multiplication principle to divide both sides by 2. Therefore, $x = 7.25$. The cat weighs 7.25 pounds.

Most of the time, real-world problems are more difficult than this one and consist of multi-step problems. The following is an example of a multi-step problem: *The sum of two consecutive page numbers is equal to 437. What are those page numbers?* First, define the unknown quantities. If x is equal to the first page number, then $x + 1$ is equal to the next page number because they are consecutive integers. Their sum is equal to 437, and this statement translates to the equation $x + x + 1 = 437$. To solve, first collect like terms to obtain $2x + 1 = 437$. Then, subtract 1 from both sides and then divide by 2. The solution to the equation is $x = 218$. Therefore, the two consecutive page numbers that satisfy the problem are 218 and 219. It is always important to make sure that answers to real-world problems make sense. For instance, if the solution to this same problem resulted in decimals, that should be a red flag indicating the need to check the work. Page numbers are whole numbers; therefore, if decimals are found to be answers, the solution process should be double-checked to see where mistakes were made.

Average Rate of Change of a Function

Linear growth involves a quantity, the **dependent variable**, increasing or decreasing at a constant rate as another quantity, the **independent variable**, increases as well. The graph of linear growth is a straight line. Linear growth is represented as the following equation: $y = mx + b$, where m is the slope of the line, also known as the **rate of change**, and b is the y-intercept. If the y-intercept is 0, then the linear growth is actually known as direct variation. If the slope is positive, the dependent variable increases as the independent variable increases, and if the slope is negative, the dependent variable decreases as the independent variable increases.

Graphs, equations, and tables are three different ways to represent linear functions. The following graph shows a linear function because the relationship between the two variables is constant. As the distance increases by 25 miles, the time lapses by 1 hour. This pattern continues for the rest of the graph. The line represents a constant rate of 25 miles per hour. This graph can also be used to solve problems involving predictions for a future time. After 8 hours of travel, the rate can be used to predict the distance covered. Eight hours of travel at 25 miles per hour covers a distance of 200 miles. The equation at the top of the graph corresponds to this rate also. The same prediction of distance in a given time can be found using the equation. For a time of 10 hours, the distance would be 250 miles, as the equation yields $d = 25 \times 10 = 250 \ miles$.

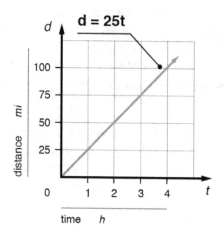

Another representation of a linear relationship can be seen in a table. The first thing to observe from the table is that the y-values increase by the same amount of 3 each time. As the x-values increase by 1, the y-values increase by 3. This pattern shows that the relationship is linear. If this table shows the money earned, y-value, for the hours worked, x-value, then it can be used to predict how much money will be earned for future hours. If 6 hours are worked, then the pay would be $19. For further hours and money to be determined, it would be helpful to have an equation that models this table of values. The equation will show the relationship between x and y. The y-value can each time be determined by multiplying the x-value by 3, then adding 1. The following equation models this relationship: $y = 3x + 1$. Now that there is an equation, any number of hours, x can be substituted into the equation to find the amount of money earned, y.

y = 3x + 1	
x	**y**
0	1
1	4
2	7
4	13
5	16

Exponential growth involves a quantity, the dependent variable, changing by a common ratio every unit increase or equal interval. The equation of exponential growth is $y = a^x$ for $a > 0, a \neq 1$. The value a is known as the **base.** Consider the exponential equation $y = 2^x$. When $x = 1, y = 2$, and when $x = 2, y = 4$.

For every unit increase in x, the value of the output variable doubles. Here is the graph of $y = 2^x$. Notice that as the dependent variable, y, gets very large, x increases slightly. This characteristic of the graph is sometimes why a quantity is said to be blowing up exponentially.

$y = 2^x$

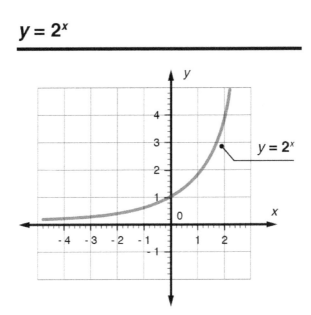

Practice Questions

1. What is $\frac{12}{60}$ converted to a percentage?
   - a. 0.20
   - b. 20%
   - c. 25%
   - d. 12%

2. Which of the following represents the correct sum of $\frac{14}{15}$ and $\frac{2}{5}$?
   - a. $\frac{20}{15}$
   - b. $\frac{4}{3}$
   - c. $\frac{16}{20}$
   - d. $\frac{4}{5}$

3. What is the product of $\frac{5}{14}$ and $\frac{7}{20}$?
   - a. $\frac{1}{8}$
   - b. $\frac{35}{280}$
   - c. $\frac{12}{34}$
   - d. $\frac{1}{2}$

4. What is the result of dividing 24 by $\frac{8}{5}$?
   - a. $\frac{5}{3}$
   - b. $\frac{3}{5}$
   - c. $\frac{120}{8}$
   - d. 15

5. Subtract $\frac{5}{14}$ from $\frac{5}{24}$. Which of the following is the correct result?

   a. $\frac{25}{168}$

   b. 0

   c. $-\frac{25}{168}$

   d. $\frac{1}{10}$

6. Which of the following is a correct mathematical statement?

   a. $\frac{1}{3} < -\frac{4}{3}$

   b. $-\frac{1}{3} > \frac{4}{3}$

   c. $\frac{1}{3} > -\frac{4}{3}$

   d. $-\frac{1}{3} \geq \frac{4}{3}$

7. Which of the following is INCORRECT?

   a. $-\frac{1}{5} < \frac{4}{5}$

   b. $\frac{4}{5} > -\frac{1}{5}$

   c. $-\frac{1}{5} > \frac{4}{5}$

   d. $\frac{1}{5} > -\frac{4}{5}$

8. How many cases of cola can Lexi purchase if each case is $3.50 and she has $40?

   a. 10
   b. 12
   c. 11.4
   d. 11

9. A car manufacturer usually makes 15,412 SUVs, 25,815 station wagons, 50,412 sedans, 8,123 trucks, and 18,312 hybrids a month. About how many cars are manufactured each month?

   a. 120,000
   b. 200,000
   c. 300,000
   d. 12,000

10. Each year, a family goes to the grocery store every week and spends $105. About how much does the family spend annually on groceries?

   a. $10,000
   b. $50,000
   c. $500
   d. $5,000

11. A grocery store sold 48 bags of apples in one day, and 9 of the bags contained Granny Smith apples. The rest contained Red Delicious apples. What is the ratio of bags of Granny Smith to bags of Red Delicious apples that were sold?

    a. 48:9
    b. 39:9
    c. 9:48
    d. 9:39

12. If Oscar's bank account totaled $4,000 in March and $4,900 in June, what was the rate of change in his bank account over those three months?

    a. $900 a month
    b. $300 a month
    c. $4,900 a month
    d. $100 a month

13. Erin and Katie work at the same ice cream shop. Together, they always work less than 21 hours a week. In a week, if Katie worked two times as many hours as Erin, how many hours did Erin work?

    a. Less than 7 hours
    b. Less than or equal to 7 hours
    c. More than 7 hours
    d. Less than 8 hours

14. Which of the following is the correct decimal form of the fraction $\frac{14}{33}$ rounded to the nearest hundredth place?

    a. 0.420
    b. 0.42
    c. 0.424
    d. 0.140

15. Gina took an algebra test last Friday. There were 35 questions, and she answered 60% of them correctly. How many correct answers did she have?

    a. 35
    b. 20
    c. 21
    d. 25

16. Paul took a written driving test, and he got 12 of the questions correct. If he answered 75% of the questions correctly, how many problems were there in the test?

    a. 25
    b. 16
    c. 20
    d. 18

17. What is the solution to the equation $3(x + 2) = 14x - 5$?

    a. $x = 1$
    b. $x = 0$
    c. All real numbers
    d. There is no solution

18. What is the solution to the equation $10 - 5x + 2 = 7x + 12 - 12x$?
    a. $x = 1$
    b. $x = 0$
    c. All real numbers
    d. There is no solution

19. Which of the following is the result when solving the equation $4(x + 5) + 6 = 2(2x + 3)$?
    a. $x = 26$
    b. $x = 6$
    c. All real numbers
    d. There is no solution

20. Two consecutive integers exist such that the sum of three times the first and two less than the second is equal to 411. What are those integers?
    a. 103 and 104
    b. 104 and 105
    c. 102 and 103
    d. 100 and 101

21. In a neighborhood, 15 out of 80 of the households have children under the age of 18. What percentage of the households have children?
    a. 0.1875%
    b. 18.75%
    c. 1.875%
    d. 15%

22. If a car is purchased for $15,395 with a 7.25% sales tax, what is the total price?
    a. $15,395.07
    b. $16,511.14
    c. $16,411.13
    d. $15,402

23. From the chart below, which two are preferred by more men than women?

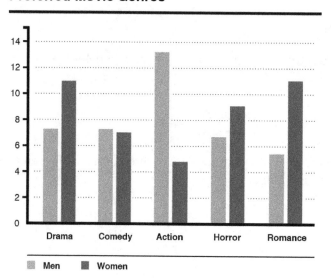

Preferred Movie Genres

   a. Comedy and Action
   b. Drama and Comedy
   c. Action and Horror
   d. Action and Romance

24. Which type of graph best represents a continuous change over a period of time?
   a. Bar graph
   b. Line graph
   c. Pie graph
   d. Histogram

25. Using the graph below, what is the mean number of visitors for the first 4 hours?

Museum Visitors

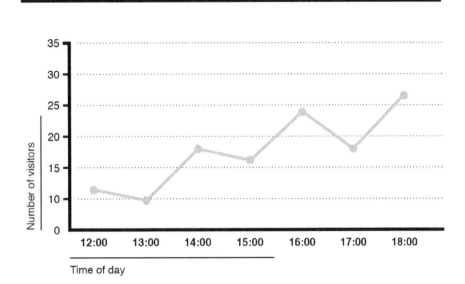

Time of day

a. 12
b. 13
c. 14
d. 15

26. What is the mode for the grades shown in the chart below?

Science Grades	
Jerry	65
Bill	95
Anna	80
Beth	95
Sara	85
Ben	72
Jordan	98

a. 65
b. 33
c. 95
d. 90

27. What type of relationship is there between age and attention span as represented in the graph below?

Attention Span

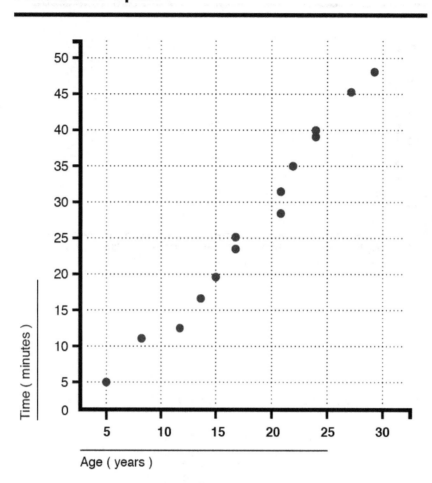

a. No correlation
b. Positive correlation
c. Negative correlation
d. Weak correlation

28. How many kiloliters are in 6 liters?
   a. 6,000
   b. 600
   c. 0.006
   d. 0.0006

29. Which of the following relations is a function?
    a. {(1, 4), (1, 3), (2, 4), (5, 6)}
    b. {(-1, -1), (-2, -2), (-3, -3), (-4, -4)}
    c. {(0, 0), (1, 0), (2, 0), (1, 1)}
    d. {(1, 0), (1, 2), (1, 3), (1, 4)}

30. Find the indicated function value: $f(5)$ for $f(x) = x^2 - 2x + 1$.
    a. 16
    b. 1
    c. 5
    d. Does not exist

31. What is the domain of $f(x) = 4x^2 + 2x - 1$?
    a. $(0, \infty)$
    b. $(-\infty, 0)$
    c. $(-\infty, \infty)$
    d. $(-1, 4)$

32. What is the range of the polynomial function $f(x) = 2x^2 + 5$?
    a. $(-\infty, \infty)$
    b. $(2, \infty)$
    c. $(0, \infty)$
    d. $[5, \infty)$

33. For which two values of x is $g(x) = 4x + 4$ equal to $g(x) = x^2 + 3x + 2$?
    a. 1, 0
    b. 2, -1
    c. -1, 2
    d. 1, 2

34. The population of coyotes in the local national forest has been declining since 2000. The population can be modeled by the function $y = -(x - 2)^2 + 1600$, where y represents number of coyotes and x represents the number of years past 2000. When will there be no more coyotes?
    a. 2020
    b. 2040
    c. 2012
    d. 2042

35. A ball is thrown up from a building that is 800 feet high. Its position s in feet above the ground is given by the function $s = -32t^2 + 90t + 800$, where t is the number of seconds since the ball was thrown. How long will it take for the ball to come back to its starting point? Round your answer to the nearest tenth of a second.
    a. 0 seconds
    b. 2.8 seconds
    c. 3 seconds
    d. 8 seconds

36. What is the domain of the following rational function?

$$f(x) = \frac{x^3 + 2x + 1}{2 - x}$$

   a. $(-\infty, -2) \cup (-2, \infty)$
   b. $(-\infty, 2) \cup (2, \infty)$
   c. $(2, \infty)$
   d. $(-2, \infty)$

37. What is the missing length x?

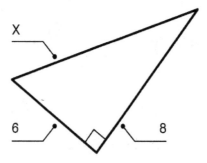

   a. 6
   b. 14
   c. 10
   d. 100

38. A study of adult drivers finds that it is likely that an adult driver wears his seatbelt. Which of the following could be the probability that an adult driver wears his seat belt?
   a. 0.90
   b. 0.05
   c. 0.25
   d. 0

39. What is the solution to the following linear inequality?

$$7 - \frac{4}{5}x < \frac{3}{5}$$

   a. $(-\infty, 8)$
   b. $(8, \infty)$
   c. $[8, \infty)$
   d. $(-\infty, 8]$

40. What is the solution to the following system of linear equations?
$$2x + y = 14$$
$$4x + 2y = -28$$

   a. (0, 0)
   b. (14, -28)
   c. All real numbers
   d. There is no solution

41. Which of the following is perpendicular to the line $4x + 7y = 23$?

 a. $y = -\frac{4}{7}x + 23$

 b. $y = \frac{7}{4}x - 12$

 c. $4x + 7y = 14$

 d. $y = -\frac{7}{4}x + 11$

42. What is the solution to the following system of equations?
$$2x - y = 6$$
$$y = 8x$$

 a. (1, 8)
 b. (-1, 8)
 c. (-1, -8)
 d. There is no solution.

43. The mass of the moon is about 7.348×10^{22} kilograms and the mass of Earth is 5.972×10^{24} kilograms. How many times GREATER is Earth's mass than the moon's mass?

 a. 8.127×10^1
 b. 8.127
 c. 812.7
 d. 8.127×10^{-1}

44. The percentage of smokers above the age of 18 in 2000 was 23.2 percent. The percentage of smokers over the age of 18 in 2015 was 15.1 percent. Find the average rate of change in the percentage of smokers over the age of 18 from 2000 to 2015.

 a. -.54 percent
 b. -54 percent
 c. -5.4 percent
 d. -15 percent

45. Triple the difference of five and a number is equal to the sum of that number and 5. What is the number?

 a. 5
 b. 2
 c. 5.5
 d. 2.5

46. In order to estimate deer population in a forest, biologists obtained a sample of deer in that forest and tagged each one of them. The sample had 300 deer in total. They returned a week later and harmlessly captured 400 deer, and 5 were tagged. Using this information, which of the following is the best estimate of the total number of deer in the forest?

 a. 24,000 deer
 b. 30,000 deer
 c. 40,000 deer
 d. 100,000 deer

47. What is the correct factorization of the following binomial?
$$2y^3 - 128$$

    a. $2(y+8)(y-8)$
    b. $2(y-4)(y^2+4y+16)$
    c. $2(y-4)(y+4)^2$
    d. $2(y-4)^3$

48. What is the simplified form of $(4y^3)^4(3y^7)^2$?
    a. $12y^{26}$
    b. $2304y^{16}$
    c. $12y^{14}$
    d. $2304y^{26}$

49. The number of members of the House of Representatives varies directly with the total population in a state. If the state of New York has 19,800,000 residents and has 27 total representatives, how many should Ohio have with a population of 11,800,000?
    a. 10
    b. 16
    c. 11
    d. 5

50. The following set represents the test scores from a university class: {35, 79, 80, 87, 87, 90, 92, 95, 95, 98, 99}. If the outlier is removed from this set, which of the following is TRUE?
    a. The mean and the median will decrease.
    b. The mean and the median will increase.
    c. The mean and the mode will increase.
    d. The mean and the mode will decrease.

51. Which of the statements below is a statistical question?
    a. What was your grade on the last test?
    b. What were the grades of the students in your class on the last test?
    c. What kind of car do you drive?
    d. What was Sam's time in the marathon?

52. Eva Jane is practicing for an upcoming 5K run. She has recorded the following times (in minutes):
    25, 18, 23, 28, 30, 22.5, 23, 33, 20
Use the above information to answer the next three questions to the closest minute. What is Eva Jane's mean time?
    a. 26 minutes
    b. 19 minutes
    c. 25 minutes
    d. 23 minutes

53. What is the mode of Eva Jane's time?
    a. 16 minutes
    b. 20 minutes
    c. 23 minutes
    d. 33 minutes

54. What is Eva Jane's median score?
    a. 23 minutes
    b. 17 minutes
    c. 28 minutes
    d. 19 minutes

55. Use the graph below entitled "Projected Temperatures for Tomorrow's Winter Storm" to answer the question.

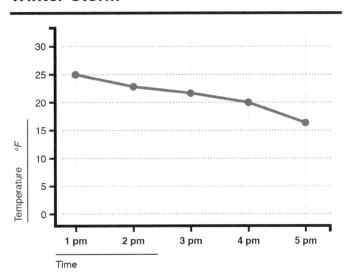

What is the expected temperature at 3:00 p.m.?
    a. 25 degrees
    b. 22 degrees
    c. 20 degrees
    d. 16 degrees

56. The function $f(x) = 3.1x + 240$ models the total U.S. population, in millions, x years after the year 1980. Use this function to answer the following question: What is the total U.S. population in 2011? Round to the nearest million.
    a. 336 people
    b. 336 million people
    c. 6,474 people
    d. 647 million people

57. What are the zeros of the following quadratic function?
$$f(x) = 2x^2 - 12x + 16$$
    a. $x = 2$ and $x = 4$
    b. $x = 8$ and $x = 2$
    c. $x = 2$ and $x = 0$
    d. $x = 0$ and $x = 4$

58. Bindee is having a barbeque on Sunday and needs 12 packets of ketchup for every 5 guests. If 60 guests are coming, how many packets of ketchup should she buy?

    a. 100

    b. 12

    c. 144

    d. 60

59. What is the volume of the cylinder below?

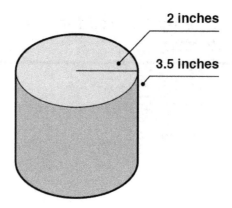

2 inches

3.5 inches

    a. 18.84 in^3

    b. 45.00 in^3

    c. 70.43 in^3

    d. 43.96 in^3

60. Given the linear function $g(x) = \frac{1}{4}x - 2$, which domain value corresponds to a range value of $\frac{1}{8}$?

    a. $\frac{17}{2}$

    b. $-\frac{63}{32}$

    c. 0

    d. $\frac{2}{17}$

61. What is the equation of the line that passes through the two points (-3, 7) and (-1, -5)?

    a. $y = 6x + 11$

    b. $y = 6x$

    c. $y = -6x - 11$

    d. $y = -6x$

62. Which of the following is the equation of a vertical line that runs through the point (1, 4)?

    a. $x = 1$

    b. $y = 1$

    c. $x = 4$

    d. $y = 4$

63. What is the area of the following figure?

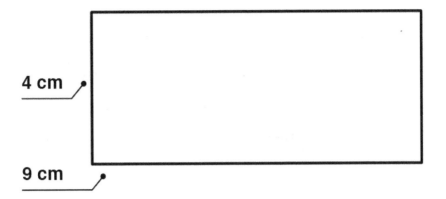

a. 26 cm
b. 36 cm
c. 13 cm^2
d. 36 cm^2

Answer Explanations

1. B: The fraction $\frac{12}{60}$ can be reduced to $\frac{1}{5}$, which puts the fraction in lowest terms. First, it must be converted to a decimal. Dividing 1 by 5 results in 0.2. Then, to convert to a percentage, move the decimal point two units to the right and add the percentage symbol. The result is 20%.

2. B: Common denominators must be used. The LCD is 15, and $\frac{2}{5} = \frac{6}{15}$. Therefore, $\frac{14}{15} + \frac{6}{15} = \frac{20}{15}$, and in lowest terms, the answer is $\frac{4}{3}$. A common factor of 5 was divided out of both the numerator and denominator.

3. A: A product is found by multiplying. Multiplying two fractions together is easier when common factors are cancelled first to avoid working with larger numbers.

$$\frac{5}{14} \times \frac{7}{20}$$

$$\frac{5}{2 \times 7} \times \frac{7}{5 \times 4}$$

$$\frac{1}{2} \times \frac{1}{4} = \frac{1}{8}$$

4. D: Division is completed by multiplying by the reciprocal. Therefore:

$$24 \div \frac{8}{5} = \frac{24}{1} \times \frac{5}{8}$$

$$\frac{3 \times 8}{1} \times \frac{5}{8}$$

$$\frac{15}{1} = 15$$

5. C: Common denominators must be used. The LCD is 168, so each fraction must be converted to have 168 as the denominator.

$$\frac{5}{24} - \frac{5}{14} = \frac{5}{24} \times \frac{7}{7} - \frac{5}{14} \times \frac{12}{12}$$

$$\frac{35}{168} - \frac{60}{168} = -\frac{25}{168}$$

6. C: The correct mathematical statement is the one in which the smaller of the two numbers is on the "less than" side of the inequality symbol. It is written in answer C that $\frac{1}{3} > -\frac{4}{3}$, which is the same as $-\frac{4}{3} < \frac{1}{3}$, a correct statement.

7. C: $-\frac{1}{5} > \frac{4}{5}$ is incorrect. The expression on the left is negative, which means that it is smaller than the expression on the right. As it is written, the inequality states that the expression on the left is greater than the expression on the right, which is not true.

8. D: This is a one-step real-world application problem. The unknown quantity is the number of cases of cola to be purchased. Let x be equal to this amount. Because each case costs $3.50, the total number of cases multiplied by $3.50 must equal $40. This translates to the mathematical equation $3.5x = 40$. Divide both sides by 3.5 to obtain $x = 11.4286$, which has been rounded to four decimal places. Because cases are sold whole (the store does not sell portions of cases), and there is not enough money to purchase 12 cases. Therefore, there is only enough money to purchase 11 cases.

9. A: Rounding can be used to find the best approximation. All of the values can be rounded to the nearest thousand. 15,412 SUVs can be rounded to 15,000. 25,815 station wagons can be rounded to 26,000. 50,412 sedans can be rounded to 50,000. 8,123 trucks can be rounded to 8,000. Finally, 18,312 hybrids can be rounded to 18,000. The sum of the rounded values is 117,000, which is closest to 120,000.

10. D: There are 52 weeks in a year, and if the family spends $105 each week, that amount is close to $100. A good approximation is $100 a week for 50 weeks, which is found through the product $50 \times 100 = \$5,000$.

11. D: There were 48 total bags of apples sold. If 9 bags were Granny Smith and the rest were Red Delicious, then 48 – 9 = 39 bags were Red Delicious. Therefore, the ratio of Granny Smith to Red Delicious is 9:39.

12. B: The average rate of change is found by calculating the difference in dollars over the elapsed time. Therefore, the rate of change is equal to $4,900−$4,000÷3 months, which is equal to $900÷3 or $300 a month.

13. A: Let x be the unknown, the number of hours Erin can work. We know Katie works $2x$, and the sum of all hours is less than 21. Therefore, $x + 2x < 21$, which simplifies into $3x < 21$. Solving this results in the inequality $x < 7$ after dividing both sides by 3. Therefore, Erin worked less than 7 hours.

14. B: If a calculator is used, divide 33 into 14 and keep two decimal places. If a calculator is not used, multiply both the numerator and denominator times 3. This results in the fraction $\frac{42}{99}$, and hence a decimal of 0.42.

15. C: Gina answered 60% of 35 questions correctly; 60% can be expressed as the decimal 0.60. Therefore, she answered $0.60 \times 35 = 21$ questions correctly.

16. B: The unknown quantity is the number of total questions on the test. Let x be equal to this unknown quantity. Therefore, $0.75x = 12$. Divide both sides by 0.75 to obtain $x = 16$.

17. A: First, the distributive property must be used on the left side. This results in $3x + 6 = 14x − 5$. The addition property is then used to add 5 to both sides, and then to subtract $3x$ from both sides, resulting in $11 = 11x$. Finally, the multiplication property is used to divide each side by 11. Therefore, $x = 1$ is the solution.

18. C: First, like terms are collected to obtain $12 − 5x = −5x + 12$. Then, if the addition principle is used to move the terms with the variable, $5x$ is added to both sides and the mathematical statement $12 = 12$ is obtained. This is always true; therefore, all real numbers satisfy the original equation.

19. D: The distributive property is used on both sides to obtain $4x + 20 + 6 = 4x + 6$. Then, like terms are collected on the left, resulting in $4x + 26 = 4x + 6$. Next, the addition principle is used to subtract $4x$ from both sides, and this results in the false statement $26 = 6$. Therefore, there is no solution.

20. A: First, the variables have to be defined. Let x be the first integer; therefore, $x + 1$ is the second integer. This is a two-step problem. The sum of three times the first and two less than the second is translated into the following expression: $3x + (x + 1 - 2)$. This expression is set equal to 411 to obtain $3x + (x + 1 - 2) = 412$. The left-hand side is simplified to obtain $4x - 1 = 411$. The addition and multiplication properties are used to solve for x. First, add 1 to both sides and then divide both sides by 4 to obtain $x = 103$. The next consecutive integer is 104.

21. B: First, the information is translated into the ratio $\frac{15}{80}$. To find the percentage, translate this fraction into a decimal by dividing 15 by 80. The corresponding decimal is 0.1875. Move the decimal point two places to the right to obtain the percentage 18.75%.

22. B: If sales tax is 7.25%, the price of the car must be multiplied by 1.0725 to account for the additional sales tax. Therefore, $15,395 \times 1.0725 = 16,511.1375$. This amount is rounded to the nearest cent, which is $16,511.14.

23. A: The chart is a bar chart showing how many men and women prefer each genre of movies. The dark gray bars represent the number of women, while the light gray bars represent the number of men. The light gray bars are higher and represent more men than women for the genres of Comedy and Action.

24. B: A line graph represents continuous change over time. The line on the graph is continuous and not broken, as on a scatter plot. A bar graph may show change but isn't necessarily continuous over time. A pie graph is better for representing percentages of a whole. Histograms are best used in grouping sets of data in bins to show the frequency of a certain variable.

25. C: The mean for the number of visitors during the first 4 hours is 14. The mean is found by calculating the average for the four hours. Adding up the total number of visitors during those hours gives:

$$12 + 10 + 18 + 16 = 56$$

Then:

$$56 \div 4 = 14$$

26. C: The mode for a set of data is the value that occurs the most. The grade that appears the most is 95. It's the only value that repeats in the set.

27. B: The relationship between age and time for attention span is a positive correlation because the general trend for the data is up and to the right. As age increases, so does attention span.

28. C: There are 0.006 kiloliters in 6 liters because 1 liter = 0.001 kiloliters. The conversion comes from the chart where the prefix kilo is found three places to the left of the base unit.

29. B: The only relation in which every x-value corresponds to exactly one y-value is the relation given in Choice B, making it a function. The other relations have the same first component paired up to different second components, which goes against the definition of a function.

30. A: To find a function value, plug in the number given for the variable and evaluate the expression, using the order of operations (parentheses, exponents, multiplication, division, addition, subtraction). The function given is a polynomial function:

$$f(5) = 5^2 - 2 \times 5 + 1$$
$$25 - 10 + 1 = 16$$

31. C: The function given is a polynomial function. Anything can be plugged into a polynomial function to get an output. Therefore, its domain is all real numbers, which is expressed in interval notation as $(-\infty, \infty)$.

32. D: This is a parabola that opens up, as the coefficient on the x^2 term is positive. The smallest number in its range occurs when plugging 0 into the function $f(0) = 5$. Any other output is a number larger than 5, even when a positive number is plugged in. When a negative number gets plugged into the function, the output is positive, and same with a positive number. Therefore, the domain is written as $[5, \infty)$ in interval notation.

33. C: First set the functions equal to one another, resulting in:

$$x^2 + 3x + 4 = 4x + 2$$

This is a quadratic equation, so the equivalent equation in standard form is:

$$x^2 - x + 2 = 0$$

This equation can be solved by factoring into:

$$(x - 2)(x + 1) = 0$$

Setting both factors equal to zero results in $x = 2$ and $x = -1$.

34. D: There will be no more coyotes when the population is 0, so set y equal to 0 and solve the quadratic equation:

$$0 = -(x - 2)^2 + 1600$$

Subtract 1600 from both sides, and divide through by -1. This results in:

$$1600 = (x - 2)^2$$

Then, take the square root of both sides. This process results in the following equation:

$$\pm 40 = x - 2$$

Adding 2 to both sides results in two solutions: $x = 42$ and $x = -38$. Because the problem involves years after 2000, the only solution that makes sense is 42. Add 42 to 2000; therefore, in 2042 there will be no more coyotes.

35. B: The ball is back at the starting point when the function is equal to 800 feet. Therefore, this results in solving the equation:

$$800 = -32t^2 + 90t + 800$$

Subtract 800 off of both sides and factor the remaining terms to obtain:

$$0 = 2t(-16 + 45t)$$

Setting both factors equal to 0 result in $t = 0$, which is when the ball was thrown up initially, and:

$$t = \frac{45}{16} = 2.8 \text{ seconds}$$

Therefore, it will take the ball 2.8 seconds to come back down to its staring point.

36. B: Given a rational function, the expression in the denominator can never be equal to 0. To find the domain, set the denominator equal to 0 and solve for x. This results in $2 - x = 0$, and its solution is $x = 2$. This value needs to be excluded from the set of all real numbers, and therefore the domain written in interval notation is $(-\infty, 2) \cup (2, \infty)$.

37. C: The Pythagorean Theorem can be used to find the missing length x because it is a right triangle. The theorem states that $6^2 + 8^2 = x^2$, which simplifies into $100 = x^2$. Taking the positive square root of both sides results in the missing value $x = 10$.

38. A: The probability of .9 is closer to 1 than any of the other answers. The closer a probability is to 1, the greater the likelihood that the event will occur. The probability of 0.05 shows that it is very unlikely that an adult driver will wear their seatbelt because it is close to zero. A zero probability means that it will not occur. The probability of 0.25 is closer to zero than to one, so it shows that it is unlikely an adult will wear their seatbelt.

39. B: The goal is to first isolate the variable. The fractions can easily be cleared by multiplying the entire inequality by 5, resulting in $35 - 4x < 3$. Then, subtract 35 from both sides and divide by -4. This results in $x > 8$. Notice the inequality symbol has been flipped because both sides were divided by a negative number. The solution set, all real numbers greater than 8, is written in interval notation as $(8, \infty)$. A parenthesis shows that 8 is not included in the solution set.

40. D: This system can be solved using the method of substitution. Solving the first equation for y results in $y = 14 - 2x$. Plugging this into the second equation gives $4x + 2(14 - 2x) = -28$, which simplifies to $28 = -28$, an untrue statement. Therefore, this system has no solution because no x value will satisfy the system.

41. B: The slopes of perpendicular lines are negative reciprocals, meaning their product is equal to -1. The slope of the line given needs to be found. Its equivalent form in slope-intercept form is $y = -\frac{4}{7}x + \frac{23}{7}$, so its slope is $-\frac{4}{7}$. The negative reciprocal of this number is $\frac{7}{4}$. The only line in the options given with this same slope is:

$$y = \frac{7}{4}x - 12$$

212

42. C: This system can be solved using substitution. Plug the second equation in for y in the first equation to obtain $2x - 8x = 6$, which simplifies to $-6x = 6$. Divide both sides by 6 to get $x = -1$, which is then back-substituted into either original equation to obtain $y = -8$.

43. A: Division can be used to solve this problem. The division necessary is:

$$\frac{5.972 \times 10^{24}}{7.348 \times 10^{22}}$$

To compute this division, divide the constants first then use algebraic laws of exponents to divide the exponential expression. This results in about 0.8127×10^2, which written in scientific notation is 8.127×10^1.

44. A: The formula for the rate of change is the same as slope: change in y over change in x. The y-value in this case is percentage of smokers and the x-value is year. The change in percentage of smokers from 2000 to 2015 was 8.1 percent. The change in x was $2000 - 2015 = -15$. Therefore:

$$8.1\%/_{-15} = -0.54\%$$

The percentage of smokers decreased 0.54 percent each year.

45. D: Let x be the unknown number. The difference indicates subtraction, and sum represents addition. To triple the difference, it is multiplied by 3. The problem can be expressed as the following equation:

$$3(5 - x) = x + 5$$

Distributing the 3 results in:

$$15 - 3x = x + 5$$

Subtract 5 from both sides, add $3x$ to both sides, and then divide both sides by 4. This results in:

$$\frac{10}{4} = \frac{5}{2} = 2.5$$

46. A: A proportion should be used to solve this problem. The ratio of tagged to total deer in each instance is set equal, and the unknown quantity is a variable x. The proportion is $\frac{300}{x} = \frac{5}{400}$. Cross-multiplying gives $120{,}000 = 5x$, and dividing through by 5 results in 24,000.

47. B: First, the common factor 2 can be factored out of both terms, resulting in $2(y^3 - 64)$. The resulting binomial is a difference of cubes that can be factored using the rule:

$$a^3 - b^3 = (a - b)(a^2 + ab + b^2)$$

with a = y and b = 4. Therefore, the result is:

$$2(y - 4)(y^2 + 4y + 16)$$

48. D: The exponential rules $(ab)^m = a^m b^m$ and $(a^m)^n = a^{mn}$ can be used to rewrite the expression as:

$$4^4 y^{12} \times 3^2 y^{14}$$

The coefficients are multiplied together and the exponential rule:

$$a^m a^n = a^{m+n}$$

Is then used to obtain the simplified form $2304y^{26}$.

49. B: The number of representatives varies directly with the population, so the equation necessary is $N = k \times P$, where N is number of representatives, k is the variation constant, and P is total population in millions. Plugging in the information for New York allows k to be solved for. This process gives $27 = k \times 20$, so $k = 1.35$. Therefore, the formula for number of representatives given total population in millions is $N = 1.35 \times P$. Plugging in $P = 11.6$ for Ohio results in $N = 15.66$, which rounds up to 16 total Representatives.

50. B: The outlier is 35. When a small outlier is removed from a data set, the mean and the median increase. The first step in this process is to identify the outlier, which is the number that lies away from the given set. Once the outlier is identified, the mean and median can be recalculated. The mean will be affected because it averages all of the numbers. The median will be affected because it finds the middle number, which is subject to change because a number is lost. The mode will most likely not change because it is the number that occurs the most, which will not be the outlier if there is only one outlier.

51. B: This is a statistical question because to determine this answer one would need to collect data from each person in the class and it is expected that the answers would vary. The other answers do not require data to be collected from multiple sources; therefore, the answers will not vary.

52. C: The mean is found by adding all the times together and dividing by the number of times recorded.

$$25 + 18 + 23 + 28 + 30 + 22.5 + 23 + 33 + 20 = 222.5$$

$$\text{divided by } 9 = 24$$

Rounding to the nearest minute, the mean is 25 minutes.

53. C: The mode is the time from the data set that occurs most often. The number 23 occurs twice in the data set, while all others occur only once, so the mode is 23 minutes.

54. A: To find the median of a data set, you must first list the numbers from smallest to largest, and then find the number in the middle. If there are two numbers in the middle, as in this data set, add the two numbers in the middle together and divide by 2. Putting this list in order from smallest to greatest yields 18, 20, 22.5, 23, 23, 25, 28, 30, and 33, where 23 is the middle number.

55. B: Look on the horizontal axis to find 3:00 p.m. Move up from 3:00 p.m. to reach the dot on the graph. Move horizontally to the left to the horizontal axis to between 20 and 25; the best answer choice is 22. The answer of 25 is too high above the projected time on the graph, and the answers of 20 and 16 degrees are too low.

56. B: The variable x represents the number of years after 1980. The year 2011 was 31 years after 1980, so plug 31 into the function to obtain:

$$f(31) = 3.1 \times 31 + 240 = 336.1$$

This value rounds to 336, and represents 336 million people.

57. A: The zeros of a polynomial function are the x-values where the graph crosses the x-axis, or where:

$$y = 0$$

Therefore, set $y = 0$ and solve the polynomial equation. This quadratic can be solved using factoring, as follows:

$$0 = 2x^2 - 12x + 16$$

$$2(x^2 - 6x + 8) = 2(x - 4)(x - 2)$$

Setting both factors equal to 0 results in the two solutions $x = 4$ and $x = 2$, which are the zeros of the original function.

58. C: This problem involves ratios and percentages. If 12 packets are needed for every 5 people, this statement is equivalent to the ratio $\frac{12}{5}$. The unknown amount x is the number of ketchup packets needed for 60 people. The proportion $\frac{12}{5} = \frac{x}{60}$ must be solved. Cross-multiply to obtain $12 \times 60 = 5x$. Therefore, $720 = 5x$. Divide each side by 5 to obtain $x = 144$.

59. D: The volume for a cylinder is found by using the formula:

$$V = \pi r^2 h$$

$$\pi(2^2) \times 3.5 = 43.96 in^3$$

60. A: The range value is given, and this is the output of the function. Therefore, the function must be set equal to $\frac{1}{8}$ and solved for x. Thus, $\frac{1}{8} = \frac{1}{4}x - 2$ needs to be solved. The fractions can be cleared by multiplying by the LCD 8. This results in $1 = 2x - 16$. Add 16 to both sides and divide by 2 to obtain:

$$x = \frac{17}{2}$$

61. C: First, the slope of the line must be found. This is equal to the change in y over the change in x, given the two points. Therefore, the slope is -6. The slope and one of the points are then plugged into the slope-intercept form of a line: $y - y_1 = m(x - x_1)$. This results in $y - 7 = -6(x + 3)$. The -6 is simplified and the equation is solved for y to obtain:

$$y = -6x - 11$$

62. A: A vertical line has the same x value for any point on the line. Other points on the line would be (1, 3), (1, 5), (1, 9,) etc. Mathematically, this is written as $x = 1$. A vertical line is always of the form $x = a$ for some constant a.

63. D: The area for a rectangle is found by multiplying the length by the width. The area is also measured in square units, so the correct answer is Choice D. The answer of 26 is the perimeter. The answer of 13 is found by adding the two dimensions instead of multiplying.

Science

Life Science

Organisms, Their Environments, and Their Life Cycles

Living Characteristics and Cells

Things are characterized as "living" due to a number of widely accepted factors. All living things must be made up of one cell (**unicellular**) or more (**multicellular**) that can perform the basic functions needed for living. These include the utilization of a nutrient source for energy to carry out other processes, respiration (which converts nutrients into necessary energy, and typically requires the presence of oxygen and carbon dioxide), the ability to move, the ability to create and remove waste, the ability to grow larger in some capacity, the ability to reproduce, and the ability to seek out and sense factors in their environment that contribute to survival and reproduction. Organisms consist of eukaryotic or prokaryotic cells. Animals, people, plants, fungi, and insects are **eukaryotes**, while bacteria are **prokaryotic** cells. Prokaryotic cells are much simpler than eukaryotic cells; they do not have a nucleus or other intra-cellular components.

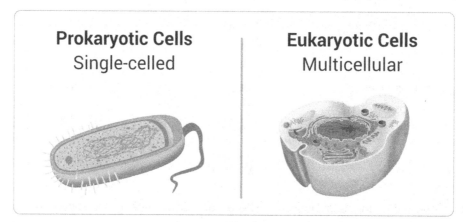

The smallest known living organism that fits these characteristics is a prokaryotic micro-bacterium; 150,000 of these bacteria could fit on the tip of a human hair. It was discovered in 2015 by University of California, Berkeley scientists who reason that there are probably many undiscovered microbes that may be as small or smaller. The largest living thing is an intact, parasitic fungus that spans approximately four miles of a forest in Oregon.

The Cell

All living organisms are made up of **cells**. They are considered the basic functional unit of organisms and the smallest unit of matter that is living. Most organisms are multicellular, which means that they are made up of more than one cell and often they are made up of a variety of different types of cells. Cells contain **organelles**, which are the little working parts of the cell, responsible for specific functions that keep the cell and organism alive.

Plant and animal cells have many of the same organelles but also have some unique traits that set them apart from each other. Plants contain a **cell wall**, while animal cells are only surrounded by a **phospholipid plasma membrane**. The cell wall is made up of strong, fibrous polysaccharides and proteins. It protects the cell from mechanical damage and maintains the cell's shape. Inside the cell wall, plant cells also have plasma membrane. The plasma membrane of both plant and animal cells is made up of two layers of phospholipids, which have a hydrophilic head and hydrophobic tails. The tails converge towards each other on the inside of the bilayer, while the heads face the interior of the cell and the exterior environment. **Microvilli** are protrusions of the cell membrane that are only found in animal cells. They increase the surface area and aid in absorption, secretion, and cellular adhesion. **Chloroplasts** are also only found in plant cells. They are responsible for **photosynthesis**, which is how plants convert sunlight into chemical energy.

The list below describes major organelles that are found in both plant and animal cells.

- **Nucleus**: The nucleus contains the DNA of the cell, which has all of the cells' hereditary information passed down from parent cells. DNA and protein are wrapped together into chromatin within the nucleus. The nucleus is surrounded by a double membrane called the nuclear envelope.

- **Endoplasmic Reticulum** (*ER*): The ER is a network of tubules and membranous sacs that are responsible for the metabolic and synthetic activities of the cell, including synthesis of membranes. Rough ER has ribosomes attached to it while smooth ER does not.

- **Mitochondrion**: The mitochondrion is essential for maintaining regular cell function and is known as the powerhouse of the cell. It is where cellular respiration occurs and where most of the cell's ATP is generated.

- **Golgi Apparatus**: The Golgi Apparatus is where cell products are synthesized, modified, sorted, and secreted out of the cell.

- **Ribosomes**: Ribosomes make up a complex that produces proteins within the cell. They can be free in the cytosol or bound to the ER.

Organ Systems

An **organ system** refers to a group of organs that work together to carry out biological functions. In organ systems, each organ is dependent on other organs to function properly. Animal species typically have a number of organ systems. These include the **respiratory system**, in which organs such as the lungs and diaphragm function to allow the organism to breathe; the **digestive system**, in which organs such as the stomach, intestines, and pancreas break down nutrients to deliver to the rest of the body or eliminate in the form of solid wastes; the **cardiovascular system**, in which organs such as the heart and blood vessels pump and carry oxygenated blood throughout the body while transporting deoxygenated blood back to the lungs; the **urinary system**, in which organs such as the kidneys and bladder balance the body's fluid and chemical levels and eliminate liquid waste; and the **integumentary system**, consisting of the skin (the body's largest organ) and other tissues. While these are all individual systems, they also work closely together within the system of the body, and the body cannot function well if one system is not working properly. For example, if the respiratory system is unable to bring in oxygen, the cardiovascular system will be unable to circulate oxygenated blood. If the digestive system is unable to break down foods into nutrients, then nutrients cannot be transported through the cardiovascular blood stream to cells that need energy.

Vascular plants are also considered to have organ systems that protect individual and bundled cells, store transport nutrients, and conduct photosynthesis. These plants also have reproductive systems that are visible in flowers or fruits.

Growth and Development

All living things grow and develop, whether through **cell division** or by **cell replication**. When cells divide, the newly created cell may have some differences from the original copy, whereas cells that replicate are identical. Cell division is often seen in periods of overall development, such as when a fetus grows in utero or a plant shoot turns into a flower. Cell replication is typically seen when a single entity becomes larger, such as when a child's bone becomes bigger as he or she ages. Growth and development are fueled by nutrition, hydration, and respiration. Animal growth requires water, macro- and micro-nutrients (carbohydrates, proteins, fats, vitamins, and minerals), and respiration to keep the body oxygenated and remove waste. Other factors that support animal growth include adequate sleep and recovery, as well as hormonal factors. Some hormonal issues may be congenital and lead to pathologies, while others may be impacted by external stressors (such as poor nutrition or sleep).

Plant growth takes place as a seedling sprouts root into soil matter and absorbs nutrients such as water, nitrogen, potassium, and phosphorus. Plants also take carbon dioxide from the air and need sunlight exposure to carry out photosynthesis, which provides their main source of energy for growth. Different plants also have different temperature requirements that aid in growth and the development of reproductive structures, such as the pistil and stamen found within flowers.

Like animals and plants, bacteria take in nutrients for physical growth. They also typically require some level of warmth and humidity to grow and reproduce. Most bacteria are unicellular and asexual; when they can no longer grow, they split through a process called **binary fission**. During this split, the unicellular organism divides into two duplicate cells containing the exact same DNA.

Basic Macromolecules in a Biological System

There are six major elements found in most biological molecules: carbon, hydrogen, oxygen, nitrogen, sulfur, and phosphorus. These elements link together to make up the basic **macromolecules** of the biological system, which are lipids, carbohydrates, nucleic acids, and proteins. Most of these molecules use carbon as their backbone because of its ability to bond four different atoms. Each type of macromolecule has a specific structure and important function for living organisms.

Lipids

Lipids are made up of hydrocarbon chains, which are large molecules with hydrogen atoms attached to a long carbon backbone. These biological molecules are characterized as hydrophobic because their structure does not allow them to form bonds with water. When mixed together, the water molecules bond to each other and exclude the lipids. There are three main types of lipids: triglycerides, phospholipids, and steroids.

Triglycerides are made up of one glycerol molecule attached to three fatty acid molecules. **Glycerols** are three-carbon atom chains with one hydroxyl group attached to each carbon atom. Fatty acids are hydrocarbon chains with a backbone of sixteen to eighteen carbon atoms and a double-bonded oxygen molecule. Triglycerides have three main functions for living organisms. They are a source of energy when carbohydrates are not available, they help with absorption of certain vitamins, and they help insulate the body and maintain normal core temperature.

Phospholipid molecules have two fatty acid molecules bonded to one glycerol molecule, with the glycerol molecules having a phosphate group attached to it. The **phosphate group** has an overall negative charge, which makes that end of the molecule hydrophilic, meaning that it can bond with water molecules. Since the fatty acid tails of phospholipids are hydrophobic, these molecules are left with the unique characteristic of having different affinities on each of their ends. When mixed with water, phospholipids create bilayers, which are double rows of molecules with the hydrophobic ends on the inside facing each other and the hydrophilic ends on the outside, shielding the hydrophobic ends from the water molecules. Cells are protected by phospholipid bilayer cell membranes. This allows them to mix with aqueous solutions while also protecting their inner contents.

Steroids have a more complex structure than the other types of lipids. They are made up of four fused carbon rings. Different types of steroids are defined by the different chemical groups that attach to the carbon rings. Steroids are often mixed into phospholipid bilayers to help maintain the structure of the cell membrane and aid in cell signaling from these positions. They are also essential for regulation of metabolism and immune responses, among other biological processes.

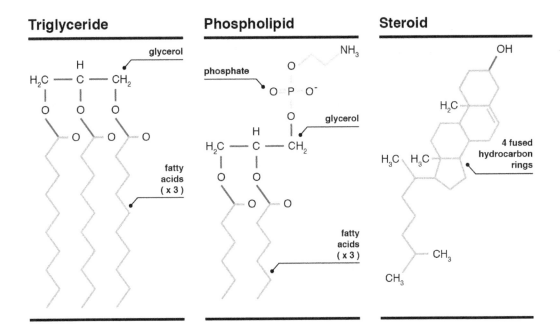

Carbohydrates
Carbohydrates are made up of sugar molecules. **Monomers** are small molecules, and **polymers** are larger molecules that consist of repeating monomers. The smallest sugar molecule, a monosaccharide, has the chemical formula of CH_2O. **Monosaccharides** can be made up of one of these small molecules or a multiple of this formula (such as $C_2H_4O_2$). **Polysaccharides** consist of repeating monosaccharides in lengths of a few hundred to a few thousand linked together. Monosaccharides are broken down by living organisms to extract energy from the sugar molecules for immediate consumption. Glucose is a common monosaccharide used by the body as a primary energy source and which can be metabolized immediately. The more complex structure of polysaccharides allows them to have a more long-term use. They can be stored and broken down later for energy. Glycogen is a molecule that consists of 1700 to 600,000 glucose units linked together. It is not soluble in water and can be stored for long periods of time. If necessary, the glycogen molecule can be broken up into single glucose molecules in order to provide energy for the body. Polysaccharides also form structurally strong materials, such as chitin, which makes up the exoskeleton of many insects, and cellulose, which is the material that surrounds plant cells.

Nucleic Acids
Nucleotides are made up of a five-carbon sugar molecule with a nitrogen-containing base and one or more phosphate groups attached to it. Nucleic acids are polymers of nucleotides, or polynucleotides. There are two main types of nucleic acids: **deoxyribonucleic acid** (*DNA*) and **ribonucleic acid** (*RNA*).

DNA is a double strand of nucleotides that are linked together and folded into a helical structure. Each strand is made up of four nucleotides, or bases: *adenine, thymine, cytosine,* and *guanine.* The adenine bases only pair with thymine on the opposite strand, and the cytosine bases only pair with guanine on the opposite strand. It is the links between these base pairs that create the helical structure of double-stranded DNA. DNA is in charge of long-term storage of genetic information that can be passed on to subsequent generations. It also contains instructions for constructing other components of the cell. RNA, on the other hand, is a single-stranded structure of nucleotides that is responsible for directing the construction of proteins within the cell. RNA is made up of three of the same nucleotides as DNA, but instead of thymine, adenine pairs with the base uracil.

Proteins

Proteins are made from a set of twenty amino acids that are linked together linearly, without branching. The amino acids have peptide bonds between them and form polypeptides. These polypeptide molecules coil up, either individually or as multiple molecules linked together, and form larger biological molecules, which are called proteins. Proteins have four distinct layers of structure. The primary structure consists of the sequence of amino acids. The secondary structure consists of the folds and coils formed by the hydrogen bonding that occurs between the atoms of the polypeptide backbone. The tertiary structure consists of the shape of the molecule, which comes from the interactions of the side chains that are linked to the polypeptide backbone. Lastly, the quaternary structure consists of the overall shape that the protein takes on when it is made up of two or more polypeptide chains. Proteins have many vital roles in living organisms. They help maintain and repair body tissue, provide a source of energy, form antibodies to aid the immune system, and are a large component in transporting molecules within the body, among many other functions.

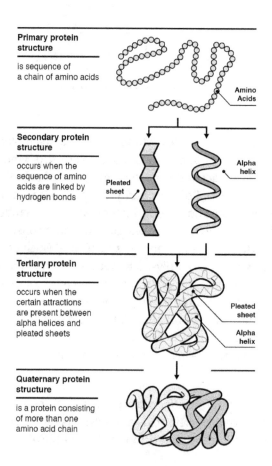

Cellular Respiration

Cellular respiration in multicellular organisms occurs in the mitochondria. It is a set of reactions that converts energy from nutrients to ATP and can either use oxygen in the process, which is called aerobic respiration, or not, which is called anaerobic respiration.

Aerobic respiration has two main parts, which are the citric acid cycle, also known as **Krebs cycle**, and **oxidative phosphorylation**. Glucose is a commonly found molecule that is used for energy production within the cell. Before the citric acid cycle can begin, the process of glycolysis converts glucose into two pyruvate molecules. Pyruvate enters the mitochondrion, is oxidized, and then is converted to a compound called acetyl CoA. There are eight steps in the citric acid cycle that start with acetyl CoA and convert it to oxaloacetate and NADH. The oxaloacetate continues in the citric acid cycle and the NADH molecule moves on to the oxidative phosphorylation part of cellular respiration. Oxidative phosphorylation has two main steps, which are the electron transport chain and chemiosmosis. The mitochondrial membrane has four protein complexes within it that help to transport electrons through the inner mitochondrial matrix. Electrons and protons are removed from NADH and $FADH_2$ and then transported along these and other membrane complexes. Protons are pumped across the inner membrane, which creates a gradient to draw electrons to the intermembrane complexes. Two mobile electron carriers, ubiquinone and cytochrome C, are also located in the inner mitochondrial membrane. At the end of these electron transport chains, the electrons are accepted by O_2 molecules and water is formed with the addition of two hydrogen atoms. Chemiosmosis occurs in an ATP synthase complex that is located next to the four electron transport complexes. As the complex pumps protons from the intermembrane space to the mitochondrial matrix, ADP molecules become phosphorylated and ATP molecules are generated. Approximately four to six ATP molecules are generated during glycolysis and the citric acid cycle and twenty-six to twenty-eight ATP molecules are generated during oxidative phosphorylation, which makes the total amount of ATP molecules generated during aerobic cellular respiration approximately thirty to thirty-two.

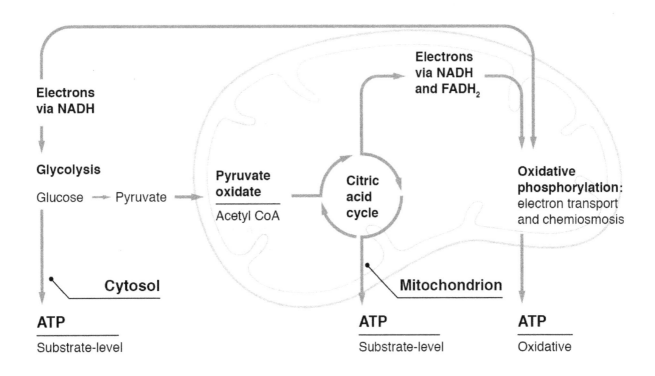

Since not all environments are oxygen-rich, some organisms must find alternate ways to extract energy from nutrients. The process of anaerobic respiration is similar to that of aerobic respiration in that protons and electrons are removed from nutrient molecules and are passed down an electron transport chain, with the end result being ATP synthesis. However, instead of the electrons being accepted by oxygen molecules at the end of the electron transport chain, they are accepted by either sulfate or nitrate molecules. Anaerobic respiration is mostly used by unicellular organisms, or prokaryotic organisms.

Photosynthesis

Photosynthesis is a set of reactions that occur to convert light energy into chemical energy. The chemical energy is then stored as sugar and other organic molecules inside the organism or plant. Within plants, the photosynthetic process takes place within chloroplasts. The two stages of photosynthesis are the light reactions and the **Calvin cycle**. Within chloroplasts, there are membranous sacs called thylakoids and within the thylakoids are a green pigment called chlorophyll. The light reactions take place in the chlorophyll. The Calvin cycle takes place in the **stroma**, or inner space, or the chloroplasts.

During the light reactions, light energy is absorbed by chlorophyll. First, a light-harvesting complex, called photosystem II (PS II), absorbs photons from light that enters the chlorophyll and then passes it onto a reaction-center complex. Once the photon enters the reaction-center complex, it causes a special pair of chlorophyll *a* molecules to release an electron. The electron is accepted by a primary electron acceptor molecule, while at the same time, a water molecule is dissociated into two hydrogen atoms, one oxygen atom, and two electrons. These electrons are transferred to the chlorophyll *a* molecules that just lost their electrons. The electrons that were released from the chlorophyll *a* molecules move down an electron transport chain using an electron carrier, called plastoquinone, a cytochrome complex, and a protein, called plastocyanin. At the end of the chain, the electrons reach another light-harvesting complex, called photosystem I (PS I). While in the cytochrome complex, the electrons cause protons to be pumped into the thylakoid space, which in turn provides energy for ATP molecules to be produced. A primary electron acceptor molecule accepts the electrons that are released from PS I and then passes them onto another electron transport chain, which includes the protein ferredoxin. At the end of the light reactions, electrons are transferred from ferredoxin to NADP+, producing NADPH. The ATP and NADPH that are produced through the light reactions are used as energy to drive the Calvin cycle forward.

The three phases of the Calvin cycle are carbon fixation, reduction, and regeneration of the CO_2 acceptor. **Carbon fixation** occurs when CO_2 is introduced into the cycle and attaches to a five-carbon sugar, called ribulose bisphosphate (RuBP). A six-carbon sugar is split into two three-carbon sugar molecules, known as 3-phosphoglycerate. Next, during the **reduction** phase, an ATP molecule loses a phosphate group and becomes ADP. The phosphate group attaches to the 3-phosphoglycerate molecule, making it 1,3-bisphosphate. Then, an NADPH molecule donates two electrons to this new molecule, causing it to lose a phosphate group and become glyceraldehyde 3-phosphate (G3P), a sugar molecule. At the end of the cycle, one G3P molecule exits the cycle and is used by the plant for energy. Five other G3P molecules continue in the cycle to **regenerate** RuBP molecules, which are the CO_2 acceptors of the cycle. When every photon has been used up, three RuBP molecules are formed from the rearrangement of five G3P molecules and wait for the cycle to start again. It takes three turns of the cycle and three CO_2 molecules entering the cycle to generate just one G3P molecule.

Refer to this flow chart to better understand the Calvin Cycle:

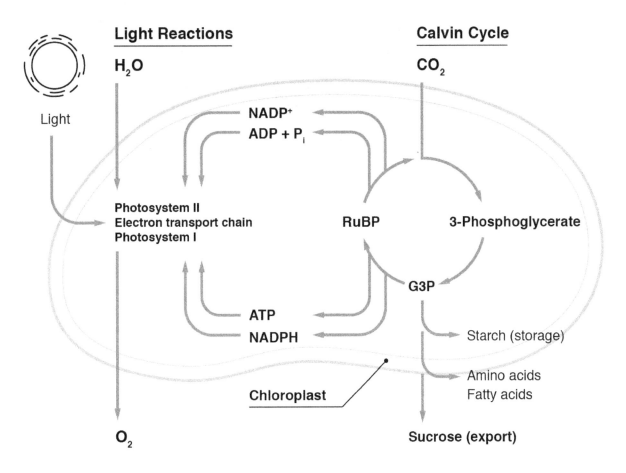

Interdependence of Organisms

Organisms on Earth are interdependent within their species and across different species. For example, all organisms are dependent on plants for food and the balance of gases in the atmosphere (i.e., producing oxygen and using carbon dioxide). However, plants benefit greatly from the presence of some fungi, which can support the nutrient composition uptake from the soil that the plant uses in photosynthesis. Humans benefit from the presence of a variety of bacteria in the gut that promote healthy digestion and nutrient absorption; in turn, the human gut provides a hospitable home for such bacteria. These types of supportive two-way interactions are called **symbiotic relationships**. Symbiotic relationships are usually divided into three different types: mutualism, parasitism, and commensalism. **Mutualism** describes the relationship between two organisms when they both benefit. In **parasitism**, one organism benefits, and the other is harmed by the interaction. **Commensalism** refers to when one organism benefits, but the other feels no effect with no benefit or harm done. **Competition** is another type of interaction between species where both groups are in pursuit of the same resource.

Consequently, it is also important to note that many organisms can be detrimental to others; for example, some fungi are toxic to plants, and many bacteria cause illnesses in larger species. **Predation** describes the relationship between predator and prey. The predator kills the prey in order to consume it.

An **ecosystem** refers to a specific physical area and the components, living and non-living, that make up the area. In an ecosystem, each component plays a functional role that can be positive or negative for the overall system. A stable, healthy ecosystem houses a precise balance of relationships between each component that allows the ecosystem to thrive. For example, a freshwater stream in a rainforest likely serves as the water source for animals and plants. The same animals graze and produce waste in the area, which fertilizes the soil for the plants. The plants contribute to the composition of the atmosphere. Additionally, a completely unique ecosystem of aquatic life likely exists within the water itself that is affected by the atmosphere. For example, a more oxygen-rich atmosphere may contribute to cleaner runoff into the water when it rains; this, in turn, will affect the type of life in the water, as well as the animals and plants that use the water. Introducing new entities into a healthy ecosystem can cause harmful shifts and disruption in the area. This can occur due to natural events (such as an earthquake or hurricane) or human activity (such as waste dumping).

Ecosystem

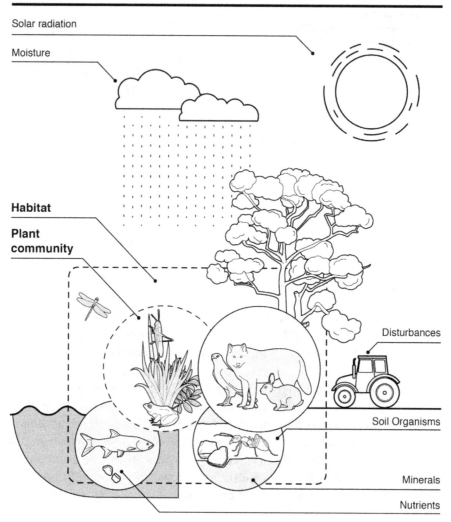

Populations of organisms typically form communities in which they can thrive off of mutually beneficial relationships. Populations refer to groups of one single species that live together or in close proximity and reproduce with one another; communities refer to members of a population that are working together toward a common goal or benefit for the species (such as procuring resources or protecting the community's physical space). Healthy populations and communities are able to work together to ensure future survival for the species.

Unity and Diversity of Life, Adaptation, and Classification

Living organisms vary in many ways, yet there are a number of similarities that help define a "living" creature. It is estimated that there are millions of distinct species; while approximately 1.5 million have been categorized, some scientists believe there could be anywhere from three to 100 million distinct species on Earth. Species are classified in a hierarchical system based on similarities in genetic composition, physical manifestations, behaviors, and other features that define how the organism survives. All species can be categorized into three domains: Bacteria, Archaea, and Eukaryotes. **Bacteria** and Archaea are single-cell and prokaryotic (having no internal structures within the cell). **Archaea** are considered the simplest of living organisms. **Eukaryotes** are far more complex; they can be single-cell or multicellular organisms. Each cell in a living organism has a variety of internal structures called organelles. *Organelles* carry out various functions that contribute to the living functions of the organism. Because different organisms evolved and adapted to external stressors in order to survive and reproduce, the need arose for a classification system to group and distinguish organisms.

After domains, organisms can further be further classified across five **kingdoms**: Bacteria, Protists, Fungi, Plants, and Animals. All organisms in the Bacteria and Archaea domains fall under the Bacteria kingdom. Protists are eukaryotic, single-celled organisms that do not qualify as bacteria or archaea. There are few commonalities among protists; these organisms simply cannot fit the classification requirements of the other five kingdoms. Fungi are eukaryotic organisms that are unique in that their cell walls are made of chitin polymers. Plants are primarily eukaryotic multicellular organisms that create energy through the process of photosynthesis. The plant kingdom is considered to be the driver for all other living organisms, as they serve as food sources, create oxygen, and balance other gases in the atmosphere. Organisms that are not classified as Bacteria, Protist, Fungi, or Plant are in the Animal kingdom. Animals are multicellular and eukaryotic organisms that are highly diverse.

After Kingdoms, organisms are categorized into a **Phylum**. There are numerous Phyla within each kingdom; new Phyla have been created to categorize new species that did not fit into the distinctions of previously established Phyla. After Phyla, species are categorized respectively into a Class, a Family, a Genus, and a Species. Distinct organisms may be referred to by their genus-species name; for example, when humans are referred to as "homo sapiens," this refers to their categorized genus ("homo") and their categorized species ("sapiens"). While a number of organisms can share Domain, Kingdom, Phyla, Class, Family, and Genus, *Species* is a distinct term (consult the chart on the following page for better understanding of this).

This chart shows how different organisms can be classified in the same ways, excluding by their species:

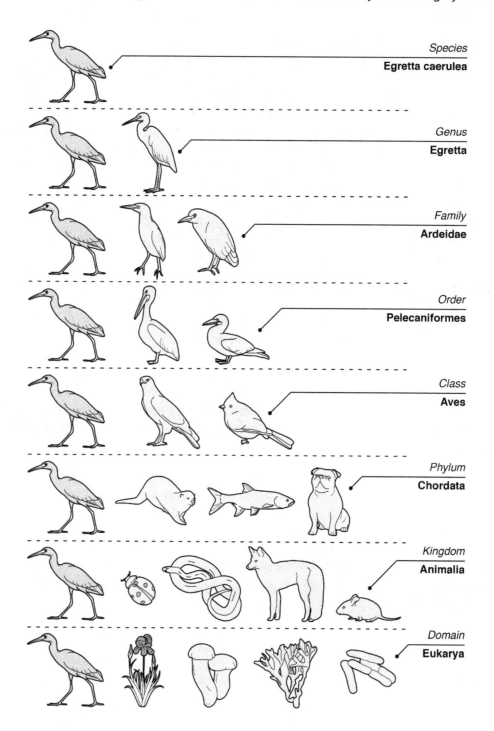

Species
Egretta caerulea

Genus
Egretta

Family
Ardeidae

Order
Pelecaniformes

Class
Aves

Phylum
Chordata

Kingdom
Animalia

Domain
Eukarya

Relationships Between Structure and Function in Living Systems

Humans are complex organisms. They have many structures and functions that work in conjunction to maintain life. The relationship of structure and function can be seen throughout living systems. For example, flagella are long whip-like structures that can move about, helping propel the cell. The small intestines contain microvilli, or small projections that significantly increase the surface area of the small intestines, which aids in absorption of nutrients. Plants have all sorts of structural adaptations that help to better serve their function. For example, root systems are far-reaching and reach underground to absorb water and minerals in the soil and anchor the plant.

There are six levels of organization that can help describe the human body. These levels, in smallest to largest size order, are chemical, cellular, tissue, organ, organ system, and organism. The **chemical level** includes atoms and molecules, which are the smallest building blocks of matter. When atoms bind together, they form molecules, which in turn make up chemicals. All body structures are made up of these small elements. **Cells** are the smallest units of living organisms. They function independently to carry out vital functions of every organism. Cells that are similar then bind together to form tissues. **Tissues** perform specific functions by having all of the cells work together. For example, muscle tissue is made up of contractile cells that help the body move. **Organs** are made up of two or more types of tissue and perform physiological functions. **Organ systems** are made up of several organs together that work to perform a major bodily function. **Organisms** include the human body as a whole and all of its structures that perform life-sustaining functions.

Human Body Systems

Respiratory System
The **respiratory system** is responsible for gas exchange between air and the blood, mainly via the act of breathing. It is divided into two sections: the upper respiratory system and the lower respiratory system. The **upper respiratory system** comprises the nose, the nasal cavity and sinuses, and the pharynx, while the **lower respiratory system** comprises the larynx (voice box), the trachea (windpipe), the small passageways leading to the lungs, and the lungs. The upper respiratory system is responsible for filtering, warming, and humidifying the air that gets passed to the lower respiratory system, protecting the lower respiratory system's more delicate tissue surfaces.

The human body has two lungs, each having its own distinct characteristics. The **right lung** is divided into three lobes (superior, middle, and inferior), and the **left lung** is divided into two lobes (superior and inferior). The left lung is smaller than the right lung, most likely because it shares space in the chest cavity with the heart. Together, the lungs contain approximately fifteen hundred miles of airway passages, which are the site of gas exchange during the act of breathing. When a breath of air is inhaled, oxygen enters the nose or mouth and passes into the sinuses, which is where the temperature and humidity of the air get regulated. The air then passes into the trachea and is filtered. From there, it travels into the bronchi and reaches the lungs. Bronchi are lined with cilia and mucus that collect dust and germs along the way. Within the lungs, oxygen and carbon dioxide are exchanged between the air in the *alveoli*, a type of airway passage in the lungs, and the blood in the **pulmonary capillaries**, a type of blood vessel in the lungs. Oxygen-rich blood returns to the heart and gets pumped through the systemic circuit. Carbon dioxide-rich air is exhaled from the body.

The respiratory system has many important functions in the body. Primarily, it is responsible for pulmonary ventilation, or the process of breathing. During pulmonary ventilation, air is inhaled through the nasal cavity or oral cavity and then moves through the pharynx, larynx, and trachea before reaching the lungs. Air is exhaled following the same pathway. When air is inhaled, the diaphragm and external intercostal muscles contract, causing the rib cage to rise and the volume of the lungs to increase. When air is exhaled, the muscles relax, and the volume of the lungs decreases. The respiratory system also provides a large area for gas exchange between the air and the circulating blood. During the process of external respiration, oxygen and carbon dioxide are exchanged within the lungs. Oxygen goes into the bloodstream, binds to hemoglobin in the red blood cells, and circulates through the body. Carbon dioxide moves from the deoxygenated blood into the alveoli of the lungs and is then exhaled from the body through the nose or mouth. Internal respiration is a process that allows gases to be exchanged between the circulating blood and body tissues. Arteries carry oxygenated blood throughout the body. When the oxygenated blood reaches narrow capillaries in the body tissue, the red blood cells release the oxygen.

The oxygen then diffuses through the capillary walls into the body tissues. At the same time, carbon dioxide diffuses from the body tissues into the blood of the narrow capillaries. Veins carry the deoxygenated blood back to the lungs for waste removal. Another function of the respiratory system is that it produces the sounds that the body makes for speaking and singing, as well as for nonverbal communication. The larynx is also known as the voice box. When air is exhaled, it passes through the larynx. The muscles of the larynx move the arytenoid cartilages, which then push the vocal cords together. When air passes through the larynx while the vocal cords are being pushed together, the vocal cords vibrate and create sound. Higher-pitched sounds are made when the vocal cords are vibrating more rapidly, and lower-pitched sounds are made when the vibrations are slower. The respiratory system also helps with an individual's sense of smell. Olfactory fibers line the nasal cavity. When air enters the nose, chemicals within the air bind to the olfactory fibers. Neurons then take the olfactory signal from the nasal cavity to the olfactory area of the brain for sensory processing. In addition to these major functions, the respiratory system also protects the delicate respiratory surfaces from environmental variations and defends them against pathogens and helps regulate blood volume, blood pressure, and body fluid pH.

Respiratory System

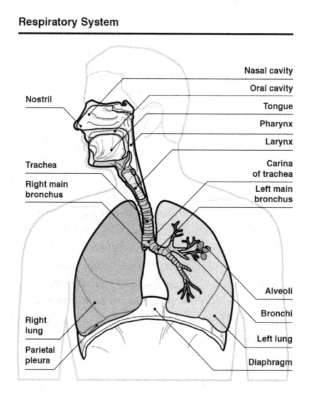

Infections of the respiratory system are common. Most upper respiratory infections are caused by viruses and include the common cold and sinusitis. Bacterial infections of the upper respiratory system are less common and include epiglottitis and croup. This part of the respiratory system gets infected by an organism entering the nasal or oral cavity and invading the mucosa that is present. The organism then begins to destroy the epithelium of the upper respiratory system. Infections of the lower respiratory system can be caused by viruses or bacteria. They include bronchitis, bronchiolitis, and pneumonia. These infections are caused by an organism entering the airway and multiplying itself in or on the epithelium in the lower respiratory system. Inflammation ensues, as well as increased mucus secretion and impaired function of the cilia lining the airways and lungs.

Cardiovascular System

The **cardiovascular system** is composed of the heart and blood vessels. It has three main functions in the human body. First, it transports nutrients, oxygen, and hormones through the blood to the body tissues and cells that need them. It also helps to remove metabolic waste, such as carbon dioxide and nitrogenous waste, through the bloodstream. Second, the cardiovascular system protects the body from attack by foreign microorganisms and toxins. The white blood cells, antibodies, and complement proteins that circulate within the blood help defend the body against these pathogens. The clotting system of the blood also helps protect the body from infection when there is blood loss following an injury. Lastly, this system helps regulate body temperature, fluid pH, and water content of the cells.

Blood Vessels

Blood circulates throughout the body in vessels called arteries, veins, and capillaries. These vessels are muscular tubes that allow gas exchange to occur. **Arteries** carry oxygen-rich blood from the heart to the other tissues of the body. **Veins** collect oxygen-depleted blood from tissues and organs and return it back to the heart. **Capillaries** are the smallest of the blood vessels and do not function individually; instead, they work together in a unit called a **capillary bed**.

Blood

Blood is an important vehicle for transport of oxygen, nutrients, and hormones throughout the body. It is composed of plasma and formed elements, which include red blood cells (RBCs), white blood cells (WBCs), and platelets. **Plasma** is the liquid matrix of the blood and contains dissolved proteins. **RBCs** contain hemoglobin, which carries oxygen through the blood. Red blood cells also transport carbon dioxide. **WBCs** are part of the immune system and help fight off diseases. **Platelets** contain enzymes and other factors that help with blood clotting.

Heart

The **heart**, which is the main organ of the cardiovascular system, acts as a pump and circulates blood throughout the body. Gases, nutrients, and waste are constantly exchanged between the circulating blood and interstitial fluid, keeping tissues and organs alive and healthy. The heart is located behind the sternum, on the left side, in the front of the chest. The heart wall is made up of three distinct layers. The outer layer, called the **epicardium**, is a serous membrane that is also known as the **visceral pericardium**. The middle layer is called the **myocardium** and contains connective tissue, blood vessels, and nerves within its layers of cardiac muscle tissue. The inner layer is called the **endocardium** and is made up of a simple squamous epithelium. It includes the heart valves and is continuous with the endothelium of the attached blood vessels. The heart has four chambers: the **right atrium**, the **right ventricle**, the **left atrium**, and the **left ventricle**. The atrium and ventricle on the same side of the heart have an opening between them that is regulated by a valve. The valve maintains blood flow in only one direction, moving from the atrium to the ventricle, and prevents backflow. The right side of the heart has a **tricuspid valve** (because it has three leaflets) between the chambers, and the left side of the heart has a **bicuspid valve** (with two leaflets) between the chambers, also called the **mitral valve**. Oxygen-poor blood from the body enters the right atrium through the superior vena cava and the inferior vena cava and is pumped into the right ventricle. The blood then enters the pulmonary trunk and flows into the pulmonary arteries, where it can become re-oxygenated. Oxygen-rich blood from the lungs then flows into the left atrium from four pulmonary veins, passes into the left ventricle, enters the aorta, and gets pumped to the rest of the body.

Cardiac Cycle

The **cardiac cycle** is the series of events that occur when the heart beats. During the cardiac cycle, blood is circulated throughout the pulmonary and systemic circuits of the body. There are two phases of the cardiac cycle—diastole and systole. During the **diastole** period, the ventricles of the heart are relaxed and are not contracting. Blood is flowing passively from the left atrium to the left ventricle and from the right atrium to the right ventricle through the atrioventricular valves. At the end of the diastole phase, both the left and right atria contract, and an additional amount of blood is pushed through to the respective ventricles. The **systole** period occurs when the left and right ventricles both contract. The aortic valve opens at the left ventricle and pushes blood through to the aorta, and the pulmonary valve opens at the right ventricle and pushes blood through to the pulmonary artery. During this phase, the atrioventricular valves are closed, and blood does not enter the ventricles from the atria.

Types of Circulation

Circulating blood carries oxygen, nutrients, and hormones throughout the body, which are vital for sustaining life. There are two types of cardiac circulation: pulmonary circulation and systemic circulation. The heart is responsible for pumping blood in both types of circulation. The **pulmonary circulatory system** carries blood between the heart and the lungs. It works in conjunction with the respiratory system to facilitate external respiration. Deoxygenated blood flows to the lungs through the vessels of the cardiovascular system to obtain oxygen and release carbon dioxide from the respiratory system. Blood that is rich with oxygen flows from the lungs back to the heart. Pulmonary circulation occurs only in the pulmonary loop. The pulmonary trunk takes the deoxygenated blood from the right ventricle to the arterioles and capillary beds of the lungs. Once the blood that is filling these spaces has been reoxygenated, it passes into the pulmonary veins and is transported to the left atrium of the heart. The **systemic circulatory system** carries blood from the heart to the rest of the body and works in conjunction with the respiratory system to facilitate internal respiration. The oxygenated blood flows out of the heart through the vessels and reaches the body tissues, while the deoxygenated blood flows through the vessels from the body back to the heart. Unlike the pulmonary loop, the systemic loop covers the whole body. Oxygen-rich blood moves out of the left ventricle into the aorta. The aorta circulates the blood to the systemic arteries and then to the arterioles and capillary beds that are present in the body tissues, where oxygen and nutrients are released into the tissues. The deoxygenated blood then moves from the capillary beds to the venules and systemic veins. The systemic veins bring the blood back to the right atrium of the heart.

Here's a visual representation of this:

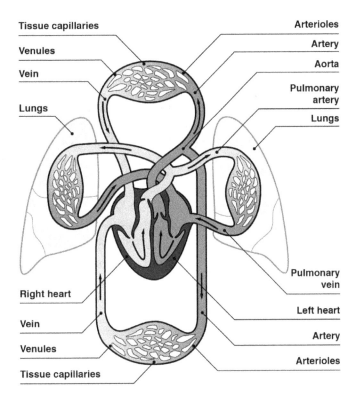

Gastrointestinal System

The **gastrointestinal system** is a group of organs that work together to fuel the body by transforming food and liquids into energy. After food is ingested, it passes through the **alimentary canal**, or **GI tract**, which comprises the mouth, pharynx, esophagus, stomach, small intestine, and large intestine. Each organ has a specific function to aid in digestion. Listed below are seven steps that incorporate the transformation of food as it travels through the gastrointestinal system.

- **Ingestion**: Food and liquids enter the alimentary canal through the mouth.

- **Mechanical processing**: Food is torn up by the teeth and swirled around by the tongue to facilitate swallowing.

- **Digestion**: Chemicals and enzymes break down complex molecules, such as sugars, lipids, and proteins, into smaller molecules that can be absorbed by the digestive epithelium.

- **Secretion**: Most of the acids, buffers, and enzymes that aid in digestion are secreted by the accessory organs, but some are provided by the digestive tract.

- **Absorption**: Vitamins, electrolytes, organic molecules, and water are absorbed by the digestive epithelium and moved to the interstitial fluid of the digestive tract.

- **Compaction**: Indigestible materials and organic wastes are dehydrated and compacted before elimination from the body.

- **Excretion**: Waste products are secreted into the digestive tract.

Mouth and Stomach

The **mouth** is the first point at which food and drink enter the gastrointestinal system. It is where food is chewed and torn apart by the teeth. Salivary glands in the mouth produce saliva, which is used to break down starches. The tongue also helps grip the food as it is being chewed and push it posteriorly toward the esophagus. The food and drink then move down the esophagus by the process of swallowing and into the **stomach**. The inferior end of the esophagus, at the stomach end, has a lower esophageal sphincter that closes off the esophagus and traps food in the stomach. The stomach stores food so that the body has time to digest large meals. It secretes enzymes and acids and also helps with mechanical processing through peristalsis or muscular contractions. The upper muscle of the stomach relaxes to let food in, and the lower muscle mixes the food with the digestive juices. The stomach secretes stomach acid to help break down proteins. Once digestion is completed in the stomach, the broken-down contents, called **chyme**, are passed into the small intestine.

Small Intestine

The **small intestine** is a thin tube that is approximately ten feet long and takes up most of the space in the abdominal cavity. The three structural parts of the small intestine are the duodenum, the jejunum, and the ileum. The **duodenum** is shaped like a C and receives the chyme from the stomach along with the digestive juices from the pancreas and the liver. The **jejunum** is the middle portion of the small intestine. It is the section with the most circular folds and villi to increase its surface area. The digestive products are absorbed into the bloodstream in the jejunum. The **ileum** follows the jejunum portion of the small intestine. It mainly absorbs bile acids and vitamin B12 and passes it into the bloodstream. The ileum connects to the large intestine. The entire small intestine secretes enzymes and has folds that increase its surface area and allow for maximum absorption of nutrients from the digested food. The small intestine digestive juice along with juices from the pancreas to liver in combination with peristalsis work to complete the digestion of starches, proteins, and carbohydrates. By the time food leaves the small intestine, approximately 90 percent of the nutrients have been absorbed. These nutrients are carried to the rest of the body. The undigested and unabsorbed food particles are passed into the large intestine.

Large Intestine

The **large intestine** is a long, thick tube that is about five feet in length, also known as the **colon**. It is the final part of the gastrointestinal tract. It has five main sections. The ascending colon connects to the small intestine and runs upward along the right side of the body. It also includes the appendix. The transverse colon runs parallel to the ground from the ascending colon to the left side of the abdominal cavity. The descending colon runs downward along the left side of the body. The sigmoid colon is an S-shaped region that connects the descending colon to the rectum. The rectum is the end of the colon and is a temporary storage place for feces. The function of the large intestine is to absorb water from the digested food and transport waste to be excreted from the body. It also contains symbiotic bacteria that break down the waste products even further, allowing for any additional nutrients to be absorbed. The final waste products are converted to stool and then stored in the rectum.

Gastrointestinal System

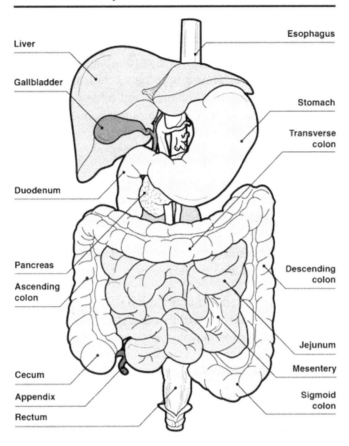

Pancreas

The **pancreas** is a large gland that is about 6 inches long. It secretes buffers and digestive enzymes into the duodenum of the small intestine. It contains specific enzymes for each type of food molecule, such as amylases for carbohydrates, lipases for lipids, and proteases for proteins. This digestive function of the pancreas is called the exocrine role of the pancreas. The cells of the pancreas are present in clusters called acini. The digestive enzymes are secreted into the middle of the acini into intralobular ducts. These ducts drain into the main pancreatic duct, which then drains into the duodenum.

In addition to these main organs, the gastrointestinal system has a few accessory organs that help break down food without having the food or liquid pass directly through them. The **liver** produces and secretes **bile**, which is important for the digestion of lipids. The bile mixes with the fat in food and helps dissolve it into the water contents within the small intestine, which helps fatty foods to be digested. The liver also plays a large role in the regulation of circulating levels of carbohydrates, amino acids, and lipids in the body. Excess nutrients are removed by the liver and deficiencies are corrected with its stored nutrients. The gallbladder is responsible for storing and concentrating bile before it gets secreted into the small intestine. It recycles the bile from the small intestine so that it can be used to digest subsequent meals.

Nervous System

The nervous system is often combined with the muscular system and considered the **neuromuscular system**, which is composed of all of the muscles in the human body and the nerves that control them. Every movement that the body makes is controlled by the brain. The nervous system and the muscular system work together to link thoughts and actions. Neurons from the nervous system can relay information from the brain to muscle tissue so fast that an individual does not even realize it is happening. Some body movements are voluntary, but others are involuntary, such as the heart beating and the lungs breathing.

The nervous system is made up of the **central nervous system** (**CNS**) and the **peripheral nervous system** (**PNS**). The CNS includes the brain and the spinal cord, while the PNS includes the rest of the neural tissue that is not included in the CNS. **Neurons**, or nerve cells, are the main cells responsible for transferring and processing information between the brain and other parts of the body. **Neuroglia** are cells that support the neurons by providing a framework around them and isolating them from the surrounding environment.

Central Nervous System

The CNS is located within the dorsal body cavity, with the brain in the cranial cavity and the spinal cord in the spinal canal. The brain is protected by the skull, and the spinal cord is protected by the vertebrae. The brain is made up of white and gray matter. The white matter contains axons and oligodendrocytes. **Axons** are the long projection ends of neurons that are responsible for transmitting signals through the nervous system, while **oligodendrocytes** act as insulators for the axons and provide support for them. The gray matter consists of neurons and fibers that are unmyelinated.

Neurons are nerve cells that receive and transmit information through electrical and chemical signals. Glial cells and astrocytes are located in both types of tissue. Different types of glial cells have different roles in the CNS. Some are immunoprotective, while others provide a scaffolding for other types of nerve cells. Astrocytes provide nutrients to neurons and clear out metabolites. The spinal cord has projections within it from the PNS. This allows the information that is received from the areas of the body that the PNS reaches to be transmitted to the brain. The CNS as a whole is responsible for processing and coordinating sensory data and motor commands. It receives information from all parts of the body, processes it, and then sends out action commands in response. Some of the reactions are conscious, while others are unconscious.

Infections of the CNS include **encephalitis**, which is an inflammation of the brain, and **poliomyelitis**, which is caused by a virus and causes muscle weakness. Other developmental neurological disorders include ADHD and autism. Some diseases of the CNS can occur later in life and affect the aging brain, such as Alzheimer's disease and Parkinson's disease. Cancers that occur in the CNS can be very serious and have high mortality rates.

Peripheral Nervous System

The PNS consists of the nerves and ganglia that are located within the body outside of the brain and spinal cord. It connects the rest of the body, organs, and limbs to the CNS. Unlike the CNS, the PNS does not have any bony structures protecting it. The PNS is responsible for relaying sensory information and motor commands between the CNS and peripheral tissues and systems. It has two subdivisions, known as the afferent, or sensory, and efferent, or motor, divisions. The **afferent division** relays sensory information to the CNS and supplies information from the skin and joints about the body's sensation and balance. It carries information away from the stimulus back to the brain. The afferent division provides the brain with sensory information about things such as taste, smell, and vision. It also monitors organs, vessels, and glands for changes in activity and can alert the brain to send out appropriate responses for bringing the body back to homeostasis. The **efferent division** transmits motor commands to muscles and glands. It sends information from the CNS to the organs and muscles to provide appropriate responses to sensations. The electrical responses from the neurons are initiated in the CNS, but the axons terminate in the organs that are part of the PNS. The efferent division consists of the autonomic nervous system (ANS), which regulates activity of smooth muscle, cardiac muscle, and glands, and allows the brain to control heart rate, blood pressure, and body temperature, and the somatic nervous system (SNS), which controls skeletal muscle contractions and allows the brain to control body movement.

Diseases of the PNS can affect single nerves or the whole system. Single nerve damage, or **mononeuropathy**, can occur when a nerve gets compressed due to trauma or a tumor. It can also be damaged as a result of being trapped under another part of the body that is increasing in size, such as in carpal tunnel syndrome. These diseases can cause pain and numbness at the affected area.

Autonomic Nervous System
The **autonomic nervous system** is made up of pathways that extend from the CNS to the organs, muscles, and glands of the body. The pathway is made up of two separate neurons. The first neuron has a cell body that is located within the CNS, for example, within the spinal cord. The axon of that first neuron synapses with the cell body of the second neuron. Part of the second neuron innervates the organ, muscle, or gland that the pathway is responsible for.

The autonomic nervous system consists of the sympathetic nervous system and the parasympathetic nervous system. The **sympathetic nervous system** is activated when mentally stressful or physically dangerous situations are faced, also known as "fight or flight" situations. Neurotransmitters are released that increase heart rate and blood flow in critical areas, such as the muscles, and decrease activity for nonessential functions, such as digestion. This system is activated unconsciously. The **parasympathetic system** has some voluntary control. It releases neurotransmitters that allow the body to function in a restful state. Heart rate and other sympathetic responses are often decreased when the parasympathetic system is activated.

Somatic Nervous System and the Reflex Arc
The **somatic nervous system** is considered the voluntary part of the PNS. It comprises motor neurons whose axons innervate skeletal muscle. However, nerve muscles and muscle cells do not come into direct contact with each other. Instead, the neurotransmitter acetylcholine transfers the signal between the nerve cell and muscle cell. These junctions are called **neuromuscular junctions**. When the muscle cell receives the signal from the nerve cell, it causes the muscle to contract.

Although most muscle contractions are voluntary, reflexes are a type of muscle contraction that is involuntary. A **reflex** is an instantaneous movement that occurs in response to a stimulus and is controlled by a neural pathway called a **reflex arc**. The reflex arc carries the sensory information from the receptor to the spinal cord and then carries the response back to the muscles. Many sensory neurons synapse in the spinal cord so that reflex actions can occur faster, without waiting for the signal to travel to the brain and back. For somatic reflexes, stimuli activate somatic receptors in the skin, muscles, and tendons. Then, afferent nerve fibers carry signals from the receptors to the spinal cord or brain stem. The signal reaches an integrating center, which is where the neurons synapse within the spinal cord or brain stem. Efferent nerve fibers then carry motor nerve signals to the muscles, and the muscles carry out the response.

Muscular System

There are approximately seven hundred muscles in the body. They are attached to the bones of the skeletal system and make up half of the body's weight. There are three types of muscle tissue in the body: skeletal muscle, smooth muscle, and cardiac muscle. An important characteristic of all types of muscle is that they are **excitable**, which means that they respond to electrical stimuli.

Skeletal Muscles

Skeletal muscle tissue is a voluntary striated muscle tissue, which means that the contractile fibers of the tissue are aligned parallel so that they appear to form stripes when viewed under a microscope. Most skeletal muscle are attached to bones by intermediary tendons. Tendons are bundles of collagen fibers. When nerve cells release acetylcholine at the neuromuscular junction, skeletal muscle contracts. The skeletal muscle tissue then pulls on the bones of the skeleton and causes body movement.

Smooth Muscles

Smooth muscle tissue is an involuntary nonstriated muscle tissue. It has greater elasticity than striated muscle but still maintains its contractile ability. Smooth muscle tissue can be of the single-unit variety or the multi-unit variety. Most smooth muscle tissues are single unit where the cells all act in unison and the whole muscle contracts or relaxes. Single-unit smooth muscle lines the blood vessels, urinary tract, and digestive tract. It helps to move fluids and solids along the digestive tract, allows for expansion of the urinary bladder as it fills, and helps move blood through the vessels. Multi-unit smooth muscle is found in the iris of the eye, large elastic arteries, and the trachea.

Cardiac Muscles

Cardiac muscle tissue is an involuntary striated muscle that is only found in the walls of the heart. The cells that make up the tissue contract together to pump blood through the veins and arteries. This tissue has high contractility and extreme endurance, since it pumps for an individual's entire lifetime without any rest. Each cardiac muscle cell has finger-like projections, called **intercalated disks**, at each end that overlap with the same projections on neighboring cells. The intercalated disks form tight junctions between the cells so that they do not separate as the heart beats and so that electrochemical signals are passed from cell to cell quickly and efficiently.

Reproductive System

The **reproductive system** is responsible for producing, storing, nourishing, and transporting functional reproductive cells, or **gametes**, in the human body. It includes the reproductive organs, also known as **gonads**, the **reproductive tract**, the accessory glands and organs that secrete fluids into the reproductive tract, and the **perineal structures**, which are the external genitalia. The human male and female reproductive systems are very different from each other.

The male gonads are called **testes**. The testes secrete androgens, mainly testosterone, and produce and store one-half billion **sperm cells**, which are the male gametes, each day. An **androgen** is a steroid hormone that controls the development and maintenance of male characteristics. Once the sperm are mature, they move through a duct system where they are mixed with additional fluids that are secreted by accessory glands, forming a mixture called **semen**. The sperm cells in semen are responsible for fertilization of the female gametes to produce offspring. The male reproductive system has a few accessory organs as well, which are located inside the body. The **prostate gland** and the **seminal vesicles** provide additional fluid that serves as nourishment to the sperm during ejaculation. The **vas deferens** is responsible for transportation of sperm to the urethra. The **bulbourethral glands** produce a lubricating fluid for the urethra that also neutralizes the residual acidity left behind by urine.

The female gonads are called **ovaries**. Ovaries generally produce one immature gamete, or **oocyte**, per month. The ovaries are also responsible for secreting the hormones estrogen and progesterone. When the oocyte is released from the ovary, it travels along the uterine tubes, or **Fallopian tubes**, and then into the **uterus**. The uterus opens into the vagina. When sperm cells enter the vagina, they swim through the uterus. If they fertilize the oocyte, they do so in the Fallopian tubes. The resulting zygote travels down the tube and implants into the uterine wall. The uterus protects and nourishes the developing embryo for nine months until it is ready for the outside environment. If the oocyte is not fertilized, it is released in the uterine, or menstrual, cycle. The **menstrual cycle** usually occurs monthly and involves the shedding of the functional part of the uterine lining. **Mammary glands** are a specialized accessory organ of the female reproductive system. The mammary glands are located in the breast tissue of females. During pregnancy, the glands begin to grow as the cells proliferate in preparation for lactation. After pregnancy, the cells begin to secrete nutrient-filled milk, which is transferred into a duct system and out through the nipple for nourishment of the baby.

Integumentary System

The **integumentary system** protects the body from damage from the outside. It consists of skin and its appendages, including hair, nails, and sweat glands. This system functions as a cushion, a waterproof layer, a temperature regulator, and a protectant of the deeper tissues within the body. It also excretes waste from the body. The skin is the largest organ of the human body, consisting of two layers called the epidermis and the dermis. The **epidermis** can be classified as thick or thin. Most of the body is covered with thin skin but areas such as the palm of the hands are covered with thick skin. The epidermis is responsible for synthesizing vitamin D when exposed to UV rays. Vitamin D is essential to the body for the processes of calcium and phosphorus absorption, which maintain healthy bones. The **dermis** lies under the epidermis and consists of a superficial papillary layer and a deeper reticular layer. The **papillary layer** is made up of loose connective tissue and contains capillaries and the axons of sensory neurons. The **reticular layer** is a meshwork of tightly packed irregular connective tissue and contains blood vessels, hair follicles, nerves, sweat glands, and sebaceous glands.

The three major functions of skin are protection, regulation, and sensation. Skin acts as a barrier and protects the body from mechanical impacts, variations in temperature, microorganisms, and chemicals. It regulates body temperature, peripheral circulation, and fluid balance by secreting sweat. It also contains a large network of nerve cells that relay changes in the external environment to the body.

Hair provides many functions for the human body. It provides sensation, protects against heat loss, and filters air that is taken in through the nose. Nails provide a hard layer of protection over soft skin. Sweat glands and sebaceous glands are two important **exocrine glands** found in the skin. **Sweat glands** regulate temperature and remove bodily waste by secreting water, nitrogenous waste, and sodium salts to the surface of the body. **Sebaceous glands** secrete sebum, which is an oily mixture of lipids and proteins. Sebum protects the skin from water loss and bacterial and fungal infections.

Endocrine System

The **endocrine system** is made up of ductless tissues and glands and is responsible for hormone secretion into either the blood or the interstitial fluid of the human body. **Hormones** are chemical substances that change the metabolic activity of tissues and organs. **Interstitial fluid** is the solution that surrounds tissue cells within the body. This system works closely with the nervous system to regulate the physiological activities of the other systems of the body in order to maintain homeostasis. While the nervous system provides quick, short-term responses to stimuli, the endocrine system acts by releasing hormones into the bloodstream, which then are distributed to the whole body. The response is slow but long lasting, ranging from a few hours to even a few weeks. While regular metabolic reactions are controlled by enzymes, hormones can change the type, activity, or quantity of the enzymes involved in the reaction. They can regulate development and growth, digestive metabolism, mood, and body temperature, among many other things. Often very small amounts of a hormone will lead to large changes in the body.

There are eight major glands in the endocrine system, each with its own specific function. They are described below.

- **Hypothalamus**: This gland is a part of the brain. It connects the nervous system to the endocrine system via the pituitary gland and plays an important role in regulating endocrine organs.

- **Pituitary gland**: This pea-sized gland is found at the bottom of the hypothalamus. It releases hormones that regulate growth, blood pressure, certain functions of the reproductive sex organs, and pain relief, among other things. It also plays an important role in regulating the function of other endocrine glands.

- **Thyroid gland**: This gland releases hormones that are important for metabolism, growth and development, temperature regulation, and brain development during infancy and childhood. Thyroid hormones also monitor the amount of circulating calcium in the body.

- **Parathyroid glands**: These are four pea-sized glands located on the posterior surface of the thyroid. The main hormone that is secreted is called **parathyroid hormone (PTH)** and helps with the thyroid's regulation of calcium in the body.

- **Thymus gland**: This gland is located in the chest cavity, embedded in connective tissue. It produces several hormones that are important for development and maintenance of normal immunological defenses.

- **Adrenal glands**: One adrenal gland is attached to the top of each kidney. Its major function is to aid in the management of stress.

- **Pancreas**: This gland produces hormones that regulate blood sugar levels in the body.

- **Pineal gland**: The pineal gland secretes the hormone melatonin, which can slow the maturation of sperm, oocytes, and reproductive organs. Melatonin also regulates the body's **circadian rhythm**, which is the natural awake-asleep cycles.

Genitourinary System

The **genitourinary system** encompasses the reproductive organs and the **urinary system**. They are often grouped together because of their proximity to each other in the body, and, for the male, because of shared pathways. The external male genitalia include the penis, urethra, and scrotum. Both urine and male gametes travel through the urethra within the penis to exit the body. In both the male and female bodies, the urinary system is made up of the kidneys, ureters, urinary bladder, and the urethra. It is the main system responsible for getting rid of the organic waste products, as well as excess water and electrolytes that are generated by the other systems of the body. Regulation of the water and electrolytes also contributes to the maintenance of blood pH.

Under normal circumstances, humans have two functioning **kidneys**. They are the main organs that are responsible for filtering waste products out of the blood and transferring them to urine. Every day, the kidneys filter approximately 120 to 150 quarts of blood and produce one to two quarts of urine. The kidneys are made up of millions of tiny filtering units called **nephrons**. Nephrons have two parts: a **glomerulus**, which is the filter, and a **tubule**. As blood enters the kidneys, the glomerulus allows for fluid and waste products to pass through it and enter the tubule. Blood cells and large molecules, such as proteins, do not pass through and remain in the blood. The filtered fluid and waste then pass through the tubule, where any final essential minerals can be sent back to the bloodstream. The final product at the end of the tubule is called urine. The urine travels through the ureters into the **urinary bladder**, which is a hollow, elastic muscular organ. As more and more urine enters the urinary bladder, its walls stretch and become thinner so that there is no significant difference in internal pressure. The urinary bladder stores the urine until the body is ready for urination, at which time its muscles contract and force the urine through the urethra and out of the body.

Immune System

The **immune system** comprises cells, tissues, and organs that work together to protect the body. They recognize millions of different microorganisms, such as parasites, bacteria, and fungi, and viruses that can invade and infect the body. When an attack on the body is sensed, immune cells are activated and communicate with each other through an elaborate network. They begin to produce chemicals and recruit other cells to defend the body at the infection site. It is important for the body's immune system to be able to distinguish between pathogens, such as viruses and microorganisms, and the body's own healthy tissue, so that the body does not attack itself. There are two types of immune systems that work to defend the body against infection: the innate system and the adaptive system.

Innate Immune System

The **innate immune system** is triggered when pattern recognition receptors recognize components of microorganisms that are the same among many different varieties, or when damaged or stressed cells send out help signals. It is always primed and ready to fight infections. This system works in a nonspecific way, which means that it works without having a memory of the pathogens it defended against previously and does not provide long-lasting immunity for the body. It does critical work in the first few hours and days of exposure to a pathogen to fight off infection by producing general immune responses, such as by producing inflammation and activating the complement cascade. During inflammation, immune cells that are already present in the injured tissue are activated, and chemicals stimulate inflammation of the tissue to create a physical barrier against the infection. The complement system is a cascade of events that the immune systems starts to help antibodies fight off pathogens. The proteins that are produced first recruit inflammatory cells and then tag the pathogens for destruction by coating them with opsonin. The proteins also put holes in the pathogen's plasma membrane and, lastly, help rid the body of the neutralized invader.

Mechanical, chemical, and surface barriers all function as part of the innate immune system. **Mechanical barriers** are the first line of defense against pathogens. Both skin and the respiratory tract are examples of mechanical barriers that provide initial protection against infection. The epithelial layer of the skin is an impermeable physical barrier that wards off many infections. When the epithelium sheds, or goes through desquamation, any microbes that are attached to the skin are removed as well. The top layers of skin are also avascular and lack a blood supply, so the environment is not ideal for survival of microbes. The respiratory tract uses a mucociliary escalator to protect against infection. Pathogens get caught in the sticky mucus that lines the respiratory tract and are moved upward toward the throat. If a mechanical barrier is passed and the pathogen enters the body, **chemical barriers** are activated. These barriers include cells that have a secondary purpose of fighting off infections. For example, the gastric acid and proteases found in the stomach act as chemical barriers against pathogens that are ingested. They create unfriendly and toxic environments for bacteria to colonize. The group of proteins called interferons helps inhibit the replication of viruses. **Biological barriers** can also help prevent infection. The bacteria that is naturally found in the gastrointestinal tract acts as a biological barrier by creating competition for nutrients and space against pathogenic bacteria. The eyes have a flushing reflex to push out pathogens that may cause infection. Tears are produced by the tear ducts and collect the microbes to be released along with them from the eye.

Adaptive Immune System

The **adaptive immune system** creates a memory of the pathogen that it fought against previously so that the body can respond again in an efficient manner the next time the pathogen is encountered. When **antigens**, or **allergens**, such as pollen, are encountered, specific antibodies are secreted to inactivate the antigen and protect the body. The response of the adaptive immune system can take several days, which is much longer than the immediate response of the innate immune system. It uses fewer types of cells to produce its immune response compared to the innate immune system as well. The adaptive immune system provides the body with a long-lasting defense mechanism and protection against recurrent infections. Vaccines use the memory of the adaptive immune system to protect individuals against diseases they have never experienced. When vaccines are administered, an active, weakened, or attenuated virus is injected into the body, which mimics the active virus. The adaptive immune system then starts an immune response and creates a memory of the antigens associated with that disease along with the antibodies to fight them off.

Lymphocytes are the specialized cells that are a part of the adaptive immune system. **B cells** and **T cells** are different types of lymphocytes that are derived from multipotent hematopoietic stem cells and that recognize specific target pathogens. B cells are formed in the bone marrow and then move into the lymphatic system and circulate through the body. These cells are called naive B cells. Naive B cells express millions of antibodies on their surfaces as well as a B-cell receptor (BCR). The BCR helps with antigen binding, internalization, and processing. It allows the B cell to then initiate signaling pathways and communicate with other immune cells. When a naive B cell encounters an antigen that fits one of its surface antibodies, it begins to mature and differentiates into a memory B cell or a plasma cell. Memory B cells retain their surface-bound antibodies. Plasma cells, on the other hand, secrete their antibodies instead of keeping them attached to the cell membrane. The freely circulating antibodies can then identify pathogens that are circulating throughout the body.

T cells are also produced in the bone marrow. They then migrate to the thymus where they mature and start to express T-cell receptors (TCRs) as well as one of two other types of receptors—CD4 or CD8. These receptors help T cells identify antigens that are bound to specific receptors on antigen-presenting cells, such as macrophages and dendritic cells. CD4 and CD8 bind to the receptor-antigen complex on these cells, which activates the T cells to produce an immune response. After maturation, the three types of T cells that are produced are helper T cells, cytotoxic T cells, and T regulatory cells. Helper T cells express CD4 and help activate B cells, cytotoxic T cells, and other immune cells. Cytotoxic T cells express CD8 and remove pathogens and infected host cells. T regulatory cells express CD4 and help distinguish between molecules that belong to the individual and foreign molecules.

Malfunctions of the Immune System

An **immunodeficiency** occurs when the body cannot respond to a pathogen because a component of the immune system is inactive. This can be caused by a genetic mutation or by behavior. For example, smoking paralyzes the mucociliary escalator of the respiratory tract, disabling a critical part of the body's innate immune system. If the immune system is not functioning properly, the body may attack its own self and develop an autoimmune disorder. When the body cannot distinguish between itself and foreign pathogens, the immune system becomes overactive and the body attacks itself unnecessarily. To diagnose an autoimmune disease, it is important to identify which antibodies an individual's body is producing and then determine which of those antibodies is attacking the individual itself. Autoantibody tests look for specific antibodies within an individual's tissues; antinuclear antibody tests are specific autoantibody tests that look for antibodies that attack the nuclei of cells; complete blood count tests count the number of red and white cells in the blood, which reveals if the body is actively fighting an infection; and C-reactive protein tests determine whether there is inflammation occurring throughout the body.

Diseases such as type I diabetes and rheumatoid arthritis are autoimmune disorders. The pancreas contains cells that produce insulin to regulate blood sugar levels. In type I diabetes, auto-aggressive T cells infiltrate the pancreas and destroy the insulin-producing beta cells population. Without the insulin-producing cells, the body cannot regulate blood sugar levels on its own and requires external sources of insulin to maintain healthy levels. Rheumatoid arthritis occurs when the body's immune system attacks the joint linings of its own body. This causes inflammation and damage in healthy tissue around the joints, leading to pain and immobility.

Skeletal System

The adult **skeletal system** consists of the 206 bones that make up the skeleton, as well as the cartilage, ligaments, and other connective tissues that stabilize them. Babies are born with roughly three hundred bones, some of which fuse together during growth. Bone is made up of collagen fibers and calcium salts. The calcium salts are strong but brittle, and the collagen fibers are weak but flexible, so the combination makes bone very resistant to shattering.

Axial Skeleton

The **axial skeleton** comprises the bones found in the head, trunk of the body, and vertebrae—a total of eighty bones. It is the central core of the body and is responsible for connecting the pelvis to the rest of the body. The axial skeleton can be divided into the following five parts:

1. The **skull bones** are made up of the cranium and the facial bones. While the skeleton gets frailer with age, the skull remains strong to protect the brain. The cranium is formed from eight flat bones that fit together without spaces in between. These meeting points are called **sutures**. The brain is held in a space inside the cranium called the **cranial vault**. The lower front part of the skull is formed by fourteen facial bones. Together with the cranium, the facial bones form spaces to hold the eyes, internal ear, nose, and mouth.

2. There are three **middle ear ossicles** in each ear, called the **malleus**, **incus**, and **stapes**, respectively, positioned between the eardrum and inner ear. They are among the smallest bones found in the human body. These bones transmit sounds from the air to the fluid-filled cochlea, and without them, individuals suffer from hearing loss.

3. The **hyoid bone** is a horseshoe-shaped bone located in the anterior midline of the neck. It is held in place by muscles and is only distantly articulated to other bones. It aids in swallowing and movement of the tongue.

4. The **rib cage** includes the **sternum**, or breastbone, and twelve pairs of **ribs**. The ribs each have one flat end and one rounded end, similar to a crescent shape. The flat ends of the upper seven ribs come together at the sternum, while their rounded ends are attached at joints to the vertebrae. The eighth, ninth, and tenth pairs of ribs connect to the ribs above with noncostal cartilage. The last two sets of ribs are floating ribs and do not attach to any other structure. The rib cage protects the heart and lungs.

5. The adult **vertebral column** comprises twenty-four vertebrae, the sacrum, and the coccyx. The sacrum is formed from five fused vertebrae, and the coccyx is formed from three to five fused vertebrae, so the vertebral column is thought to have thirty-three bones total. The seven cervical vertebrae are the uppermost vertebrae and connect the cranium. Below the cervical vertebrae are twelve thoracic vertebrae, followed by five lumbar vertebrae.

Appendicular Skeleton

The appendicular skeleton comprises 126 bones that support the body's appendages. It helps with movement of the body as well as with manipulation of external objects and interactions with the environment. It is divided into the following four parts:

1. The **pectoral girdles** include four bones, which are the right and left clavicle and two scapula bones. The scapulae help keep the shoulders in place.

2. The **upper limbs** comprise the arms and the hands. The arms each consist of three bones. The humerus is in the upper part of the arm, and the ulna and radius are below the elbow, making up the forearm. The arm bones provide attachment points for muscles that allow for movement of the arms and wrist. Each hand has twenty-seven bones.

3. The **pelvis** comprises two coxal bones that provide an attachment point for the lower limbs and the axial skeleton.

4. The **lower limbs** comprise the legs, ankles, and feet. The legs are each made up of four bones. The femur is located in the thigh. The patella covers the knee. The tibia and fibula are in the lower leg below the knee. These bones have muscles attached to them that allow for movement of the leg. The foot and ankle of each leg have twenty-six bones total.

Functions of the Skeletal System

One of the major functions of the skeletal system is to provide structural support for the entire body. It provides a framework for the soft tissues and organs to attach to. The skeletal system also provides a reserve of important nutrients, such as calcium and lipids. Normal concentrations of calcium and phosphate in body fluids are partly maintained by the calcium salts stored in bone. Lipids that are stored in yellow bone marrow can be used as a source of energy. Red bone marrow produces red blood cells, white blood cells, and platelets that circulate in the blood. Certain groups of bones form protective barriers around delicate organs. The ribs, for example, protect the heart and lungs, the skull encloses the brain, and the vertebrae cover the spinal cord.

Compact and Spongy Bone

There are two types of bone: compact and spongy. **Compact bone** is dense with a matrix filled with organic substances and inorganic salts. There are only tiny spaces left between these materials for the **osteocytes**, or bone cells, to fit into. **Spongy bone**, in contrast to compact bone, is lightweight and porous. It covers the outside of the bone and gives it a shiny, white appearance. It has a branching network of parallel lamellae called **trabeculae**. Although spongy bone forms an open framework around the compact bone, it is still quite strong. Different bones have different ratios of compact to spongy bone depending on their function. The outside of the bone is covered by a **periosteum**, which has four major functions. It isolates and protects bones from the surrounding tissue, provides a place for attachment of the circulatory and nervous system structures, participates in growth and repair of the bone, and attaches the bone to the deep fascia. An **endosteum** is found inside the bone; it covers the trabeculae of the spongy bone and lines the inner surfaces of the central canals.

Physical Science

Physical Properties such as Volume, Mass, Color, and Temperature

The universe is composed completely of matter. **Matter** is any material or object that takes up space and has a mass. Certain properties of matter can be readily observed and measured. There are certain properties of matter that are observable and measurable such as volume, mass, color, and temperature.

Physical properties can be detected without changing the substance's identity or composition. If a substance is subject to a physical change, its appearance changes, but its composition does not change at all. Common physical properties include color, odor, density, and hardness. Physical properties can also be classified as extensive or intensive. Extensive physical properties depend on the amount of substance that is present. These include characteristics such as mass and volume. Intensive properties do not depend on how much of the substance is present and remain the same regardless of quantity. These include characteristics such as density and color.

It is often important to quantify the properties that are observed in a substance. Quantitative measurements are associated with a number, and it is imperative to include a unit of measure with each of those measurements. Without the unit of measure, the number may be meaningless. If the weight of a substance is noted as 100 without a unit of measure, it is unknown whether the substance is equivalent to 100 of something very light, such as milligram measurements, or 100 of something very heavy, such as kilogram measurements. There are many other quantitative measurements of matter that can be taken, including length, time, and temperature, among many others. Contrastingly, there are many properties that are not associated with a number; these properties can be measured qualitatively. The measurements of qualitative properties use a person's senses to observe and describe the characteristics. Since the descriptions are developed by individual persons, they can be very subjective. These properties can include odor, color, and texture, among many others. They can also include comparisons between two substances without using a number; for example, one substance may have a stronger odor or be lighter in color than another substance.

As those familiar with word roots can determine, thermometers measure heat or temperature (*thermo-* is from Greek for heat and *meter* from Greek for measure). Some thermometers measure outdoor or indoor air temperature, and others measure body temperature in humans or animals. Meat thermometers measure temperature in the center of cooked meat to ensure sufficient cooking to kill bacteria. Refrigerator and freezer thermometers measure internal temperatures to ensure sufficient coldness. Although digital thermometers eliminate the task of reading a thermometer scale by displaying a specific numerical reading, many thermometers still use visual scales. In these, increments are typically two-tenths of a degree. However, basal thermometers, which are more sensitive and accurate, and are frequently used to track female ovulation cycles for measuring and planning fertility, display increments of one-tenth of a degree. Whole degrees are marked by their numbers; tenths of degrees are the smallest marks in between.

Position and Motion of Objects

Cultures have been studying the movement of objects since ancient times. These studies have been prompted by curiosity and sometimes by necessity. On Earth, items move according to guidelines and have motion that is fairly predictable. To understand why an object moves along its path, it is important to understand what role forces have on influencing its movements. The term **force** describes an outside influence on an object. Force does not have to refer to something imparted by another object. Forces can act upon objects by touching them with a push or a pull, by friction, or without touch like a magnetic force or even gravity. Forces can affect the motion of an object.

To study an object's motion, it must be located and described. When locating an object's position, it can help to pinpoint its location relative to another known object. Comparing an object with respect to a known object is referred to as establishing a frame of reference. If the placement of one object is known, it is easier to locate another object with respect to the position of the original object.

Motion can be described by following specific guidelines called **kinematics**. Kinematics use mechanics to describe motion without regard to the forces that are causing such motions. Specific equations can be used when describing motions; these equations use time as a frame of reference. The equations are based on the change of an object's position (represented by x), over a change in time (represented by Δt).

This describes an object's **velocity**, which is measured in meters/second (m/s) and described by the following equation:

$$v = \frac{\Delta x}{\Delta t} = \frac{x_f - x_i}{\Delta t}$$

Velocity is a vector quantity, meaning it measures the magnitude (how much) and the direction that the object is moving. Both of these components are essential to understanding and predicting the motion of objects. The scientist Isaac Newton did extensive studies on the motion of objects on Earth and came up with three primary laws to describe motion:

Law 1: An object in motion tends to stay in motion unless acted upon by an outside force. An object at rest tends to stay at rest unless acted upon by an outside force (also known as the **law of inertia**).

For example, if a book is placed on a table, it will stay there until it is moved by an outside force.

Law 2: The force acting upon an object is equal to the object's mass multiplied by its acceleration (also known as F = ma).

For example, the amount of force acting on a bug being swatted by a flyswatter can be calculated if the mass of the flyswatter and its acceleration are known. If the mass of the flyswatter is 0.3 kg and the acceleration of its swing is 2.0 m/s^2, the force of its swing can be calculated as follows:

$$m = 0.3\,kg$$
$$a = 2.0\,m/s^2$$
$$F = m \times a$$
$$F = (0.3) \times (2.0)$$
$$F = 0.6\,N$$

Law 3: For every action, there is an equal and opposite reaction.

For example, when a person claps their hands together, the right hand feels the same force as the left hand, as the force is equal and opposite.

Another example is if a car and a truck run head-on into each other, the force experienced by the truck is equal and opposite to the force experienced by the car, regardless of their respective masses or velocities. The ability to withstand this amount of force is what varies between the vehicles and creates a difference in the amount of damage sustained.

Newton used these laws to describe motion and derive additional equations for motion that could predict the position, velocity, acceleration, or time for objects in motion in one and two dimensions. Since all of Newton's work was done on Earth, he primarily used Earth's gravity and the behavior of falling objects to design experiments and studies in free fall (an object subject to Earth's gravity while in flight). On Earth, the acceleration due to the force of gravity is measured at 9.8 meters per second² (m/s²). This value is the same for anything on the Earth or within Earth's atmosphere.

Acceleration
Acceleration is the change in velocity over the change in time. It is given by the following equation:

$$a = \frac{\Delta v}{\Delta t} = \frac{v_f - v_i}{\Delta t}$$

Since velocity is the change in position (displacement) over a change in time, it is necessary for calculating an acceleration. Both of these are vector quantities, meaning they have magnitude and direction (or some amount in some direction). Acceleration is measured in units of distance over time² (meters/second² or m/s² in metric units).

For example, what is the acceleration of a vehicle that has an initial velocity of 35 m/s and a final velocity of 10 m/s over 5.0 s?

Using the givens and the equation:

$$a = \frac{\Delta v}{\Delta t} = \frac{v_f - v_i}{\Delta t}$$

V_f = 10 m/s

V_i = 35 m/s

Δt = 5.0 s

$$a = \frac{10 - 35}{5.0} = \frac{-25}{5.0} = -5.0 \, m/s^2$$

The vehicle is decelerating at –5.0 m/s².

If an object is moving with a constant velocity, its velocity does not change over time. Therefore, it has no (or 0) acceleration.

It is common to misuse vector terms in spoken language. For example, people frequently use the term "speed" in situations where the correct term would be "velocity." However, the difference between velocity and speed is not just that velocity must have a direction component with it. Average velocity and average speed actually are looking at two different distances as well. Average speed is calculated simply by dividing the total distance covered by the time it took to travel that distance. If someone runs four miles along a straight road north and then makes a 90-degree turn to the right and runs another three miles down that straight road east (such that the runner's route of seven miles makes up two sides of a rectangle) in seventy minutes, the runner's average speed was 6 miles per hour (one mile covered every ten minutes). Using the same course, the runner's average velocity would be about 4.29 miles per hour northeast.

Why is the magnitude less in the case of velocity? Velocity measures the change in position, or **displacement**, which is the shortest line between the starting point and ending point. Even if the path between these two points is serpentine or meanders all over the place racking up a great distance, the displacement is still just the shortest straight path between the change in the position of the object. In the case of the runner, the "distance" used to calculate velocity (the displacement) is the hypotenuse of the triangle that would connect the two side lengths at right angles to one another. Using basic trigonometric ratios, we know this distance is 5 miles (since the lengths of the other two legs are 3 miles and 4 miles and the Pythagorean Theorem says that $a^2 + b^2 = c^2$). Therefore, although the distance the runner covered was seven miles, his or her displacement was only five miles. Average velocity is thus calculated by taking the total time (70 minutes) and dividing it by the displacement (5 miles northeast). Therefore, to calculate average velocity of the runner, 70 minutes is divided by 5 miles, so each mile of displacement to the northeast was covered in 14 minutes. To find this rate in miles per hour, 60 minutes is divided by 14, to get 4.29 miles per hour northeast.

Another misconception is if something has a negative acceleration, it must be slowing down. If the change in position of the moving object is in a negative direction, it could have a negative velocity. If the acceleration is in the same direction as this negative velocity, it would be increasing the velocity in the negative direction, thus resulting in the object actually increasing in velocity.

For example, if west is designated to be a negative direction, a car increasing in speed to the west would have a negative velocity. Since it is increasing in speed, it would be accelerating in the negative direction, resulting in a negative acceleration.

Another common misconception is if a person is running around an oval track at a constant velocity, they would have no (or 0) acceleration because there is no change in the runner's velocity. This idea is incorrect because the person is changing direction the entire time they are running around the track, so there would be a change in their velocity. Therefore, the runner would have an acceleration.

One final point regarding acceleration is that it can result from the force a rotating body exerts toward its center. For planets and other massive bodies, it is called **gravity**. This type of acceleration can also be utilized to separate substances, as in a centrifuge.

Projectile Motion

When objects are launched or thrown into the air, they exhibit what is called **projectile motion**. This motion takes a parabolic (or arced) path as the object rises and/or falls with the effect of gravity. In sports, if a ball is thrown across a field, it will follow a path of projectile motion. Whatever angle the object leaves the horizon is the same angle with which it will return to the horizon. The height the object achieves is referred to as the *y*-**component**, and the distance along the horizon the object travels is referred to as the *x*-**component**. To maximize the horizontal distance an object travels, it should be launched at a 45-

degree angle relative to the horizon. The following shows the range, or x-distance, an object can travel when launched at various angles:

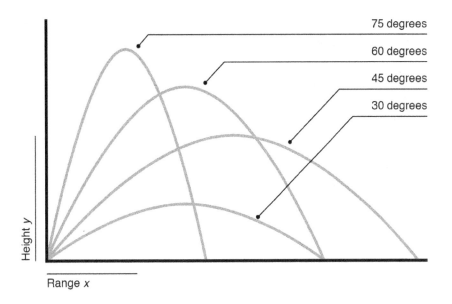

If something is traveling through the air without an internal source of power or any extra external forces acting upon it, it will follow these paths. All projectiles experience the effects of gravity while on Earth; therefore, they will experience a constant acceleration of 9.8 m/s^2 in a downward direction. While on its path, a projectile will have both a horizontal and vertical component to its motion and velocity. At the launch, the object has both vertical and horizontal velocity. As the object increases in height, the y-component of its velocity will diminish until the very peak of the object's path. At the peak, the y-component of the velocity is zero. Since the object still has a horizontal, or x-component, to its velocity, it would continue its motion. There is also a constant acceleration of 9.8 m/s^2 acting in the downward direction on the object the entire time, so at the peak of the object's height, the velocity would then switch from zero to being pulled down with the acceleration. The entire path of the object takes a specific amount of time, based on the initial launch angle and velocity. The time and distance traveled can be calculated using the kinematic equations of motion. The time it takes the object to reach its maximum height is exactly half of the time for the object's entire flight.

Similar motion is exhibited if an object is thrown from atop a building, bridge, cliff, etc. Since it would be starting at its maximum height, the motion of the object would be comparable to the second half of the path in the above diagram.

Principles of Light, Heat, Electricity, and Magnetism

Light
The movement of light is described like the movement of waves. **Light** travels with a wave front and has an **amplitude** (a height measured from the neutral), a cycle or wavelength, a period, and energy. Light travels at approximately 3.00 × 10^{8}m/s and is faster than anything created by humans.

Light is commonly referred to by its measured wavelengths, or the length for it to complete one cycle. Types of light with the longest wavelengths include radio, TV, micro, and infrared waves. The next set of wavelengths are detectable by the human eye and make up the visible spectrum. The visible spectrum has wavelengths of 10^{-7} m, and the colors seen are red, orange, yellow, green, blue, indigo, and violet. Beyond the visible spectrum are even shorter wavelengths (also called the **electromagnetic spectrum**) containing ultraviolet light, x-rays, and gamma rays. The wavelengths outside of the visible light range can be harmful to humans if they are directly exposed, especially for long periods of time.

For example, the light from the sun has a small percentage of ultraviolet (UV) light, which is mostly absorbed by the UV layer of the Earth's atmosphere. When this layer does not filter out the UV rays, the exposure to the wavelengths can be harmful to human skin. When there is an extra layer of pollutant and the light from the sun is trapped by repeated reflection back to the ground, unable to bounce back into space, it creates another harmful condition for the planet called the **greenhouse effect**. This is an overexposure to the sun's light and contributes to global warming by increasing the temperatures on Earth.

When a wave crosses a boundary or travels from one medium to another, certain actions take place. If the wave travels through one medium into another, it experiences **refraction**, which is the bending of the wave from one medium's density to another, altering the speed of the wave.

For example, a side view of a pencil in half a glass of water appears as though it is bent at the water level. What the viewer is seeing is the refraction of light waves traveling from the air into the water. Since the wave speed is slowed in water, the change makes the pencil appear bent.

When a wave hits a medium that it cannot pass through, it is bounced back in an action called **reflection**. For example, when light waves hit a mirror, they are reflected, or bounced off, the back of the mirror. This can cause it to seem like there is more light in the room due to the doubling back of the initial wave. This is also how people can see their reflection in a mirror.

When a wave travels through a slit or around an obstacle, it is known as **diffraction**. A light wave will bend around an obstacle or through a slit and cause a diffraction pattern. When the waves bend around an obstacle, it causes the addition of waves and the spreading of light on the other side of the opening.

Electricity

Electrostatics is the study of electric charges at rest. A balanced atom has a neutral charge from its number of electrons and protons. If the charge from its electrons is greater than or less than the charge of its protons, the atom has a charge. If the atom has a greater charge from the number of electrons than protons, it has a negative charge. If the atom has a lesser charge from the number of electrons than protons, it has a positive charge. Opposite charges attract each other, while like charges repel each other, so a negative attracts a positive, and a negative repels a negative. Similarly, a positive charge repels a positive charge. Just as energy cannot be created or destroyed, neither can charge; charge can only be transferred. The transfer of charge can occur through touch, or the transfer of electrons. Once electrons have transferred from one object to another, the charge has been transferred.

For example, if a person wears socks and scuffs their feet across a carpeted floor, the person is transferring electrons to the carpeting through the friction from his or her feet. Additionally, if that person then touches a light switch, they he or she receive a small shock. This "shock" is the person feeling the electrons transferring from the switch to their hand. Since the person lost electrons to the carpet, that person now has fewer negative charges, resulting in a net positive charge. Therefore, the electrons from the light switch are attracted to the person for the transfer. The shock felt is the electrons moving from the switch to the person's finger.

Another method of charging an object is through induction. **Induction** occurs when a charged object is brought near two touching stationary objects. The electrons in the objects will attract and cluster near another positively-charged object and repel away from a negatively-charged object held nearby. The stationary objects will redistribute their electrons to allow the charges to reposition themselves closer or farther away. This redistribution will cause one of the touching stationary objects to be negatively charged and the other to be positively charged. The overall charges contained in the stationary objects remain the same but are repositioned between the two objects.

Another way to charge an object is through **polarization**. Polarization can occur simply by the reconfiguration of the electrons within a single object.

For example, if a girl at a birthday party rubs a balloon on her hair, the balloon could then cling to a wall if it were brought close enough. This would be because rubbing the balloon causes it to become negatively charged. When the balloon is held against a neutrally-charged wall, the negatively charged balloon repels all the wall's electrons, causing a positively-charged surface on the wall. This type of charge is temporary, due to the massive size of the wall, and the charges will quickly redistribute.

An electric current is produced when electrons carry charge across a length. To make electrons move so they can carry this charge, a change in voltage must be present. On a small scale, this is demonstrated through the electrons traveling from the light switch to a person's finger in the example where the person had run their socks on a carpet. The difference between the charge in the switch and the charge in the finger causes the electrons to move. On a larger and more sustained scale, this movement would need to be more controlled. This can be achieved through batteries/cells and generators. Batteries or cells have a chemical reaction that takes place inside, causing energy to be released and charges to move freely. Generators convert mechanical energy into electric energy for use after the reaction.

For example, if a wire runs touching the end of a battery to the end of a lightbulb, and then another wire runs touching the base of the lightbulb to the opposite end of the original battery, the lightbulb will light up. This is due to a complete circuit being formed with the battery and the electrons being carried across the voltage drop (the two ends of the battery). The appearance of the light from the bulb is the visible presence of the electrons in the filament of the bulb.

Electric energy can be derived from a number of sources, including coal, wind, sun, and nuclear reactions. Electricity has numerous applications, including being transferable into light, sound, heat, or magnetic forces.

Magnetism

Magnetic forces occur naturally in specific types of materials and can be imparted to other types of materials. If two straight iron rods are observed, they will naturally have a negative end (pole) and a positive end (pole). These charged poles follow the rules of any charged item: Opposite charges attract, and like charges repel. When set up positive to negative, they will attract each other, but if one rod is turned around, the two rods will repel each other due to the alignment of negative to negative poles and positive to positive poles. When poles are identified, magnetic fields are observed between them. If small iron filings (a material with natural magnetic properties) are sprinkled over a sheet of paper resting on top of a bar magnet, the field lines from the poles can be seen in the alignment of the iron filings, as pictured below:

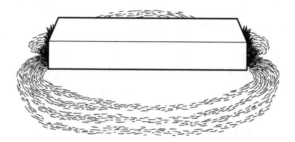

These fields naturally occur in materials with magnetic properties. There is a distinct pole at each end of such a material. If materials are not shaped with definitive ends, the fields will still be observed through the alignment of poles in the material.

For example, a circular magnet does not have ends but still has a magnetic field associated with its shape, as pictured below:

Magnetic forces can also be generated and amplified by using an electric current. For example, if an electric current is sent through a length of wire, it creates an electromagnetic field around the wire from the charge of the current. This force is from the moving of negatively-charged electrons from one end of the wire to the other. This is maintained as long as the flow of electricity is sustained. The magnetic field can also be used to attract and repel other items with magnetic properties. A smaller or larger magnetic force can be generated around this wire, depending on the strength of the current in the wire. As soon as the current is stopped, the magnetic force also stops.

Magnetic energy can be harnessed, or manipulated, from natural sources or from a generated source (a wire carrying electric current). When a core with magnetic properties (such as iron) has a wire wrapped around it in circular coils, it can be used to create a strong, non-permanent electromagnet. If current is run through the wrapped wire, it generates a magnetic field by polarizing the ends of the metal core, as described above, by moving the negative charge from one end to the other. If the direction of the current is reversed, so is the direction of the magnetic field due to the poles of the core being reversed. The term **non-permanent** refers to the fact that the magnetic field is generated only when the current is present, but not when the current is stopped.

The following is a picture of a small electromagnet made from an iron nail, a wire, and a battery:

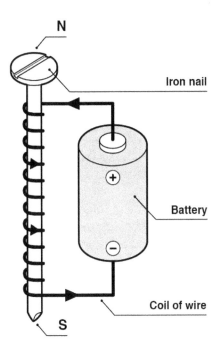

This type of electromagnetic field can be generated on a larger scale using more sizable components. This type of device is useful in the way it can be controlled. Rather than having to attempt to block a permanent magnetic field, the current to the system can simply be stopped, thus stopping the magnetic field. This provides the basis for many computer-related instruments and magnetic resonance imaging (MRI) technology. Magnetic forces are used in many modern applications, including the creation of super-speed transportation. Super magnets are used in rail systems and supply a cleaner form of energy than coal or gasoline. Another example of the use of super-magnets is seen in medical equipment, specifically MRI. These machines are highly sophisticated and useful in imaging the internal workings of the human body. For super-magnets to be useful, they often must be cooled down to extremely low temperatures to dissipate the amount of heat generated from their extended usage. This can be done by flooding the magnet with a super-cooled gas such as helium or liquid nitrogen. Much research is continuously done in this field to find new ceramic–metallic hybrid materials that have structures that can maintain their charge and temperature within specific guidelines for extended use.

Principles of Matter and Atomic Structure

Matter
The universe is composed completely of matter. **Matter** is any material or object that takes up space and has a mass. Although there is an endless variety of items found in the universe, there are only about one hundred elements, or individual substances, that make up all matter. These **elements** are different types of atoms and are the smallest units that anything can be broken down into while still retaining the properties of the original substance. Different elements can link together to form compounds, or

molecules. Hydrogen and oxygen are two examples of elements, and when they bond together, they form water molecules. Matter can be found in three different states: gas, liquid, or solid.

Gases

Gases have three main distinct properties. The first is that they are easy to compress. When a gas is compressed, the space between the molecules decreases, and the frequency of collisions between them increases. The second property is that they do not have a fixed volume or shape. They expand to fill large containers or compress down to fit into smaller containers. When they are in large containers, the gas molecules can float around at high speeds and collide with each other, which allows them to fill the entire container uniformly. Therefore, the volume of a gas is generally equal to the volume of its container. The third distinct property of a gas is that it occupies more space than the liquid or solid from which it was formed. One gram of solid CO_2, also known as dry ice, has a volume of 0.641 milliliters. The same amount of CO_2 in a gaseous state has a volume of 556 milliliters. Steam engines use water in this capacity to do work. When water boils inside the steam engine, it becomes steam, or water vapor. As the steam increases in volume and escapes its container, it is used to make the engine run.

Liquids

A **liquid** is an intermediate state between gases and solids. It has an exact volume due to the attraction of its molecules to each other and molds to the shape of the part of the container that it is in. Although liquid molecules are closer together than gas molecules, they still move quickly within the container they are in. Liquids cannot be compressed, but their molecules slide over each other easily when poured out of a container. The attraction between liquid molecules, known as **cohesion**, also causes liquids to have surface tension. They stick together and form a thin skin of particles with an extra strong bond between them. As long as these bonds remain undisturbed, the surface becomes quite strong and can even support the weight of an insect such as a water skipper. Another property of liquids is adhesion, which is when different types of particles are attracted to each other. When liquids are in a container, they are drawn up above the surface level of the liquid around the edges. The liquid molecules that are in contact with the container are pulled up by their extra attraction to the particles of the container.

Solids

Unlike gases and liquids, **solids** have a definitive shape. They are similar to liquids in that they also have a definitive volume and cannot be compressed. The molecules are packed together tightly, which does not allow for movement within the substance. There are two types of solids: crystalline and amorphous. **Crystalline solids** have atoms or molecules arranged in a specific order or symmetrical pattern throughout the entire solid. This symmetry makes all of the bonds within the crystal of equal strength, and when they are broken apart, the pieces have straight edges. Minerals are all crystalline solids. **Amorphous solids**, on the other hand, do not have repeating structures or symmetry. Their components are heterogeneous, so they often melt gradually over a range of temperatures. They do not break evenly and often have curved edges. Examples of amorphous solids are glass, rubber, and most plastics.

Matter can change between a gas, liquid, and solid. When these changes occur, the change is physical and does not affect the chemical properties or makeup of the substance. Environmental changes, such as temperature or pressure changes, can cause one state of matter to convert to another state of matter. For example, in very hot temperatures, solids can melt and become a liquid, such as when ice melts into liquid water, or sublimate and become a gas, such as when dry ice becomes gaseous carbon dioxide. Liquids can evaporate and become a gas, such as when liquid water turns into water vapor. In very cold temperatures, gases can depose and become a solid, such as when water vapor becomes icy frost on a windshield, or condense and become a liquid, such as when water vapor becomes dew on grass. Liquids can freeze and become a solid, such as when liquid water freezes and becomes ice.

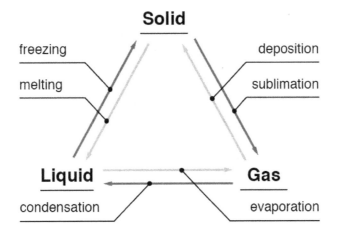

Basic Atomic Structure

The **Bohr model** of the atom is the one most commonly used in science today. It was proposed by physicist Niels Bohr and caused some controversy with its configuration of the electrons and their location within the atom. The Bohr model of the atom consists of the nucleus, or core, which is made up of positively-charged protons and neutrally-charged neutrons. The neutrons are theorized to be in the nucleus with the protons. This pairing of protons and neutrons provides a "balance" at the center of the atom. The nucleus of the atom consists of more than 99 percent of the mass of an atom. Surrounding the nucleus are orbitals containing negatively-charged electrons. The entire structure of an atom is too small to be seen with the unaided eye, which has contributed to the differing ideas about its structure. Most research has been focused around recording the reactions of the atom or the energy emitted from the electrons to test any theories about its structure.

Below is an illustration of Bohr's Model:

Bohr's Model

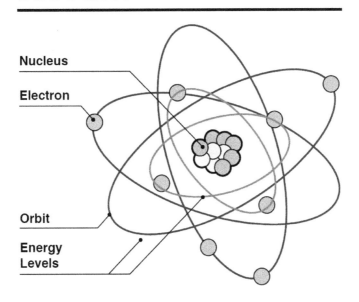

Protons

Protons are found in the nucleus of an atom and have a positive electric charge. At least one proton is found in the nucleus of all atoms. Protons have a mass of about one atomic mass unit. The number of protons in an element is referred to as the element's atomic number. Each element has a unique number of protons and therefore its own unique atomic number. Hydrogen ions are unique in that they only contain one proton and do not contain any electrons. Since they are free protons, they have a very short lifespan and immediately become attracted to any free electrons in the environment. In a solution with water, they bind to a water molecule, H_2O, and turn it into H_3O+.

Neutrons

Neutrons are also found in the nucleus of atoms. These subatomic particles have a neutral charge, meaning that they do not have a positive or negative electric charge. Their mass is slightly larger than that of a proton. Together with protons, they are referred to as the nucleons of an atom. Interestingly, although neutrons are not affected by electric fields because of their neutral charge, they are affected by magnetic fields and are said to have magnetic moments.

Electrons

Electrons have a negative charge and are the smallest of the subatomic particles. They are located outside the nucleus in orbitals, which are shells that surround the nucleus. If an atom has an overall neutral charge, it has an equal number of electrons and protons. If it has more protons than electrons or vice versa, it becomes an ion. When there are more protons than electrons, the atom is a positively charged ion, or cation. When there are more electrons than protons, the atom is a negatively charged ion, or anion.

The location of electrons within an atom is more complicated than the locations of protons and neutrons. Within the orbitals, electrons are always moving. They can spin very fast and move upward, downward, and sideways. There are many different levels of orbitals around the atomic nucleus, and each orbital has a different capacity for electrons. The electrons in the orbitals closest to the nucleus are more tightly bound to the nucleus of the atom. There are three main characteristics that describe each orbital. The first is the principle quantum number, which describes the size of the orbital. The second is the angular momentum quantum number, which describes the shape of the orbital. The third is the magnetic quantum number, which describes the orientation of the orbital in space.

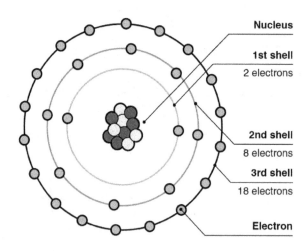

Although atoms are very small in size, they have many properties that can be quantitatively measured. The **atomic mass** of an atom is equal to the sum of the protons and neutrons in the atom. The **ionization energy** of an atom is the amount of energy that is needed to remove one electron from an individual atom. The greater the ionization energy, the harder it is to remove the electron, meaning the greater the bond between the electron and that atom's nucleus. Ionization energies are always positive because atoms require extra energy to let go of an atom. On the other hand, **electron affinity** is the change in energy that occurs when an electron is added to an individual atom. Electron affinity is always negative because atoms release energy when an electron is added. The **effective nuclear charge** of an atom is noted as Z_{eff} and accounts for the attraction of electrons to the nucleus of an atom as well as the repulsion of electrons to other neighboring electrons in the same atom. The **covalent radius** of an atom is the radius of an atom when it is bonded to another atom. It is equal to one-half the distance between the two atomic nuclei. The **van der Waal's radius** is the radius of an atom when it is only colliding with another atom in solution or as a gas and is not bonding to the other atom.

Periodic Table and Periodicity

The **periodic table** is a chart that describes all the known elements. Currently, there are 118 elements listed on the periodic table. Each element has its own cell on the table. Within the cell, the element's abbreviation is written in the center, with its full name written directly below. The atomic number is recorded in the top left corner, and the atomic mass is recorded at the bottom center of the cell in atomic mass units, or **amus**. The first ninety-four elements are naturally occurring, whereas the last twenty-four can only be made in laboratory environments.

The rows of the table are called **periods**, and the columns are called **groups**. Elements with similar properties are grouped together. The periodic table can be divided and described in many ways according to similarities of the grouped elements. When it is divided into blocks, each block is named in accordance with the subshell in which the final electron sits. The blocks are named alkali metals, alkaline earth metals, transition metals, basic metals, semimetals, nonmetals, halogens, noble gases, lanthanides, and actinides. If the periodic table is divided according to shared physical and chemical properties, the elements are often classified as metals, metalloids, and nonmetals.

Use this graphic to help with understanding the Periodic Table of Elements:

The Periodic Table of the Elements

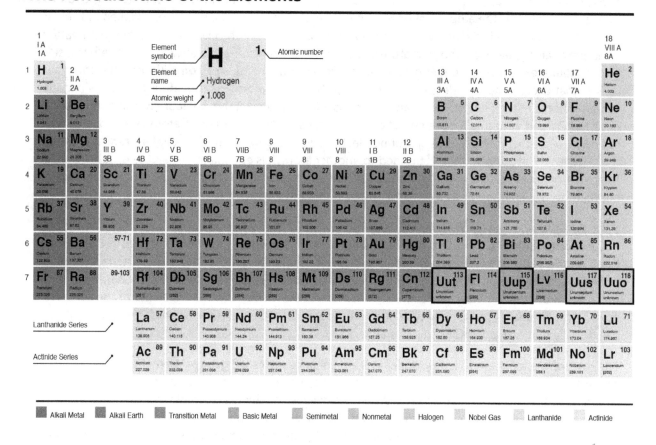

Periodicity

Periodicity refers to the trends that are seen in the periodic table among the elements. It is a fundamental aspect of how the elements are arranged within the table. These trends allow scientists to learn the properties of families of elements together instead of learning about each element only individually. They also elucidate the relationships among the elements. There are many important characteristics that can be seen by examining the periodicity of the table, such as ionization energy, electron affinity, length of atomic radius, and metallic characteristics. Moving from left to right and from bottom to top on the periodic table, elements have increasing ionization energy and electron affinity. Moving in the opposite directions, from top to bottom and from right to left, the elements have increasing length of atomic radii. Moving from the bottom left corner to the top right corner, the elements have increasing nonmetallic properties, and moving in the opposite direction from the top right corner to the bottom left corner, the elements have increasing metallic properties.

Principles of Chemical Reactions

A **chemical reaction** is a process that involves a change in the molecular arrangement of a substance. Generally, one set of chemical substances, called the reactants, is rearranged into a different set of chemical substances, called the products, by the breaking and re-forming of bonds between atoms. In a chemical reaction, it is important to realize that no new atoms or molecules are introduced. The products are formed solely from the atoms and molecules that are present in the reactants. These can involve a change in state of matter as well. Making glass, burning fuel, and brewing beer are all examples of chemical reactions.

Types of Chemical Reactions

Generally, chemical reactions are thought to involve changes in positions of electrons with the breaking and re-forming of chemical bonds, without changes to the nucleus of the atoms. The three main types of chemical reactions are combination, decomposition, and combustion.

Combination

In **combination reactions**, two or more reactants are combined to form one more complex, larger product. The bonds of the reactants are broken, the elements rearranged, and then new bonds are formed between all of the elements to form the product. It can be written as A + B → C, where A and B are the reactants and C is the product. An example of a combination reaction is the creation of iron (II) sulfide from iron and sulfur, which is written as $8Fe + S_8 → 8FeS$.

Decomposition

Decomposition reactions are almost the opposite of combination reactions. They occur when one substance is broken down into two or more products. The bonds of the first substance are broken, the elements rearranged, and then the elements bonded together in new configurations to make two or more molecules. These reactions can be written as C → B + A, where C is the reactant and A and B are the products. An example of a decomposition reaction is the electrolysis of water to make oxygen and hydrogen gas, which is written as $2H_2O → 2H_2 + O_2$.

Combustion

Combustion reactions are a specific type of chemical reaction that involves oxygen gas as a reactant. This mostly involves the burning of a substance. The combustion of hexane in air is one example of a combustion reaction. The hexane gas combines with oxygen in the air to form carbon dioxide and water. The reaction can be written as:

$$2C_6H_{14} + 17O_2 → 12CO_2 + 14H_2O$$

Balancing Chemical Reactions

The way the hexane combustion reaction is written above ($2C6H14 + 17O2 → 12CO2 + 14H2O$) is an example of a chemical equation. Chemical equations describe how the molecules are changed when the chemical reaction occurs. The "+" sign on the left side of the equation indicates that those molecules are reacting with each other, and the arrow, "→," in the middle of the equation indicates that the reactants are producing something else. The coefficient before a molecule indicates the quantity of that specific molecule that is present for the reaction. The subscript next to an element indicates the quantity of that element in each molecule. In order for the chemical equation to be balanced, the quantity of each element on both sides of the equation should be equal. For example, in the hexane equation above, there are twelve carbon elements, twenty-eight hydrogen elements, and thirty-four oxygen elements on each side of the equation. Even though they are part of different molecules on each side, the overall quantity is the same. The state of matter of the reactants and products can also be included in a chemical equation and would be written in parentheses next to each element as follows: gas (g), liquid (l), solid (s), and dissolved in water, or aqueous (aq).

Catalysts

The rate of a chemical reaction can be increased by adding a catalyst to the reaction. **Catalysts** are substances that lower the activation energy required to go from the reactants to the products of the reaction but are not consumed in the process. The activation energy of a reaction is the minimum amount of energy that is required to make the reaction move forward and change the reactants into the products. When catalysts are present, less energy is required to complete the reaction. For example, hydrogen peroxide will eventually decompose into two water molecules and one oxygen molecule. If potassium permanganate is added to the reaction, the decomposition happens at a much faster rate. Similarly, increasing the temperature or pressure in the environment of the reaction can increase the rate of the reaction. Higher temperatures increase the number of high-energy collisions that lead to the products. The same happens when increasing pressure for gaseous reactants, but not with solid or liquid reactants.

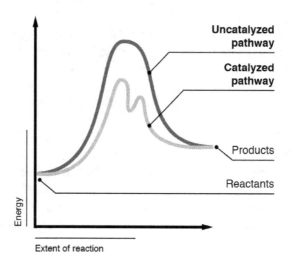

Enzymes

Enzymes are a type of biological catalyst that accelerate chemical reactions. They work like other catalysts by lowering the activation energy of the reaction to increase the reaction rate. They bind to substrates and change them into the products of the reaction without being consumed by the reaction. Enzymes differ from other catalysts by their increased specificity to their substrate. They can also be affected by other molecules. Enzyme inhibitors can decrease their activity, whereas enzyme activators increase their activity. Their activity also decreases dramatically when the environment is not in their optimal pH and temperature range.

An enzyme can encounter several different types of inhibitors. In biological systems, inhibitors are many times expressed as part of a feedback loop where the enzyme is producing too much of a substance too quickly and the reaction needs to be stopped. Competitive inhibitors and substrates cannot bind to the enzyme at the same time. These inhibitors often resemble the substrate of the enzyme. Noncompetitive inhibitors bind to the enzyme at a different location than the substrate. The substrate can still bind with the same affinity, but the efficiency of the enzyme is decreased. Uncompetitive inhibitors cannot bind to the enzyme alone and can only bind to the enzyme–substrate complex. Once an uncompetitive inhibitor binds to the complex, the complex becomes inactive. An irreversible inhibitor binds to the enzyme and permanently inactivates it. It usually does so by forming a covalent bond with the enzyme.

pH

In chemistry, **pH** stands for the potential of hydrogen. It is a numeric scale ranging from 0 to 14 that determines the acidity and basicity of an aqueous solution. Solutions with a pH of greater than 7 are considered basic, those with a pH of less than 7 are considered acidic, and solutions with a pH equal to 7 are considered neutral. Pure water is considered neutral with a pH equal to 7. It is important to remember that the pH of solutions can change at different temperatures. While the pH of pure water is 7 at 25°C, at 0°C it is 7.47, and at 100°C it is 6.14. pH is calculated as the negative of the base 10 logarithm of the activity of the hydrogen ions in the solution and can be written as:

$$pH = -\log_{10}(a_H+) = \log_{10}(1/a_H+)$$

One of the simplest ways to measure the pH of a solution is by performing a litmus test. Litmus paper has a mixture of dyes on it that react with the solution and display whether the solution is acidic or basic depending on the resulting color. Red litmus paper turns blue when it reacts with a base. Blue litmus paper turns red when it reacts with an acid. Litmus tests are generally crude measurements of pH and not very precise. A more precise way to measure the pH of a solution is to use a pH meter. These specialized meters act like voltmeters and measure the electrical potential of a solution. Acids have a lot of positively charged hydrogen ions in solutions that have greater potential than bases to produce an electric current. The voltage measurement is then compared with a solution of known pH and voltage. The difference in voltage is translated into the difference in pH.

Acids and Bases

Acids and bases are two types of substances with differing properties. In general, when acids are dissolved in water, they can conduct electricity, have a sour taste, react with metals to free hydrogen ions, and react with bases to neutralize their properties. Bases, in general, when dissolved in water, conduct electricity, have a slippery feel, and react with acids to neutralize their properties.

The Brønsted-Lowry acid–base theory is a commonly accepted theory about how acid–base reactions work. **Acids** are defined as substances that dissociate in aqueous solutions and form hydrogen ions. **Bases** are defined as substances that dissociate in aqueous solutions and form hydroxide (OH^-) ions. The basic idea behind the theory is that when an acid and base react with each other, a proton is exchanged, and the acid forms its conjugate base, while the base forms its conjugate acid. The acid is the proton donor, and the base is the proton acceptor. The reaction is written as $HA + B \leftrightarrow A^- + HB^+$, where HA is the acid, B is the base, A^- is the conjugated base, and HB^+ is the conjugated acid. Since the reverse of the reaction is also possible, the reactants and products are always in equilibrium. The equilibrium of acids and bases in a chemical reaction can be determined by finding the acid dissociation constant, K_a, and the base dissociation constant, K_b. These dissociation constants determine how strong the acid or base is in the aqueous solution. Strong acids and bases dissociate quickly to create the products of the reaction.

In other cases, acids and bases can react to form a neutral salt. This happens when the hydrogen ion from the acid combines with the hydroxide ion of the base to form water, and the remaining ions combine to form a neutral salt. For example, when hydrochloric acid and sodium hydroxide combine, they form salt and water, which can be written as:

$$HCl + NaOH \rightarrow NaCl + H_2O$$

Earth Science

Properties of Earth Materials

The Earth is a sphere comprised of four distinct layers. The center of the Earth, known as the **inner core**, makes up the innermost layer. This layer is solid (made primarily of iron and nickel) and approximately 760 miles across. Around the inner core is a layer known as the **outer core**, which is similar in elemental make-up but consisting of liquid matter. The next layer is known as the **mantle**; this layer makes up most of the Earth's total volume. It is made of silicate particles and considered a solid layer; the deepest part of the mantle is solid rock. The more superficial part of the mantle transitions into the consistency of thick, slow-flowing magma. In fact, this is like the magma seen in volcanic eruptions and along tectonic fault lines. Finally, the outermost layer of the Earth is called the **crust.** The crust is the area where all living creatures reside. It is relatively thin in relation to the core and the mantle. The crust is a rocky surface. **Continental crust** refers to landmasses, while **oceanic crust** refers to areas under the oceans. Approximately 75% of the crust is oceanic. The **hydrosphere** refers to all areas of the planet's surface that contain water (in any form of matter), while the **atmosphere** refers to the types of gases present on the planet's surface. The hydrosphere and atmosphere directly influence each other; the unique water cycle on Earth results from their interaction. As water in the hydrosphere is heated by the Sun, it turns into gas and becomes part of the atmosphere. As the atmospheric temperature cools, the gaseous water vapor turns into cloud matter. Eventually, atmospheric pressure will lead to precipitation in the form of rain or snow, returning the once-evaporated water to the hydrosphere.

Minerals, rocks, soil, and water are important Earth materials. Minerals occur naturally, are inorganic, and their defined structure and composition is crystalline in nature. Rocks are amalgamations of different minerals and substances. There are three types of rocks: sedimentary, igneous, and metamorphic and a rock can be transformed between different types via time, pressure, and heat.

Earth's Systems, Processes, Geologic Structures, and Time

Processes of the Earth System
The Earth system is complex, made of interrelated parts; process variation in one arena can significantly impact other components. The planet can be thought of as consisting of four main parts: the hydrosphere, the atmosphere; the geosphere (referring to any physical matter that makes up the Earth); and the biosphere (referring to all living organisms across the planet).

Different chemical, physical, and energetic cycles give rise to the activities that take place across the Earth system. These include processes such as evaporation/precipitation (which drive how water is dispersed over the planet) and oceanic convection currents (which influence weather patterns and animal migration patterns, amongst others). Most of these cycles are initially triggered by the radiation provided by the Sun. For example, **photosynthesis** refers to the process of plants converting sunlight to chemical energy that plants can better utilize. This process is also the primary basis of converting carbon dioxide to oxygen, a critical component for the biosphere (as animals need oxygen to survive), the atmosphere (as imbalanced levels of carbon dioxide or oxygen can cause repercussions such as global warming and climate change), and the geosphere (as plants also need to take water and other nutrients from their surrounding soil for the full functions of photosynthesis to take place). However, other events may be triggered by processes within the Earth itself. For example, **plate tectonics** refers to how plates that make up the surface of the Earth's crust move. When plates move, they can create earthquakes, volcanic eruptions, mountain ranges, or deep trenches. These greatly impact the physical makeup of the geosphere, as well as the existence of a number of organisms in the biosphere.

Earth History
The Earth is believed to be approximately 4.5 billion years old, resulting from solar nebula remnants that dispersed after the Sun was formed. Consequently, the planet was originally hot and gaseous, but with no oxygen. Most of the activity that took place on the face of the "Early Earth" was frequent volcanic eruptions. Over a period of approximately half a billion years, the Earth cooled to where a solid layer was able to form. It was during this time that liquid water appeared on the surface and original life began. Once the crust solidified, rock and fossil records were able to appear. As rocks show physical changes due to heat, chemicals, and erosion, they can provide some insight into the type of environment that existed in the past. Fossils, as organic matter, can provide a more accurate timeline of events than rock records. It is believed that living organisms first appeared approximately 1 billion years ago. These early organisms did not need much oxygen. Plants that photosynthesized appeared approximately between 2.5 to 3 billion years ago; their appearance contributed substantially more oxygen to the Earth's atmosphere.

The periods of time on Earth can be divided chronologically into the **Hadean Eon** (the period before the crust solidified), the **Archean Eon**, the **Proterozoic Eon**, and the **Phanerozoic Eon**. The Phanerozoic Eon is further distinguished into eras that define when complex life forms came to be. Between the Proterozoic and Phanerozoic Eons, a phenomenon known as the **Cambrian Explosion** took place (approximately 540 million years ago). During this time, there was major diversification and growth of organisms, which quickly evolved from unicellular entities to complex multicellular creatures. The **Paleozoic Era** saw the evolution of arthropods and fish. The **Mesozoic Era** saw the arrival and extinction of dinosaurs. The last era, the **Cenozoic era**, introduced mammals. Ancestral humans are believed to have evolved from apes that appeared during the Cenozoic Era. The original human relative dates to approximately 2 million years ago.

Earth's Movements and Position in the Solar System

Within galaxies, the next largest system is a **solar system**. These systems contain of one or more stars with a massive gravitational pull. Various planets, moons, and debris orbit the star or stars as a result. Stars form because of the gravitational collapse of explosive elements (primarily hydrogen and helium). The center of the star continues in a state of thermonuclear fusion that creates the star's internal gravitational pull. Over many thousands of years, the central thermonuclear activity will slowly cease until the star collapses, resulting in the absorption of orbiting bodies. The Earth is part of a solar system in the Milky Way galaxy. Its solar system consists of one sun, eight planets, their moons, and other cosmic bodies such as meteors and asteroids (including an asteroid belt that orbits the Sun in a full ring). In order of proximity to the Sun, the planets in this solar system are Mercury, Venus, Earth, Mars, Jupiter, Saturn, Uranus, and Neptune. Beyond Neptune is a dwarf planet named Pluto, and at least four other known dwarf planets. While the Earth has only a single moon, most other planets have multiple moons (although Mercury and Venus have none). Based on their mass, moons, and position from the Sun, each planet takes a different period to orbit the Sun and rotate upon its axis. The Earth takes 24 hours to rotate on its axis and 365 days to orbit the Sun. Solar eclipses occur on Earth when the moon's orbit appears to cross between the Earth and the Sun, while lunar eclipses occur when the Earth's orbit crosses between the Sun and the moon. Additionally, the moon's gravitational pull on the Earth impacts the water on the planet, causing high and low tides throughout the day. The difference in tidal height varies based on the location of the moon in its orbit around the Earth (as well as different features of the water body and shoreline).

Earth Patterns, Cycles, and Change

Several patterns and cycles take place on Earth that create and influence the most unique feature of the planet: its ability to host life. One feature that influences these patterns and cycles is Earth's position relative to the Sun. The Earth is located approximately 93 million miles from the Sun; variations in this distance would likely make the planet too hot or too cold to support life. As the Earth tilts closer to or further from the Sun, different parts of the planet experience different patterns of weather known as **seasons**. As the top hemisphere of the Earth tilts toward the Sun, those geographical areas experience spring and summer, while the bottom hemisphere, which has tilted away from the Sun, experiences fall and winter. As the Earth tilts the opposite way, the seasons across each northern and southern hemisphere flip. It takes approximately one year for an area far from the equator (the equidistant point between the north and south poles of the Earth) to experience four seasons. Areas that are close to the equator have approximately the same temperature year-round but may experience differences in climate, such as rainier or drier seasons. These are impacted by ocean currents that travel around the Earth and can be warm or cool based on the Earth's positioning, natural variances (such as the El Nino and La Nina phenomena), and human activities that contribute to temperature changes.

Important cycles on the Earth include the water cycle, which involves the evaporation of surface water into the atmosphere that later returns as precipitation, and the carbon cycle, a complex system that creates the atmospheric conditions on the planet. Carbon is an element found in all organic chemical compounds and is therefore found in the air, living organisms, the ocean, rocks and other solid matter, and in the Earth's deeper layers. Carbon is exchanged across these entities through processes such as photosynthesis, volcanic eruptions, oceanic currents, and the water cycle. Without human activities that contribute to excess carbon production, the carbon present on Earth stays approximately the same through these exchanges, except for a few distinct periods in history (such as ice ages). Human activities such as burning fossil fuels, concentrated animal feeding operations, and deforestation have contributed to a steady increase in the levels of carbon and carbon dioxide present in the atmosphere. In turn, this traps more heat near the Earth's surface and deposits more carbon into the hydrosphere through acid rain. An ocean temperature increase of just two degrees Celsius causes shifts in ocean ecosystems and convection currents that can disrupt weather and climate patterns, such as abnormally hot or cool temperatures, increased precipitation or aridity, and the timing, intensity, and location of natural disasters such as hurricanes.

The Sun, Other Stars, and the Solar System

The **universe** is defined as the largest entity made by space and time, and it includes all of the smaller entities contained within it. The largest known systems within the universe are known as **galaxies**. It is believed there are over 100 billion galaxies within the universe. Galaxies can be made up of different gases, cosmic debris, stars, planets, moons, black holes, and other bits of matter. Billions of these entities can make up a single galaxy. They primarily appear in either spiral or disk shapes. Galaxies that cannot be categorized as these shapes are known as irregularly shaped. These shapes are made based on the masses of the various entities within the galaxy. The gravitational pull of these objects upon one another creates not only the shape of the galaxy, but also influences how other systems operate within the galaxy. Galaxies can be hundreds of thousands to millions of light years in size across.

Within galaxies, the next largest system is a **solar system**. These systems contain of one or more stars with a massive gravitational pull. Various planets, moons, and debris orbit the star or stars as a result. Stars form because of the gravitational collapse of explosive elements (primarily hydrogen and helium). The center of the star continues in a state of thermonuclear fusion that creates the star's internal gravitational pull. Over many thousands of years, the central thermonuclear activity will slowly cease until the star collapses, resulting in the absorption of orbiting bodies. The Earth is part of a solar system in the Milky Way galaxy. Its solar system consists of one sun, eight planets, their moons, and other cosmic bodies, such as meteors and asteroids (including an asteroid belt that orbits the Sun in a full ring). In order of proximity to the Sun, the planets in this solar system are Mercury, Venus, Earth, Mars, Jupiter, Saturn, Uranus, and Neptune. Beyond Neptune is a dwarf planet named Pluto, and at least four other known dwarf planets. While the Earth has only a single moon, most other planets have multiple moons (although Mercury and Venus have none). Based on their mass, moons, and position from the Sun, each planet takes a different period to orbit the Sun and rotate upon its axis. The Earth takes 24 hours to rotate on its axis and 365 days to orbit the Sun.

Solar eclipses occur on Earth when the moon's orbit appears to cross between the Earth and the Sun, while lunar eclipses occur when the Earth's orbit crosses between the Sun and the moon. Additionally, the moon's gravitational pull on the Earth impacts the water on the planet, causing high and low tides throughout the day. The difference in tidal height varies based on the location of the moon in its orbit around the Earth (as well as different features of the water body and shoreline).

The positioning of the moon between the Earth and the Sun at different positions in the Earth's orbit results in the phases of the moon. While a common misconception is that the phases are simply caused from the shadow the Earth casts on the moon, the different phases we see are actually due to the position of the moon relative to the Sun. A portion of the moon that is not visible or shadowed is turned away from the Sun. At all times, half of the moon is illuminated while half is shadowed, yet our perception of different phases is caused by the moon's position relative to us on Earth.

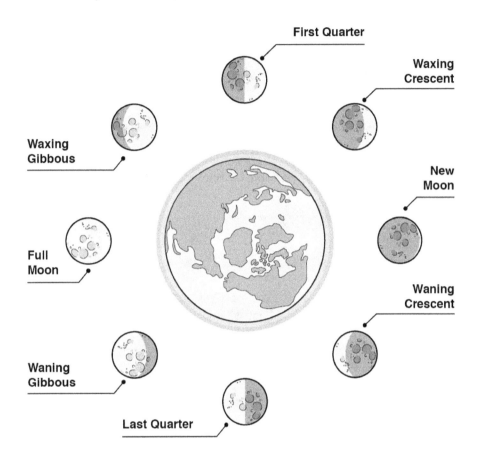

Practice Questions

1. If a vehicle increases speed from 20 m/s to 40 m/s over a time period of 20 seconds, what is the vehicle's rate of acceleration?
   a. 2 m/s^2
   b. 1 m/s^2
   c. 10 m/s^2
   d. 20 m/s^2

2. Under which situation does an object have a negative acceleration?
   a. It is increasing velocity in a positive direction.
   b. It is at a complete stop.
   c. It is increasing velocity in a negative direction.
   d. It is moving at a constant velocity.

3. What part of the respiratory system is responsible for regulating the temperature and humidity of the air that comes into the body?
   a. Larynx
   b. Lungs
   c. Trachea
   d. Sinuses

4. Which of the following is true regarding the circulatory system?
   a. Carbon dioxide-rich blood returns to the heart to get pumped through systemic circulation.
   b. The mitral valve prevents backflow of blood from the right ventricle back into the right atrium.
   c. Oxygen-rich blood leave the right ventricle and enters pulmonary circulation.
   d. The bicuspid valve closes after the left ventricle has filled with blood.

5. Which organ is responsible for gas exchange between air and circulating blood?
   a. Nose
   b. Larynx
   c. Lungs
   d. Stomach

6. Which layer of the heart wall contains nerves?
   a. Epicardium
   b. Myocardium
   c. Endocardium
   d. Sternum

7. What is another name for the mitral valve?
   a. Bicuspid valve
   b. Pulmonary valve
   c. Aortic valve
   d. Tricuspid valve

8. Which component of blood helps to fight off diseases?
   a. Red blood cells
   b. White blood cells
   c. Plasma
   d. Platelets

9. Which system consists of a group of organs that work together to transform food and liquids into fuel for the body?
   a. Respiratory system
   b. Immune system
   c. Genitourinary system
   d. Gastrointestinal system

10. During which step performed by the gastrointestinal system do chemicals and enzymes break down complex food molecules into smaller molecules?
   a. Digestion
   b. Absorption
   c. Compaction
   d. Ingestion

Questions 11–15 are based on the following information:

The annual wildland fire statistics are compiled and reported by the National Interagency Coordination Center at NIFC for federal and state agencies. Information is gathered through situation reports from many decades.

The table below lists the number of fires during the years since 2007 and the acres those fires covered:

Total Wildland Fires and Acres (1926 - 2017)

The National Interagency Coordination Center at NIFC compiles annual wildland fire statistics for federal and state agencies. This information is provided through Situation Reports, which have been in use for several decades. Prior to 1983, sources of these figures are not known, or cannot be confirmed, and were not derived from the current situation reporting process. As a result, the figures prior to 1983 should not be compared to later data.

Source: National Interagency Coordination Center

Year	Fires	Acres
2017	71,499	10,026,086
2016	67,743	5,509,995
2015	68,151	10,125,149
2014	63,312	3,595,613
2013	47,579	4,319,546
2012	67,774	9,326,238
2011	74,126	8,711,367
2010	71,971	3,422,724
2009	78,792	5,921,786
2008	78,979	5,292,468
2007	85,795	9,328,045

The chart below lists temperature anomalies (both land and ocean) beginning in 1880 in degrees Celsius and degrees Fahrenheit:

Global Land and Ocean Temperature Anomalies, March

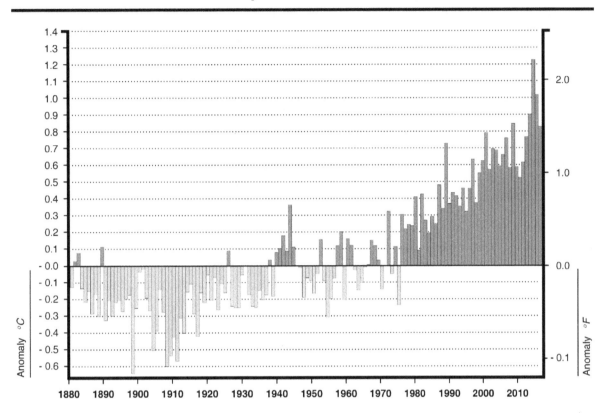

11. Approximately which year shows the highest temperature anomaly?
    a. 1900
    b. 1944
    c. 1990
    d. 2016

12. What was the average number of fires from 2007 to 2012?
    a. 67,774
    b. 76,224.5
    c. 70,512
    d. 78,792

13. What could be inferred about the years with the highest number of acres covered by its fires?
    a. There was a misreporting of the number of fires.
    b. The most acres reflected the greatest number of fires.
    c. The most acres reflected larger-sized fires.
    d. No conclusion can be drawn.

14. What could be inferred about the trend in temperature anomalies?
    a. Temperatures are increasing.
    b. Temperatures are decreasing.
    c. Temperatures have remained steady.
    d. Nothing can be inferred about the temperature anomalies.

15. Overall, what could NOT have caused the number of total fires to decrease?
    a. Better policing of camp grounds
    b. Improved fire-fighting technologies
    c. Increased fire-fighting awareness
    d. Decrease in climate temperatures

16. How many gametes do the ovaries produce every month?
    a. One million
    b. One
    c. One-half billion
    d. Two

17. In what part of the female reproductive system do sperm fertilize an oocyte?
    a. Ovary
    b. Mammary gland
    c. Vagina
    d. Fallopian tubes

18. What is the largest organ in the human body?
    a. Brain
    b. Large intestine
    c. Skin
    d. Liver

19. Which is a function of the skin?
    a. Temperature regulation
    b. Breathing
    c. Ingestion
    d. Gas exchange

20. Which accessory organ of the integumentary system provides a hard layer of protection over the skin?
    a. Hair
    b. Nails
    c. Sweat glands
    d. Sebaceous glands

Questions 21–25 are based on the following information:

Our solar system is made up of eight planets. In order from the Sun moving outward, they are Mercury, Venus, Earth, Mars, Jupiter, Saturn, Uranus, and Neptune. The first four planets are the smallest and have some similarities in characteristics, while the outer four planets are larger and also have some comparable traits. Kepler studied the movements of the planets and devised three laws for planetary motion. Some basic information regarding the planets is listed in the tables below:

Planet Name	Distance from Sun (km)	Mass (10^{24} kg)	Gravity (m/s^2)
Mercury	57.9 million	0.330	3.7
Venus	108.2 million	4.87	8.9
Earth	149.6 million	5.97	9.8
Mars	227.9 million	0.642	3.7
Jupiter	778.6 million	1898	23.1
Saturn	1,433.5 million	568	9.0
Uranus	2,872.5 million	86.8	8.7
Neptune	4,495.1 million	102	11.0

Planet	Mercury	Venus	Earth	Mars	Jupiter	Saturn	Uranus	Neptune
Length of Orbit (days)	88.0	224.7	365.2	687	4,331	10,747	30,589	59,800
Number of moons	0	0	1	2	67	62	27	14

21. The mass of Venus is approximately what percentage of the Earth's mass?
    a. 8.16 percent
    b. 81.6 percent
    c. 1.22 percent
    d. 12.2 percent

22. What would NOT be a likely reason for the variations in gravity on each planet?
    a. The pull of the moons on a planet
    b. The speed of rotation of a planet
    c. The make-up of the core of a planet
    d. The mass of a planet

23. What is the average length of orbit, in days, of the planets listed?
    a. 365.2
    b. 13,354
    c. 29,944
    d. 59,800

24. How many times farther is Neptune from the Sun than Mercury, and how many times longer is Neptune's orbit than Earth's?
    a. 78 times as far, and 164 times as long
    b. 78 times as far, and 679 times as long
    c. 30 times as far, and 679 times as long
    d. 30 times as far, and 164 times as long

25. Which distance is the shortest?
    a. Distance from Mercury to the Sun
    b. Distance from Mercury to Venus
    c. Distance from Venus to Earth
    d. Distance from Mars to Earth

Questions 26 and 27 are based on the following information:

Soil samples were collected from various locations and analyzed for their composition. These minerals were sand, silt, and clay. Three types of minerals were identified and measured by percent in each sample. Particle size ranges for each mineral were also measured and recorded for the soil samples.

Soil Sample	Sand (%)	Clay (%)	Silt (%)
1	75	5	20
2	5	80	15
3	20	35	45
4	70	15	15
5	55	25	20

Type of mineral particle	Size range of particles (mm)
Sand	3.0-0.07
Silt	0.07-0.003
Clay	Less than 0.003

26. Which minerals mainly comprised Sample 3?
    a. Sand and silt
    b. Silt and clay
    c. Sand and clay
    d. Sand

27. Which soil sample would MOST likely have particle sizes around 1.7 mm?
    a. Sample 2
    b. Sample 3
    c. Sample 4
    d. Sample 5

28. In the Periodic Table, what similarity do the elements in columns have with each other?
    a. Have the same atomic number
    b. Similar chemical properties
    c. Similar electron valence configurations
    d. Start with the same first letter

29. Which is NOT an example of how to charge an object?
    a. Transferring electrons
    b. Induction
    c. Polarization
    d. Refraction

30. Which two magnetic poles would attract each other?
    a. Two positive poles
    b. Two negative poles
    c. One positive pole and one negative pole
    d. Two circular positive magnets

31. Which of the following is an example of a physical property of substances?
    a. Odor
    b. Reactivity
    c. Flammability
    d. Toxicity

32. Which of the following is one main type of transfer that occurs in redox reactions?
    a. Transfer of carbon
    b. Transfer of hydrogen
    c. Transfer of nitrogen
    d. Transfer of sulfur

33. Which of the following descriptions characterizes an oxidative agent in a redox reaction?
    a. When it gains a hydrogen
    b. When it loses an oxygen
    c. When it gains one or more electrons
    d. When it gains a carbon

34. What is the sum of the oxidation numbers of all the atoms in a neutral compound?
    a. 1
    b. −1
    c. 2
    d. 0

35. What color does red litmus paper turn if the solution is a base?
    a. Blue
    b. Green
    c. Stays red
    d. Purple

36. What number on the pH scale indicates a neutral solution?
    a. 13
    b. 8
    c. 7
    d. 2

37. Which of the following compounds is an example of a strong base?
    a. HCl
    b. HNO_3
    c. HBr
    d. NaOH

38. What is the primary difference between the Earth's inner core and outer core?
    a. The inner core is made of nickel, while the outer core is made of iron
    b. The inner core is solid, while the outer core is molten
    c. The inner core is inaccessible to people, while the outer core is accessible to people
    d. The inner core is hot, while the outer core is cool

39. The separation of a single radioactive element to create a powerful force is known as what type of process?
    a. Nuclear fission
    b. Nuclear meltdown
    c. Thermodynamics
    d. Proton pump inhibitor

40. Combining two or more atoms of a single radioactive element to create a powerful force is known as what type of process?
    a. Nuclear fission
    b. Nuclear meltdown
    c. Nuclear fusion
    d. Nuclear dynamic

Questions 41–45 are based on the following information:

Capacitors can be used for storing energy. Capacitors quickly discharge energy, unlike batteries. Among their applications, capacitors can be used in devices that need a sudden amount of energy at once, such as flash bulbs and defibrillators. When charged particles pass through a capacitor, it can change the direction of the charged particle. This technology is used in printers and televisions.

Most capacitors are made up of two parallel plates placed a slight distance apart. A complete circuit is made by connecting the positive end of a battery to one of the plates, and a negative end of the battery to the other plate.

Several capacitors are attached in a circuit with a 12 V battery, in an attempt to determine factors that could affect the capacitance, charge, and energy of each. The capacitors are different sizes, the plates are different distances (measured in mm) apart, and the plate sizes are different (measured in mm^2). Results are seen below:

Capacitor	Distance between plates (mm)	Area of plates (cm²)	Capacitance (pF)	Charge (nC)
1	0.5	10	178	2.11
2	0.5	20	710	8.50
3	1.0	10	89.5	1.10
4	1.0	20	355	4.22
5	1.5	10	58	0.72
6	1.5	20	234	2.84

Capacitor 3 from the above chart is used to test the effects of dielectrics on capacitors. A dielectric is placed in between the two plates of a capacitor and multiplies the overall capacitance. The stronger the dielectric, the larger the increase in capacitance. Different materials were used as a dielectric and the following measurements for capacitance and charge were recorded in the chart below:

Dielectric material	Capacitance (pF)	Charge (nC)
Air (no dielectric)	89.5	1.10
Glass	687.0	8.21
Paper	266.5	3.20
Polystyrene	229.0	2.70
Quartz	335.4	4.04
Teflon	188.8	2.25

41. From the information in Chart 1, what type of capacitor would have the highest capacitance?
    a. Large distance between plates and large area
    b. Large distance between plates and small area
    c. Small distance between plates and large area
    d. Small distance between plates and small area

42. If another capacitor were added to Chart 1, with twice the distance between the plates and twice the area of Capacitor 6, what would be the charge on this new capacitor?
    a. 0.37 nC
    b. 2.66 nC
    c. 5.68 nC
    d. 11.31 nC

43. Which of the following correctly ranks the dielectrics in order of decreasing strength?
    a. Polystyrene, paper, quartz, glass
    b. Polystyrene, quartz, paper, glass
    c. Glass, paper, quartz, Teflon
    d. Glass, quartz, paper, Teflon

44. If another capacitor with plates of area 10 cm² and a distance between the plates of 1.25 mm were tested, which would be closest to the most likely resulting capacitance?
    a. 31 pF
    b. 58 pF
    c. 73.8 pF
    d. 89.5 pF

45. Which combination would result in the highest charge?
    a. Capacitor 1 with a paper dielectric
    b. Capacitor 2 with a glass dielectric
    c. Capacitor 3 with a Teflon dielectric
    d. Capacitor 4 with a quartz dielectric

46. In what type of chemical reaction is heat absorbed from the surrounding environment?
    a. Exothermic
    b. Endothermic
    c. Rusting
    d. Freezing

47. Which of the following is an example of friction?
    a. A rubber band that is pulled taut then snapped
    b. Jumping on the end of a diving board
    c. Throwing a baseball at an angle
    d. Applying a car's brakes at a red light

48. Isaac is reducing strawberry jam to make a dessert sauce. He takes the jam out of the refrigerator and spoons some into a saucepan. The jam is a cold, thick glob and takes a few minutes to slowly fall off the spoon and into the pan. Isaac turns the stove top on medium heat and the jam begins to turn runny and bubble. What factor of the jam did the heat affect?
    a. Its flavor
    b. Its mass
    c. Its viscosity
    d. Its caloric energy

49. Objects in which molecules are closest together can be described as which of the following in comparison to objects in which molecules are farther apart?
    a. High vibrational
    b. Dense
    c. Forceful
    d. Pressurized

50. The process in which a gas becomes a liquid is known as which of the following?
    a. Melting
    b. Sublimation
    c. Vaporization
    d. Condensation

51. Which of the following is an example of a homogenous mixture?
    a. Distilled water
    b. Sand
    c. Shallow tidal pools
    d. Deep ocean water

52. In chemical reactions, what is the state in which reactants and products no longer change formally known as?
    a. End state
    b. Equilibrium
    c. Finality
    d. Termination

53. Tropical clownfish are often found living among sea anemone. The clownfish eat algae and dead material off the sea anemone, which keeps the anemones clean and the clownfish fed. The sea anemone also provides a protective environment for the clownfish. What is this an example of?
    a. Aquatic friendship
    b. Evolution
    c. Symbiotic relationship
    d. Conservation

54. What is one way to mitigate transmission of communicable disease?
    a. Vaccination
    b. Traveling by car versus traveling by plane
    c. Avoiding people who have traveled abroad in the last 12 months
    d. Following a diet high in biotin

55. Which is the defining feature of organisms found in the Archaea and Bacteria domains?
    a. They are eukaryotes
    b. They are prokaryotes
    c. They contain the most recently evolved organisms
    d. They are all multicellular

56. Scientists believe there may be up to how many distinct species on Earth?
    a. 100 million
    b. 500 million
    c. 1 billion
    d. 5 billion

57. Which layer of the Earth is approximately 760 miles wide and comprised primarily of nickel and iron?
    a. The crust
    b. The atmosphere
    c. The mantle
    d. The inner core

58. Earthquakes and the formation of mountain ranges are two natural events influenced directly by which of the following geological processes?
    a. Underground sulfur springs
    b. Lunar cycles
    c. Plate tectonics
    d. Atmospheric pressure shifts

59. The interaction between the Earth's atmosphere and the Earth's hydrosphere creates which of the following cycles?
   a. The carbon cycle
   b. The water cycle
   c. The Krebs cycle
   d. The life cycle

60. Which of the following terms refers to all living organisms on Earth, regardless of species?
   a. Biosphere
   b. Geosphere
   c. Biodome
   d. Binomial nomenclature

Answer Explanations

1. B: The equation for calculating acceleration is:
$$a = \frac{\Delta v}{\Delta t} = \frac{v_f - v_i}{\Delta t}$$
In this case, V_f = 40 m/s, V_i = 20 m/s, and Δt = 20 s. Plugging in the numbers gives an acceleration rate of 1 m/s^2, Choice *B*.

2. C: If an object is moving in a negative direction, it has a negative velocity. If the velocity is increasing in that negative direction, it is becoming increasingly more negative and would therefore have a negative acceleration. In Choice *A*, the object would have a positive acceleration. Acceleration is zero in both Choices *B* and *D* because the velocity of the object is not changing.

3. D: After air enters the nose or mouth, it gets passed on to the sinuses. The sinuses regulate temperature and humidity before passing the air on to the rest of the body. Volume of air can change with varying temperatures and humidity levels, so it is important for the air to be a constant temperature and humidity before being processed by the lungs. The larynx is the voice box of the body, making Choice *A* incorrect. The lungs are responsible for oxygen and carbon dioxide exchange between the air that is breathed in and the blood that is circulating the body, making Choice *B* incorrect. The trachea takes the temperature- and humidity-regulated air from the sinuses to the lungs, making Choice *C* incorrect.

4. D: The bicuspid valve, also called the mitral valve, is on the left side of the heart between the left atrium and left ventricle. It closes after the left ventricle has filled with blood to prevent blood from flowing backward into the atrium. Choice *B* is incorrect because this valve is on the left, not right, side of the heart. Choices *A* and *C* are incorrect because they state the opposite in terms of oxygenation of the blood for the circulation route mentioned. Oxygen-depleted blood from the body enters the right atrium and is pumped into the right ventricle. This blood, which has a high carbon dioxide content then enters pulmonary circulation where it can become re-oxygenated. Oxygen-rich blood from the lungs then flows into the left atrium, passes into the left ventricle, and enters the aorta for systemic circulation.

5. C: The main function of the lungs is to provide oxygen to the blood circulating in the body and to remove carbon dioxide from blood that has already circulated the body. The passageways within the lungs are responsible for this gas exchange between the air and the blood. The nose, Choice *A*, is the first place where air is breathed in. The larynx, Choice *B*, is the voice box of the body. The stomach, Choice *D*, is responsible for getting nutrients out of food and drink that are ingested, not from air.

6. B: The myocardium is the middle layer of the heart wall and contains connective tissue, blood vessels, and nerves. The epicardium, Choice *A*, is the outer layer of the heart wall and is solely made up of a serous membrane without any nerves or blood vessels. The endocardium, Choice *C*, is the inner layer of the heart wall and is made up of a simple squamous epithelium. The sternum, Choice *D*, is the chest bone and protects the heart.

7. A: The mitral valve is also known as the bicuspid valve because it has two leaflets. It is located on the left side of the heart between the atrium and the ventricle. The pulmonary valve, Choice *B*, is located between the right ventricle and the pulmonary artery and has three cusps. The aortic valve, Choice *C*, is located between the left ventricle and the aorta and also has three cusps. The tricuspid valve, Choice *D*, has three leaflets and is located on the right side of the heart between the atrium and the ventricle.

8. B: White blood cells are part of the immune system and help fight off diseases. Red blood cells contain hemoglobin, which carries oxygen through the blood. Plasma is the liquid matrix of the blood. Platelets help with the clotting of the blood.

9. D: The gastrointestinal system consists of the stomach and intestines, which help process food and liquid so that the body can absorb nutrients for fuel. The respiratory system is involved with the exchange of oxygen and carbon dioxide between the air and blood. The immune system helps the body fight against pathogens and diseases. The genitourinary system helps the body to excrete waste.

10. A: During digestion, complex food molecules are broken down into smaller molecules so that nutrients can be isolated and absorbed by the body. Absorption, Choice *B*, is when vitamins, electrolytes, organic molecules, and water are absorbed by the digestive epithelium. Compaction, Choice *C*, occurs when waste products are dehydrated and compacted. Ingestion, Choice *D*, is when food and liquids first enter the body through the mouth.

11. D: For the years listed on Chart 2, Choice *D* shows an anomaly of approximately 2.22 degrees Fahrenheit, Choice *A* shows an anomaly of approximately -1.22 degrees Fahrenheit, Choice *B* shows an anomaly of approximately 0.7 degrees Fahrenheit, and Choice *C* shows an anomaly of approximately 1.46 degrees Fahrenheit. Overall, Choice *D* is the highest.

12. B: Looking at Table 1, the number listed for 2012 is Choice *A*, 67,774. The number listed for 2008 is Choice *D*, 78,792. Neither of these represents the average. In order to calculate the average, the following formula should be used:

$$\frac{\Sigma \text{ fires}}{Number\ of\ values} = \frac{457,347}{6} = 76,224.5$$

Choice *B* lists the correct value for the average number of fires and Choice *C* is the average for 2007 to 2017, which is erroneous because the question only requests the average of number of fires from 2007 to 2012.

13. C: Because the highest number of acres that are listed are not paired with the largest numbers of fires, another factor would have to contribute to the size of the fires. Many of the larger numbers in acres are after a year with a higher temperature anomaly. This potentially indicates possible drought conditions, which would make it more difficult to contain or extinguish even a small number of fires.

14. A: Chart 2 shows an increase in temperatures, thus causing an increase in anomalies. There have not been any decreases in temperatures significant enough to cause an anomaly since the 1970s; thus, Choice *B* is not correct. The chart displays multiple anomalies, so Choices *C* and *D* are also not correct.

15. D: There is no way to confirm the assertions of Choices *A, B,* or *C,* but all would contribute to the reduction of accidental fires and fires in general, while Choice *D* is not displayed through the information represented in Chart 2. Therefore, the only verifiable thing not contributing to the overall decrease in the number of fires.

16. B: The ovaries produce only one mature gamete each month. If it is fertilized, the result is a zygote that develops into an embryo. The male reproductive system produces one-half billion sperm cells each day.

17. D: Once sperm enter the vagina, they travel through the uterus to the Fallopian tubes to fertilize a mature oocyte. The ovaries are responsible for producing the mature oocyte. The mammary glands produce nutrient-filled milk to nourish babies after birth.

18. C: The skin is the largest organ of the body, covering every external surface of the body and protecting the body's deeper tissues. The brain performs many complex functions, and the large intestine is part of the gastrointestinal system, but neither of these is largest organ in the body. The liver is the second largest organ of the body, weighing roughly three pounds in the average adult.

19. A: The skin has three major functions: protection, regulation, and sensation. It has a large supply of blood vessels that can dilate to allow heat loss when the body is too hot and constrict in order to retain heat when the body is cold. The organs of respiratory system are responsible for breathing and work together with the circulatory system for gas exchange, and the mouth is responsible for ingestion.

20. B: The nails on a person's hands and feet provide a hard layer of protection over the soft skin underneath. The hair, Choice *A*, helps to protect against heat loss, provides sensation, and filters air that enters the nose. The sweat glands, Choice *C*, help to regulate temperature. The sebaceous glands, Choice *D*, secrete sebum, which protects the skin from water loss and bacterial and fungal infections.

21. B: Choice *B* correctly calculates the relative percentage that the mass of Venus is relative to that of Earth mass by utilizing the following:

$$mass\ of\ Venus = x\% \times Earth's\ mass$$

$$4.87 = x\% \times 5.97$$

$$x = 81.6\%$$

Choice *A* does not move the decimal correctly to convert to a percent; Choice *C* reverses the values for the mass of Venus and the mass of Earth in the calculations; and Choice *D* reverses the values for the mass of Venus and the mass of Earth in the calculations, and it also does not move the decimal correctly to convert to a percent.

22. A: Not all planets in the solar system have moons, so Choice *A* would not be a cause for variation in the gravity of a planet. Choices *B, C,* and *D* all have to do with characteristics that could potentially influence a gravitational pull on a planet.

23. B: Choice *B* is the only option that correctly calculates the average orbit, in days, using the following:

$$\frac{\Sigma\ orbits}{number\ of\ values} = \frac{106,831.9}{8} = 13,354$$

Choice *A* is the orbit of Earth, Choice *C* is an incorrect calculation of the average, and Choice *D* is the orbit of Neptune.

24. A: Choice *A* is the only correct combination. To calculate the distance of Neptune from the Sun relative to that of Mercury to the Sun, the following should be used:

$$\frac{4495.1}{57.9} = 78 \; times \; as \; far$$

In order to calculate the orbit time of Neptune relative to that of Earth, the following should be used:

$$\frac{59800}{365.2} = 164 \; times \; as \; long$$

25. C: Choice *C* is the closest at 41.4 million km apart, while Choice *B* is next at 50.3 million km apart. Next is Choice *A* at 57.9 million km apart, and the furthest is Choice *D* at 78.3 million km apart.

26. B: According to Table 1, Sample 3 is made up of 45 percent silt and 35 percent clay. Sand comprised only 20 percent of the sample. Therefore Choices *A, C,* and *D* are incorrect.

27. C: According to Table 2, sand particles can have a size of 1.7 mm. As sample 4 is the only sample listed from Table 1 that has mostly sand as its percent composition, Choice *C* is the best option. Choice *D,* Sample 5, also has a higher percent of sand, but still less than Sample 4.

28. B: The elements are arranged such that the elements in columns have similar chemical properties, such as appearance and reactivity. Each element has a unique atomic number, not the same one, so Choice *A* is incorrect Elements are arranged in rows, no columns, with similar electron valence configurations, so Choice *C* is incorrect. The names of the elements are not arranged alphabetically in columns, Choice *D*.

29. D: Refraction involves the bending of light through different mediums and does not affect the charge of an object. Electrons can be transferred from one object to another, giving an object a negative charge, so Choice *A* is incorrect. Induction, Choice *B,* involves the redistribution of charge by bringing a charged object close to two stationary objects. If the charged object is negatively charged, the electrons of the stationary objects will be repelled away and one of the stationary objects will take on a positive charge while the other one will become negatively charged. Polarization, Choice *C,* occurs when electrons in an object are reconfigured temporarily.

30. C: Magnets follow the same rules of charge as other items. Positive and negative poles attract each other. Two poles with the same charge, positive or negative, would repel each other; therefore, Choices *A* and *B* are incorrect. Although circular magnets do not have ends, they still have poles and follow the same rules of charge. So, circular positive magnets would repel each other, making Choice *D* incorrect.

31. A: Physical properties of substances are those that can be observed without changing that substance's chemical composition, such as odor, color, density, and hardness. Reactivity, flammability, and toxicity are all chemical properties of substances. They describe the way in which a substance may change into a different substance. They cannot be observed from the surface.

32. B: Redox reactions are chemical reactions that involve one species of the reactants being oxidized and another species of the reactants being reduced. The transferring of hydrogen between reactants is one type of redox reaction. Another type involves the transferring of oxygen between reactants, and the third type involves the transferring of electrons between the reactants. Carbon, nitrogen, and sulfur, Choices *A, C,* and *D,* do not drive redox reactions.

33. C: A reactant is an oxidative agent when it gains electrons, gains an oxygen, or loses a hydrogen. Choices *A* and *B* describe the opposite situations. Carbon is not involved in the transfer of electrons, or when hydrogen acts as a proton, in redox reactions. Therefore, Choice *D* is incorrect.

34. D: The oxidation number of a compound is the charge that the compound would have if it was composed of ions. Neutral compounds do not have a charge, so their oxidation number is 0. The oxidation number of hydrogen is 1, but it is –1 when it is combined with elements that are not as electronegative. Group 2 elements in a compound have an oxidation number of 2. Thus, Choices *A*, *B*, and *C* are incorrect.

35. A: Red litmus paper will turn blue if the solution being tested is a base. If the solution is neutral or acidic, the paper will stay red, Choice *C*. Red litmus paper does not turn green or purple, so Choices *B* and *D* do not apply in this situation.

36. C: The pH scale goes from 0 to 14. A neutral solution falls right in the middle of the scale at 7. Choice *A*, a value of 13, indicates a strong base. Choice *B*, a value of 8, indicates a solution that is weakly basic. Choice *D*, a value of 2, indicates a strong acid.

37. D: A strong base dissociates completely and forms the anion OH^-. All bases include a hydroxide group (OH) in their formula, but not all basic compounds will dissociate completely; only strong bases dissociate completely into ions. Choices *A*, *B*, and *C* all represent strong acids and have a hydrogen atom, which is always present in acidic compounds.

38. B: The inner and outer core are both made up of the same elements (iron and nickel), are extremely hot, and both are inaccessible to humans. However, the inner core is solid while the outer core is molten.

39. A: Nuclear fission occurs when the nucleus of a radioactive atomic element is split by force. This causes an explosive release of energy. A nuclear meltdown refers to any accident that occurs at a core reactor in a power plant; these are catastrophic to the surrounding geographic areas, as extreme amounts of radiation are released. Thermodynamics is the physical science of energetic relationships. Proton pump inhibitors are a class of drugs that limit intracellular pumping activity, typically to manage digestive issues.

40. C: Nuclear fusion occurs when lower-weight radioactive elements fuse and release energy rather than splitting, which occurs in the nuclear fission process. Nuclear meltdowns refer to catastrophic events at nuclear power plants, while nuclear dynamic is not a true physical science term.

41. C: Looking at Table 1, a comparison of Capacitor 1 to Capacitor 3 shows that the area between the plates is the same, but the distance between the plates varies. The distance between the plates of Capacitor 1 is smaller than the distance between the plates of Capacitor 3, and Capacitor 1 has a greater capacitance. A comparison of Capacitor 2 to Capacitor 1 shows that the distance between the plates is the same, but the area of the plates varies. The area of Capacitor 2's plates is larger with a larger capacitance. This shows that a smaller distance and a larger area will result in a larger capacitance.

42. C: Looking at Table 1, a comparison of the capacitance of Capacitors 1 and 4 shows the relationship of doubling the area and the distance, which results in a doubling of the charge. Therefore, if another capacitor were added with twice the area and twice the distance of Capacitor 6, it would double the charge, which would be 5.68 nC.

43. D: The information states that a stronger dielectric would result in a larger capacitance. Looking at Chart 2 from the question, the dielectrics in order from largest capacitance to smallest capacitance are glass, quartz, paper, and Teflon.

44. C: Using Table 1, Capacitors 3 and 5 both have the same area for their plates. Capacitor 3 has a distance of 1.0 mm and a capacitance of 89.5 pF, while Capacitor 5 has a distance of 1.5 mm and a capacitance of 58 pF. Therefore, a capacitor with a distance between its plates halfway between Capacitor 3 and Capacitor 5 would have a capacitance approximately halfway between their capacitances, which would be 73.8 pF.

45. B: Combining the charges from the listed capacitors in Table 1 with the charge from the respective dielectrics in Table 2 results in the largest combination being that of Capacitor 2 (at 8.50 nC) and glass (at 8.21 nC).

46. B: Endothermic reactions consume heat from the surrounding environment to complete the chemical reaction and create a product. For example, holding an ice cube in one's hand and allowing it to melt into water (as a result of body heat) is an example of an endothermic reaction. Exothermic reactions release heat into the environment (i.e., igniting wood to create a fire). Rusting occurs when iron reacts to oxygen over a period of time. Freezing is a type of exothermic reaction.

47. D: Friction refers to any force that works opposite to an initial force, typically to slow down or stop the initial force. In this example, the brakes are working in the opposite direction from the way the wheels are rolling. Pulling a rubber band, jumping, and throwing a ball are adding more force to an event, rather than working in the opposite manner.

48. C: Viscosity refers to a liquid material's friction or resistance in relation to another surface. Heating a viscous substance makes it less viscous. In this case, the jam was viscous when it was cold and in a more solid state; it resisted moving off the spoon and onto the pan. As Isaac heated the jam, it became less viscous and was able to easily move around the pan. Low temperature is unlikely to quickly change the taste, caloric content, or flavor of a food item.

49. B: An object's density refers to how close its molecules physically are to one another in space. Typically, as objects cool and solidify (such as when water freezes to ice), density increases and volume decreases. As objects heat up and become more liquid or gaseous, volume increases and density decreases.

50. D: Condensation refers to the process of a gas becoming a liquid. For example, surface water on blades of grass evaporates when ground and surface temperatures are warmer in the daytime; as night time temperatures cool, the water vapor condenses into droplets. Melting is the process of turning a solid into a liquid. Vaporization is the process of turning a liquid into a gas. Sublimation is the process of turning a solid into a gas.

51. A: A homogenous mixture refers to any substance composed of two or more elements. These elements can be metal or non-metal in nature. Homogenous mixtures cannot be perceived to have various components within them with the five senses. Distilled water is the purest type of water, consisting of just hydrogen and oxygen atoms. Other types of water, such as tidal pool water and deep ocean water, will contain compounds such as salt or sea debris. Sand is also made up of various materials, such as rocks and shell. Mixtures that can be perceived to have multiple components are known as heterogeneous mixtures.

52. B: When chemical reactions reach a state of chemical equilibrium, the reactants (substances that cause the reaction) and the products (the end result of the reaction) remain in unchanging states and concentrations.

53. C: A symbiotic relationship is between two entities in which the presence of one party benefits the other, and vice versa. Aquatic friendship is not a scientific term. Evolution refers to the way a specific organism changes over a long period of time. Conservation refers to the preservation of natural areas.

54. A: Vaccinations introduce and encourage immunity to specific contagious diseases. When large populations are vaccinated and develop immunity, this mitigates the presence of specific communicable diseases within that population. It protects the few who may not have been vaccinated through the practice of herd immunity (the larger vaccinated group is able to prevent introduction and spread of the specific disease for which the group has been vaccinated). Car travel and avoiding persons who travel do not necessarily limit interactions that would prevent the spread of communicable disease. Biotin is a vitamin that promotes hair, nail, and skin strength; it is not involved in immunity or disease prevention.

55. B: All species in the Archaea and Bacteria domains are single-celled prokaryotes. The oldest and most primitive organisms are found in these domains, while multicellular, complex, and more recently evolved species are found in the third domain, Eukaryotes.

56. A: Scientists estimate there may be up to 100 million distinct species of life, although only about 1.5 million have been categorized so far. As new species continue to be discovered, the number of Phyla in the taxonomic classification system have greatly expanded to accommodate organisms that do not fit currently stablished categorical groups.

57. D: The inner core of the Earth is a solid sphere made of nickel and iron. Scientists believe that while the Earth initially formed as a molten, gaseous ball, it cooled down and its four main layers were created. While the outer three layers are made of a variety of elements and consist of different forms of matter, the inner core is very hot, yet solid.

58. C: Plate tectonics refers to a scientific theory that describes the Earth's crust as consisting of numerous plates that shift against one another. The tremendous friction and pressure that consequently take place result in earthquakes (as plates violently slide against or away from one another) or contribute to the formation of mountain ranges (as plates push into one another and create elevation where they meet). Underground bodies of water, lunar cycles, and atmospheric changes do not directly influence earthquakes or mountain formation.

59. B: The atmosphere and hydrosphere interact to create the water cycle, a unique cycle that allows Earth to have life. The water cycle is initiated from solar heat, which causes water sources on the ground to evaporate into the atmosphere. Cooler atmospheric temperature triggers precipitation, which returns water vapor to the ground. The carbon cycle primarily takes place between the atmosphere and the geosphere, while the Krebs cycle takes place within living cells as a means of energy production. The life cycle refers to an organism's life processes from birth to death.

60. A: Biosphere is the all-encompassing term for living organisms on a planet. Geosphere refers to the physical components of a planet, while biodome is not a real term. Binomial nomenclature is a method of naming an organism by using its categorized genus and species.

Social Studies

History

Analyzing Historical Sources and Recognizing Perspectives

Every scholar, educator, and student of history must be able to analyze historical sources. These sources are usually categorized into two main groups: primary sources and secondary sources. **Primary sources** are the texts or artefacts that are representative of a particular event, moment, or era in the past because they were written during (or directly after) that event, moment, or era. Primary sources really capture the past. They help convey the effects of every event, the might of every moment, and the essence of every era. Primary sources are what scholars, educators, and students of history use to create secondary sources. **Secondary sources** recount or analyze events that happened in the distant past rather than the proximal present. While an example of a primary source would be a personal diary, journal, or letter, an example of a secondary source would be a textbook, scholarly article, or monograph. Secondary sources weave together excerpts from primary sources, synthesizing these facts with historical theories. There are also tertiary sources that serve as compendia of information derived from secondary sources.

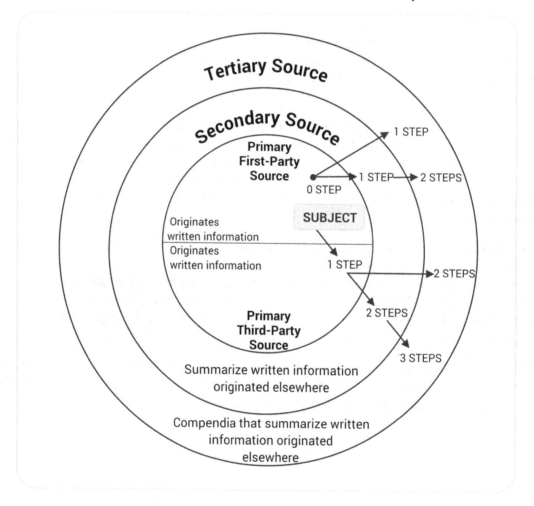

Regardless of whether the source is primary, secondary, or tertiary, it must be analyzed with a keen eye for unique perspectives. Some historians refer to this process as "deconstruction"—it is an analysis that breaks down the ideas constructed by a text or artefact. Deconstruction is essential for recognizing the biases embedded within every source. Every unique perspective in a text has its own set of biases, which are tied to the personal beliefs, preferences, blind spots, and values of the author. History centers on understanding every sources perspective and biases in order to make sense of what is fact and what is opinion. At times, the line separating fact and opinion becomes blurred, so scholars, educators, and students of history must try to assess the historical validity of a source. This means these stakeholders of historical analysis must infuse the old facts with new theories, breathing life into history by making unique inferences.

Throughout history, primary sources have naturally captured different personal perspectives. As soon as the ink hits the paper, reality becomes filtered through the biases of others. All history is an inevitable process of conscious selection. Authors of primary sources choose to tell about some details; they also choose to ignore others. All primary sources fail to capture the entire essence of history. They are entirely driven by point of view, or perspective. Sometimes these perspectives are shaped radically by ideals. History, in this sense, becomes a political platform. Below are two examples of primary sources that view desegregation from two radically different perspectives.

Governor George Wallace's Inaugural Address as Governor of Alabama: "Segregation Now, Segregation Forever" Speech (1963)	Dr. Martin Luther King, Jr.'s "Address at the Freedom Rally in Cobo Hall" (1963)
"Today I have stood, where once Jefferson Davis stood, and took an oath to my people. It is very appropriate then that from this Cradle of the Confederacy, this very Heart of the Great Anglo-Saxon Southland, that today we sound the drum for freedom as have our generations of forebears before us done, time and time again through history. Let us rise to the call of freedom-loving blood that is in us and send our answer to the tyranny that clanks its chains upon the South. In the name of the greatest people that have ever trod this earth, I draw the line in the dust and toss the gauntlet before the feet of tyranny . . . and I say . . . segregation today . . . segregation tomorrow . . . segregation forever."	*"For Birmingham tells us something in glaring terms. It says first that the Negro is no longer willing to accept racial segregation in any of its dimensions. For we have come to see that segregation is not only sociologically untenable, it is not only politically unsound, it is morally wrong and sinful. Segregation is a cancer in the body politic, which must be removed before our democratic health can be realized. Segregation is wrong because it is nothing but a new form of slavery covered up with certain niceties of complexity. Segregation is wrong because it is a system of adultery perpetuated by an illicit intercourse between injustice and immorality. And in Birmingham, Alabama, and all over the South and all over the nation, we are simply saying that we will no longer sell our birthright of freedom for a mess of segregated pottage. In a real sense, we are through with segregation now, henceforth, and forevermore."*

Notice how both men—Governor George Wallace and Dr. Martin Luther King, Jr.—are talking about the same topic (segregation) in the same year (1963), but they are viewing their historical context through two entirely different lenses, or perspectives. Governor Wallace is pro-segregation. Dr. King is antisegregation. Two men, from two different cultures, with two different sets of biases, are analyzing the same historical context in two radically different ways.

Secondary sources function in the same manner. Scholars often take different stances about particular historical events or contacts. Below is a chart showing the ways in which different historians have analyzed the Great Depression.

Historian	Key Quotes	Summary of Perspective
FRIEDMAN AND SCHWARTZ	"The Depression, they claimed, was a result of a drastic contraction of the currency." "The Federal Reserve Board in Washington came to dominate monetary policy, with disastrous results."	The Depression was caused by a failure to control the supply of money. Federal Reserve was to blame for the Great Depression because of its monetary policy.
TEMIN	"The New Deal never ended the Depression because it did not spend enough."	The government did not intervene and spend enough money.
POWELL	"Even if [some events had happened differently], there would have been a Depression because of the severe monetary contraction."	The Depression was caused by monetary contraction.
BADGER	"Lack of demand was not offset in the early years of the Depression by any compensatory spending."	The US government did not spend enough money to stop the Depression.
KINDELBERGER	"The world economic system was unstable unless some country stabilized it."	The USA failed to take an economic lead in the world.
KENNEDY	"American banks were rotten."	American banks were unregulated and were to blame to for the Depression.

As this chart demonstrates, historians are inclined to take different stances on how they interpret history. Not every historian agrees about the root causes of the Great Depression. Each has his or her own perspective.

Different perspectives exist in history because history is a very human process. History is as diverse as human identities. So how can students of history figure out which perspective is right? This is a difficult question to answer. The "right" answer is the one that is the most objective. Objectivity is a fancy word for truthfulness—it is the quest to capture reality and convey that reality to others. Fenwick English (1992) once stated: "[One who strives for objectivity] is more interested in truthfulness than truth. The former is established by a continued pursuit never culminated." Like an asymptote on a mathematical graph, objectivity (truthfulness) is something you can get infinitesimally closer to, but it is something that may never be fully realized. Nevertheless, this reality should not stop students from searching for truthfulness. The "right" perspective is usually the one that is repeated most often by primary and secondary sources. Thus, students should continuously cross-reference relevant sources to get closer to historical objectivity.

Interconnections Among the Past, Present, and Future

Historical analysis is actually a reflection of both the past and present—it combines old primary sources with new secondary perspectives. Additionally, many scholars, educators, and students of history introduce future implications into their research—they try to predict how historical themes, past and present, may play out in the future. The art and craft of historical analysis thus rests at the praxis of the past, present, and future. This is often overlooked by students. They believe history is something that is dead. But history is very much alive. It shapes the past, informs the present, and paves a path to the future.

In this sense, history has its own agency. It cannot fully be sequestered to the past because, as a sociological phenomenon, it impacts the present and future. In any given moment in time, we can be studying history (past), living it (present), and making it (future). This is a very difficult concept to understand, but the course of history has its own momentum. It is an inescapable force that moves both with, against, and without humanity. Below are a few anecdotes that illustrate different interconnections that can exist between the past, present, and future.

Example of how the PAST influences the PRESENT:

> Every year, Americans join together to memorialize D-Day on June 6. D-Day is a past moment. It occurred on June 6, 1944. It is the moment Allied forces—led by the Americans—landed on the beaches of Normandy and began marching eastward to defeat the Germans. It is a past moment that continues to have present-day effects.

This is an example of the past influencing the present because it is a historical moment that has contemporary memorials.

Example of how the PRESENT influences the PAST:

> Noticing the heartache brought on by the Great Recession (2007-2009), a young scholar writes an article in 2009 that compares the Great Recession to the Great Depression. The article changes the way historians understand the Great Depression, adding to the canon of historical reflections on the topic.

This is an example of how present moments can illuminate the past.

Example of how the PAST and PRESENT influence the FUTURE:

> Protesters decide to topple a Confederate statue on a college campus, breaking it into pieces; the damage is irreparable, and the crowd decides to erect a new statue of an African-American leader.

This is a present event that destroys the past (a statue memorializing the Confederacy), which inevitably changes the future (both physically and philosophically).

Example how the FUTURE might connect to the PRESENT:

> Barring any imminent demise of humanity or its cognitive functions, our current moment in history will likely be reviewed by future historians. Since we have moved to a digital age, future historians may have to be skilled computer forensics experts who can unearth the past digital interactions of humanity.

This is a future scenario that will likely be shaped by present interactions.

Additionally, in history classrooms, **causation** and **correlation** are key to understanding the relationships (or lack of relationships) between the past, present, and future. Causation is more easily understood than correlation. For instance, events cause certain changes that have an effect on present actions, which, in turn, affect visions of the future and the future itself. But causation need not be linear or progressive when it comes to history; it is not always a timeline like in the textbooks. Future prospects can cause us to re-envision past occurrences, which, in turn, affect the way we view the present. However, most phenomena in history are not causally related—there is no direct line of cause and effect. Since history is chock full of mysteries and silences, it typically seems like a giant blob of ideas and actions. The events, people, and values that exist within this amorphous blob of details usually have loose connections that are called correlations. We know that these events, people, and values may be correlated in some way, but not directly or causally.

Notable Eras in U.S. and World History

Classical Civilizations
Classical civilizations developed many innovations that contributed to our postmodern world—metal tools and weapons, written languages, calendars, representative government, professional militaries, urban planning, and large-scale farming, among others. These innovations allowed classical civilizations to conquer huge territories and establish historic empires. Examples of classical civilizations include Egypt, Greece, Persia, China, and Rome.

Ancient Egypt
Ancient Egypt's first ruling dynasty was established in the 32nd century B.C., and over the next three thousand years, Egyptian innovations were adopted by many other classical civilizations. In addition, Egypt was conquered and controlled by three of those civilizations—Greece, Persia, and Rome.

Developed out of necessity to limit the Nile River's annual flooding, Egyptian irrigation techniques changed the world. Where floods once washed away crops and settlements, Egyptian irrigation left a rich delta that could grow enough food to support an empire. To govern their empire, the Egyptians independently developed a writing system (hieroglyphs) and created a professional bureaucracy of clerks and writers (scribes). Egypt also signed the earliest formal peace treaty during a war with the Hittites. Egyptian pharaohs built enormous burial structures, such as the Great Pyramid. The demand for these massive public works incentivized the invention of new quarrying, surveying, and construction techniques.

Greece
Greek city-states, such as Sparta and Athens, started to develop in the 8th century B.C. Some of the city-states united to fend off the Persian Empire, and in 4th century B.C., Alexander the Great conquered a then-unprecedented empire that included Greece, Egypt, Persia, Syria, Mesopotamia, central Asia, and northern India. Alexander the Great's empire marks the start of the Hellenistic period, which lasted until the Roman Republic conquered the eastern Mediterranean and turned Greece into a Roman province.

Greek political systems and philosophy have had a major influence on Western civilization. The Greek city-states invented the principle of self-government and *demokratia* (direct democracy). Although Greece is considered the birthplace of democracy, the city-states were slave societies, and only the wealthy elite could participate politically. Still, Greece served as a model for Rome and early modern Europe on how to govern, collect taxes, and militarize.

The three most famous Greek scholars are **Socrates**, **Plato**, and **Aristotle**. All three are considered the fathers of Western philosophy, and their work continues to be studied in classrooms across the world. Plato's *Republic* and *Laws* are foundational texts in the field of political philosophy, and Aristotle's work in the physical sciences and logic were the basis for European scholarship throughout the Middle Ages. In addition to philosophy, Aristarchus was the first person to theorize that the Earth revolved around the Sun; Archimedes discovered pi (π); and Euclid's *Elements* is a collection of mathematical definitions, theorems, and proofs that went unchallenged until the 19th century.

Persia

Cyrus the Great founded the first Persian Empire, the Achaemenid Empire, in the 6th century B.C., effectively turning a group of nomadic shepherds into one of the largest empires in history. His successor, Darius I, further expanded the empire from Eastern Europe to Central and South Asia. Alexander the Great overthrew Darius III and conquered the Persian Empire in 330 B.C.

The Persian Empire was multicultural, consisting of multiple civilizations with varying religions, languages, and ethnicities. Persian rulers took a "carrot or stick" approach to diplomacy. If a civilization willingly joined the Empire and paid taxes, they would receive some degree of self-government and enjoy better economic relations. If the civilization resisted, they were invaded and enslaved.

The Persian Empire was organized under a federal system, much like the American model of government. Persian rulers functioned as the central government, and they appointed satraps (governors) and records-keepers to every region. The satraps allowed for limited local self-government, depending on the region's relationship with the Persian Empire. In exercising control over their vast empire, Persians developed innovative roadway and postal systems that the Romans later expanded upon.

China

Ancient China was ruled under successive dynasties, beginning with the Xia dynasty (21st century B.C.) and ending with the Qing dynasty (1911). The dynasties legitimized their power through the Mandate of Heaven—the belief that a higher power selected the ruler. The Mandate of Heaven contributed to the concept of legitimization in political philosophy.

The Han dynasty (206 B.C.—220 A.D.) was a golden era for Chinese trade, technological innovation, and scientific advancement. Han rulers opened the Silk Road, a series of trade routes connecting China with the rest of the Asian continent and Roman territory. The Silk Road linked the Far East and West for the first time. To facilitate this multicultural trade, Han rulers issued one of the world's first uniform currencies. Han Chinese technological innovations included papermaking, wheelbarrows, steering rudders, mapmaking (grids and raised-relief), and the seismometer (measure earthquakes). In science and math, Han scholars discovered herbal remedies, square roots, cube roots, the Pythagorean theorem, negative numbers, and advanced pi (π) calculations.

Rome

The city of **Rome** was founded in the 8th century B.C. and initially ruled under a monarchical government. Rome transitioned into a republic in the latter half of the 6th century B.C. In the 1st century B.C., a series of civil wars destabilized the Roman Republic, which culminated with Gaius Julius Caesar seizing absolute power. Caesar's victory caused Rome's transition from a republic to an empire. The Roman Empire would become the largest in history. Political corruption, religious conflict, economic challenges, and repeated invasions by German tribes led to the empire splitting into Western and Eastern halves in 395 A.D. The Eastern half, known as the Byzantine Empire, outlasted the Western Empire, surviving until 1453.

Rome is often referred to as the birthplace of Western civilization, and its influence on world history cannot be overstated. James Madison used the Roman Republic's separation of powers as a model for the United States Constitution. Rome spread Greek philosophy and culture across its many territories, including advancements in architecture and urban planning. Roman law influenced the development of many legal practices, like trials by juries, contracts, wills, corporations, and civil rights. All Romantic languages evolved from Latin—French, Italian, Spanish, and Portuguese. Roman numerals are still commonly used. Rome adopted and mastered Persian theories related to bureaucracy, civil engineering, "carrot and stick" diplomacy, and multiculturalism. To build their large infrastructure projects, the Romans invented a superior form of concrete. The largest religion in the world, Christianity, also started in a Roman province (Judea), spread across the empire, and eventually became Rome's state-sanctioned religion.

European Exploration and Colonization

In what's commonly referred to as the "Age of Exploration," Europeans began exploring overseas to regions they'd never before seen. The Age of Exploration started in the early 15th century and lasted until the late 18th century, and in practice, it represents the earliest stage of globalization. Technological innovation greatly facilitated European exploration, like the **caravel**—a smaller and more maneuverable ship first used by the Portuguese to explore West Africa—and the *astrolabe* and *quadrant,* which increased navigators' accuracy.

European monarchies hired explorers primarily to expand trade routes, particularly to increase access to the Indian spice markets. There were other motivating factors, however, such as the desire to spread Christianity and pursue glory.

As exploration increased, European powers adopted the policies of mercantilism and colonialism. **Mercantilism** is an economic policy that prioritizes the wealth, trade, and accumulation of resources for the sole benefit of the nation. **Colonization** was an outgrowth of mercantilism that involved establishing control over foreign people and territories. Unsurprisingly, mercantilism and colonialism were a major source of conflict between native people, colonists, and colonizers.

In 1492, Christopher Columbus arrived in the Caribbean, though he initially, and mistakenly, believed he had landed in India. Over the next two centuries, European powers established several colonies in North America to extract the land's resources and prevent their rivals from doing the same. In 1565, Spain established the first European colony in North America at St. Augustine in present-day Florida. Along with building a South and Central American empire, Spain colonized Mexico and much of the present-day American southwest, southeast, and heartland (Louisiana Territory). Sweden colonized the mid-Atlantic region of the present-day United States in 1638, but the Netherlands conquered these colonies in 1655. France colonized Canada, Hudson's Bay, Acadia (near present-day Maine), and Newfoundland. In 1717, France assumed control over the Louisiana Territory from Spain.

In 1607, England established her first colony at Jamestown. Six years later, Jamestown colonists imported the first slaves to North America. During the rest of the 17th century, British separatist Puritans (Pilgrims) arrived on the *Mayflower* in Cape Cod, Massachusetts. Great Britain conquered New Netherlands (New York) in 1674, effectively driving the Dutch out of North America.

England's North American holdings expanded into the Thirteen Colonies—Connecticut, Delaware, Georgia, Maryland, Massachusetts, New Hampshire, New Jersey, New York, North Carolina, Pennsylvania, Rhode Island, South Carolina, and Virginia. Some colonies were royal colonies, while others were chartered to business corporations or proprietary local governments. The chartered colonies generally allowed for more self-government, but the Crown withdrew the charters and placed all thirteen colonies under direct royal control by the second half of the 18th century.

The American Revolution

The aftermath of the French and Indian War (1756-1763) set the stage for a conflict between the American colonists and Great Britain. After defeating France and her Native American allies, England passed a series of controversial laws. First, the Proclamation of 1763 barred the colonists from settling west of the Appalachian Mountains in an effort to appease Native Americans. Second, the Quebec Act of 1774 granted protections to their recently acquired French-Canadian colonies. Third, England passed a series of taxes on the colonists to pay their war debt, including the Sugar Act (1764), Quartering Act (1765), Stamp Act (1765), Townshend Acts (1767), Tea Act (1773), and the Intolerable Acts (1774).

To the colonists, it appeared as if England was rewarding the combatants of the French and Indian War and punishing the colonists who fought for England. As a result, protests erupted, especially in New England. The most significant were those surrounding the Boston Massacre (1770) and the Boston Tea Party (1773). Colonists organized the First Continental Congress in 1774 to request the repeal of the Intolerable Acts and affirm their loyalty to the Crown. When King George III refused, the crisis escalated in April 1775 at the Battle of Lexington and Concord, the first armed conflict between British troops and colonial militias. A Second Continental Congress then met in Philadelphia, and the delegates issued Thomas Jefferson's Declaration of Independence on July 4, 1776. The Declaration of Independence declared that all people enjoyed basic rights, specifically the right to life, liberty, and the pursuit of happiness, and accused King George III of violating colonists' rights, which justified the American revolution.

The Continental Army lost its first major offensive campaign in Quebec City in December 1775, but victories at Trenton and Saratoga boosted morale. The tides of war turned against England when the Americans and France signed the Treaty of Alliance in 1778. France sent resources and troops to support the Americans, and the Marquis de Lafayette served as a combat commander at the final decisive battle at Yorktown. On September 3, 1783, the Treaty of Paris secured American independence.

Founding of the United States

Enacted during the American Revolution, the **Articles of Confederation** was the original governing document of the United States. However, the Articles of Confederation was ineffective due to its weak central government. The government didn't include a president or judiciary branch, and Congress didn't have the power to tax or raise money for an army. The final straw was the Articles of Confederation's failure to handle the Shays' Rebellion (1786–1787). In May 1787, the Founding Fathers convened the **Constitutional Convention** in Philadelphia. George Washington served as the Convention's president, and James Madison wrote the draft that was the basis for the Constitution.

Slavery challenged the Constitutional Convention from the outset, foreshadowing the American Civil War. The South wanted slaves to count for representation, even though slaves couldn't vote, but not taxes, while the North advocated for the opposite. The **Three-Fifths Compromise** settled the issue by counting slaves as three-fifths of a person for taxation and representation. In addition, the delegates agreed to a compromise that allowed slave owners to capture escaped slaves in exchange for ending the international slave trade by 1808.

The delegates also debated representation in Congress. **The New Jersey Plan** proposed a single legislative body with one vote per state. In contrast, **the Virginia Plan** proposed two legislative bodies with representation decided by the states' populations. The delegates agreed to the Connecticut Compromise—two legislative bodies with one house based on population (House of Representatives) and the other granting each state two votes (Senate). Two other branches, the judiciary and executive, were also included in the final document, and a series of checks and balances divided power between all three branches. The Constitution also expressly addressed issues from the Articles of Confederation by providing for a significantly stronger central government.

The Constitution, including its interpretation by the Supreme Court of the United States, divided the powers as follows:

Exclusive federal government powers
- Coin money
- Declare war
- Establish lower federal courts
- Sign foreign treaties
- Expand territories and admit new states
- Regulate immigration
- Regulate interstate commerce

Exclusive state government powers
- Establish local government
- Hold and regulate elections
- Implement welfare and benefit programs
- Create and oversee public education
- Establish licensing requirements
- Regulate state corporations
- Regulate commerce within the state

Concurrent powers (shared)
- Levy taxes
- Borrow money
- Charter corporations

Nine of the thirteen states needed to ratify the Constitution before it became law, and a heated debate over the Constitution spread throughout the nation. Those in favor of the Constitution, called the Federalists, produced and distributed the *Federalist Papers*, which James Madison, Alexander Hamilton, and John Jay wrote under the pseudonym "Publius." Thomas Jefferson and Patrick Henry led the Anti-Federalists and called for the inclusion of a bill of rights. The Constitution was ratified on June 21, 1778, and three years later, the Bill of Rights was added. The Bill of Rights is the first ten amendments to the Constitution ratified together in December 1791.

Bill of Rights
- **First Amendment**: freedom of speech, freedom of press, free exercise of religion, and the right to assemble peacefully and petition the government

- **Second Amendment**: the right to bear arms

- **Third Amendment**: the right to refuse to quarter (house) soldiers

- **Fourth Amendment**: prohibits unreasonable searches and seizures and requires a warrant based on probable cause

- **Fifth Amendment**: protects due process, requires a grand jury indictment for certain felonies, protects against the government seizing property without compensation, protects against self-incrimination, and prohibits double jeopardy

- **Sixth Amendment**: the right to a fair and speedy criminal trial, the right to view criminal accusations, the right to present witnesses, and the right to counsel in criminal trials

- **Seventh Amendment**: the right to a trial by jury in civil cases

- **Eighth Amendment**: prohibits cruel and unusual punishment

- **Ninth Amendment**: establishes the existence of unnamed rights and grants them to citizens

- **Tenth Amendment**: reserves all non-specified powers to the states or people

The Founding Fathers created a decentralized federal system to protect against tyranny. The Tenth Amendment is what enshrines the principle of **federalism**—the separation of power between federal, state, and local government. In general, the federal government can limit or prohibit the states from enacting certain policies, and state governments exercise exclusive control over local government.

Growth and Expansion of the United States
Shortly after gaining independence, the United States rapidly expanded based on **manifest destiny**—the belief that Americans hold special virtues, America has a duty to spread those virtues westward, and success is a certainty.

In 1803, President Thomas Jefferson purchased the Louisiana territory from France. **The Louisiana Purchase** included 828,000 square miles of land west of the Mississippi River. Unsure of what he had purchased, Jefferson organized several expeditions to explore the new territory, including the famous **Lewis and Clark Expedition**. To consolidate the eastern seaboard under American control, President James Monroe purchased New Spain (present-day Florida) in the *Adams-Onís Treaty* of 1819.

Conflict with Native Americans was constant and brutal. Great Britain allied with Native American tribes in the present-day Midwest, using them as a buffer to protect her Canadian colonies. Britain's support for Native American raids on American colonies ignited tensions and led to the War of 1812. Two years later, Great Britain and the United States signed the Treaty of Ghent, ending the war with a neutral resolution, and British support for the Native Americans evaporated. The Supreme Court's decision in *Worcester v. Georgia* (1832) later established the concept of tribal sovereignty; however, President Andrew Jackson refused to enforce the Court's decision. Consequently, Americans continued to colonize Native American land at will. President Jackson also passed the first of several Indian Removal Acts, forcing native tribes westward on the Trail of Tears.

Tensions on the American-Mexican border worsened after President John Tyler annexed Texas in 1845. That same year, James K. Polk succeeded Tyler after winning the election of 1844 on a platform of manifest destiny. President Polk ordered General Zachary Taylor to march his army into the disputed territory, which ignited the *Mexican-American War*. The United States dominated the conflict and annexed the present-day American southwest and California through the 1848 Treaty of Guadalupe Hidalgo. Polk also settled the Oregon Country dispute with Britain, establishing the British-American boundary at the 49th parallel.

The United States purchased Alaska from Russia in 1867, and Hawaii was annexed in 1898 to complete what would become the fifty states.

The migration of settlers into these new territories was facilitated by technological innovation and legislation. Steamships allowed the settlers to navigate the nation's many winding rivers, and railroads exponentially increased the speed of travel and transportation of supplies to build settlements. Canals were also important for connecting the eastern seaboard to the Midwest. Thus, the fastest growing settlements were typically located near a major body of water or a railroad. Starting in 1862, Congress incentivized Americans to travel westward and populate the territories with a series of Homestead acts, which gave away public lands, called "homesteads," for free. In total, the United States gave away 270 million acres of public land to support the country's expansion.

Civil War

Congress repeatedly attempted to compromise on slavery, especially as to whether it would be expanded into new territories. The first such attempt was the **Missouri Compromise of 1820**, which included three parts. First, Missouri was admitted as a slave state. Second, Maine was admitted as a free state to maintain the balance between free and slave states. Third, slavery was prohibited north of the 36°30' parallel in new territories, but the Supreme Court overturned this in *Dred Scott v. Sandford* (1837).

The **Compromise of 1850** admitted California as a free state, allowed popular sovereignty to determine if Utah and New Mexico would be slave states, banned slavery in Washington D.C., and enforced a harsh **Fugitive Slave Law**. The **Kansas-Nebraska Act** *of 1854* started a mini civil war, as slave owners and abolitionists rushed into the new territories. Slavery eventually caused the collapse of the Whigs, the dominant political party of the early and middle 19th century. The Northern Whigs created the anti-slavery Republican Party.

Republican Abraham Lincoln won the election of 1860. Before his inauguration, seven Southern states seceded from the United States and established the **Confederate States of America** to protect slavery. On April 12, 1861, the first shots of the **Civil War** were fired by Confederate artillery at Fort Sumter, South Carolina. On January 1, 1863, after the particularly bloody Battle of Antietam, President Lincoln issued the **Emancipation Proclamation**, freeing the Confederacy's 3 million slaves; however, slavery continued in Union-controlled states and territories.

Later that year, the Battle of Gettysburg gave the **Union Army** a decisive victory, turning back the Confederacy's advances into the North. President Lincoln's famous **Gettysburg Address** called for the preservation of the Union and equality for all citizens. After a series of Union victories, Confederate General Robert E. Lee surrendered at Appomattox Court House and ended the Civil War in April 1865. Less than one week after Lee's surrender, John Wilkes Booth assassinated President Lincoln during a play at Ford's Theater.

Vice President Andrew Johnson assumed the presidency and battled the Radical Republicans in Congress over the Reconstruction Amendments. Federal troops were deployed across the South to enforce the new amendments, but Reconstruction concluded in 1877. Immediately after the troops withdrew, the Southern stated enacted Jim Crow laws to rollback African Americans' right to vote and enforce segregationist policies.

Industrial Revolution

The First Industrial Revolution exploded in the United States during the War of 1812, as more American industrialists adopted British textile innovations. New manufacturing firms utilized the assembly line to increase production speed and accuracy. Each worker completed one individualized task on the assembly line, and the team of workers collectively produced the goods. New England manufacturers adopted the Waltham-Lowell labor and production system to maximize the use of new technologies such as the spinning jenny, spinning mule, and water frame. Eli Whitney's cotton gin was similarly revolutionary for the cotton industry.

The Second Industrial Revolution began after the Civil War and lasted until World War I. Technology again drove the economic changes, particularly in factory machinery and steel production. This era transformed the United States from an agricultural and textile economy into an industrial powerhouse. The invention of the telegraph and innovations in transportation like railways, canals, and steamboats further connected the economy.

Industrialization drove urbanization, as most new jobs were in booming city centers. To accommodate new arrivals from rural areas, cities had to adopt urban planning strategies to create a sewer system and facilitate the supply of gas and water. Economic opportunity also attracted waves of immigrants to American cities. Before 1880, most immigrants were English and Irish, but from 1880 to 1920, more than 20 million immigrants came from Central, Eastern, and Southern Europe.

Despite the explosion in wealth creation, the Second Industrial Revolution caused crises in political corruption, working conditions, and wealth inequality. Political machines, like Tammany Hall in New York, consolidated political power in major American cities under a party boss who doled out special favors to his constituents. A lack of government regulation also created unsafe labor conditions. Most laborers worked twelve-hour shifts, with minimal breaks, six or seven days every week, and child labor was rampant. The long hours, dangerous machines, and young workers led to workplace accidents, and death rates soared. Wages were also low, and income inequality exploded, which is why the Second Industrial Revolution is often referred to as the "Gilded Age."

Labor began to organize in response to dangerous working conditions and meager pay. Between 1881 and 1905, labor unions organized 37,000 strikes across every major industry. Anarchists, populists, and socialists challenged the political establishment, demanding reform for the working poor. These movements led to the rise of the Progressive Era (the 1890s to the 1920s). Progressives passed the direct election of U.S. Senators (17th Amendment), anti-corruption laws, anti-trust laws, women's suffrage, and the prohibition of alcohol.

Early American Foreign Policy
In the late 19th century, a public debate raged over the United States' international role. Part of the public wished to heed George Washington's famous farewell address in which he warned against intervening in foreign conflicts. Another faction supported military intervention to free colonies, as France had done for the United States. The final group called for following the European model of imperialism.

Issued in 1823, the **Monroe Doctrine** promised that the United States wouldn't meddle in existing European colonies, but it also vowed that the U.S. would consider any future European intervention to be a hostile act. President Theodore Roosevelt would later strengthen the Monroe Doctrine during his State of the Union address in 1904, explicitly threatening unilateral military intervention in Latin America whenever it suited American interests.

In 1898, the American battleship *USS Maine* exploded in Havana Harbor, and President McKinley declared war on Spain. In less than five months, the United States had defeated Spain and assumed control over her first overseas colonies—Puerto Rico, Guam, and the Philippines.

The outbreak of **World War I** in 1914 presented a serious challenge to the United States. President Woodrow Wilson attempted to keep the nation neutral, but his hand was ultimately forced. In 1915, German U-boats (submarines) torpedoed a British passenger ship *RMS Lusitania* carrying American passengers, and two years later, Britain intercepted the Zimmerman Telegram, which showed that Germany was attempting to conspire with Mexico to invade the United States. President Wilson had run out of options, and the United States declared war on Germany and joined the Triple Entente (Britain, France, and Russia). More than 116,000 Americans lost their lives serving in World War I, but the conflict was a turning point in American history, marking the arrival of the United States as a global power.

In the interwar period, known as the Roaring Twenties, there was rampant stock market speculation. On "Black Tuesday," October 29, 1929, the stock market crashed, sending the global economy off a cliff into the Great Depression. Immediately after winning the 1932 presidential election, Franklin D. Roosevelt launched the New Deal to increase employment, stimulate demand, and increase regulation over capital. Much of the current social welfare state—Social Security, unemployment insurance, disability, labor laws, and housing and food subsidies—date back to the New Deal. President Lyndon B. Johnson later introduced a series of legislation modeled after the New Deal collectively referred to as the War on Poverty or the Great Society. His aim was to alleviate poverty by increasing access to education, health care, and housing.

After **World War II** broke out in Europe, the United States again tried to remain neutral, but on December 7, 1941, Japan bombed Pearl Harbor. In response, President Roosevelt declared war on Japan. Germany preemptively declared war on the United States, who did the same in return. On June 6, 1944, commonly referred to as "D-Day," American forces landed on the beaches of Normandy, France, and by May 1945, American and Soviet forces had conquered Germany. However, the Japanese continued to fight in the Pacific theater. To avoid what would've been a costly invasion of Japan's mainland, the United States dropped nuclear bombs on Nagasaki and Hiroshima, ending World War II.

20th Century Developments and Transformations

The major themes of the 20th century are industrialization, imperialism, nationalism, global conflict, independence movements, technological advancement, and globalization.

The United States and Europe completed the transition from agrarian to industrial economies in the early 20th century, and by the end of the 20th century, nearly the entire world had at least started industrializing. **Industrialization** created global wealth, exponentially improving the global standard of living, but it also caused explosive population growth and environmental destruction. Over the 20th century, the global population increased from 1.6 billion people to 6 billion people. This unprecedented population growth has been driving rapid deforestation, as more land is cleared for agriculture and settlements. In addition, industrialization has polluted the Earth's atmosphere, land, and oceans. As a result, temperatures are increasing, sea levels are rising, and biodiversity is declining. The 21st century will face dire consequences, like mass migration and global instability, if these trends aren't reversed.

The 20th century opened with much of Asia and Africa divided into European colonies. **Nationalism** was on the rise and used to justify European **imperialism** and **militarism**. Nationalism, imperialism, and militarism also markedly increased in the United States and Japan. The overwhelming majority of present-day countries didn't achieve independence until after the World Wars or the Cold War.

Early 20th century nation-states were entangled in a complex web of military alliances, and when Serbian nationalists assassinated Archduke Franz Ferdinand of Austria (1914), those alliances triggered a global conflict, World War I. _Russia_ joined the **Triple Entente alliance** with Great Britain and France, but in 1917, Vladimir Lenin's Bolsheviks (communists) led an armed insurrection against the Russian Tsar, Nicholas II. Following the Russian Revolution, the **Bolsheviks** won a victory against the European-backed counter-revolutionaries, paving the way for the creation of the **Union of Soviet Socialist Republics** (_USSR_).

The Treaty of Versailles ended World War I in June 1919, and it included several of American President Woodrow Wilson's **Fourteen Points**. Wilson was an advocate for countries' right to self-determination, and his Fourteen Points helped discredit colonialism and imperialism as legitimate foreign policy goals. Part I of the Treaty established Wilson's League of Nations, but Germany was prohibited from joining until 1926. The United States also refused to join, further undermining the League's legitimacy. The Treaty of Versailles also forced Germany to claim total responsibility for causing World War I and pay extensive reparations. This amounted to a national humiliation for Germany, fueling a reactionary right-wing movement that culminated in Adolf Hitler's rise to power in the 1930s.

The global women's rights movement gained support because of World War I. When the men left to fight overseas, women joined the workforce and contributed to the war effort. British Parliament granted women the right to vote in 1918, and Germany did the same in the following year. President Woodrow Wilson also reversed his previous position and declared his support for the 19th Amendment (1920), establishing women's suffrage in the United States.

The Great Depression occurred between the World Wars, lasting from 1929 until the late-1930s. A stock market crash triggered a depression in the United States that quickly spread around the world. The Great Depression put enormous stress on Western governments to alleviate poverty and lessen the appeal of communist revolution. Some governments, like the United States and Great Britain, increased their public investment to increase employments and social services. Other governments, like Germany and Italy, transitioned from militarized nationalism into outright far-right fascism.

Hitler's **Nazi Party** exploited global instability in his efforts to right the perceived injustices inflicted on Germany during World War I. First, Nazi Germany remilitarized the Rhineland and stopped paying reparations under the Treaty of Versailles. Second, Hitler enforced collective punishments on German Jews as retribution for their alleged disloyalty to the state, previewing what would become the Holocaust. Third, Hitler supported Italy's invasion of Ethiopia (1935), and German forces fought a proxy war to assist the fascist Francisco Franco during the Spanish Civil War (1936-1939). Fourth, Hitler annexed two territories in 1938—Austria and the Sudetenland in Czechoslovakia—under the pretext of protecting ethnic Germans. Much of Europe was horrified by these events, but the leadership mostly followed a policy of appeasement, hoping to avoid a second global conflict.

One year later, in September 1939, the Nazis invaded Poland to clear *lebensraum* (living space) for Germans. Two days later, France and Britain declared war on Germany, igniting World War II. Germany, Italy, and Japan formed the **Axis alliance**. Japan immediately conquered China and the Korean Peninsula, and Germany added the conquest of mainland Europe and North Africa. Germany also signed a nonaggression pact with Joseph Stalin's Soviet Union, but Hitler broke the pact and invaded the Soviet Union in June 1941. This decision was disastrous, as it forced the Nazis into an expensive and bloody war of attrition. Hitler also committed significant resources to the Holocaust—his plan to exterminate Jews, gypsies, homosexuals, Afro-Europeans, and disabled people.

The United States entered World War II in 1941, changing the tides of war. In 1942 alone, the U.S. Navy defeated Japan at the Battle of Midway; the Allies routed Italy and Germany in North Africa; and the Soviet Union decisively defeated Germany at Stalingrad. Nazi Germany agreed to an unconditional surrender on May 8, 1945, and an American nuclear attack forced Japan to surrender in September 1945.

With most of Europe and Asia lying in ruins, the United States and the Soviet Union were the only global superpowers to survive World War II. The Soviet Union's resilience was remarkable, considering the intense suffering Soviets endured, but they did benefit from consolidating power in Eastern Europe under Soviet-controlled satellite governments. The two superpowers engaged in a series of proxy wars, but the threat of nuclear war ultimately deterred a direct conflict. However, there were several close calls, any of which would have likely ended life on Earth.

The Cuban Missile Crisis (1962) began after the failed American **Bay of Pigs** invasion angered Cuba and the Soviet Union. The Soviets also objected to American ballistic missiles in Turkey, and they responded by covertly sending ballistic missiles to Cuba. American generals urged President John F. Kennedy to preemptively launch a nuclear strike against the Soviet Union, but an agreement was reached at the last minute. The Soviets dismantled their Cuban missile system, and the United States publicly promised not to invade Cuba. In addition, the United States secretly promised to dismantle her Turkish missile system.

At the end of the Cold War, the United States was the world's only remaining superpower. Without formal military opposition, the United States has increasingly used trade as a foreign policy tool. Until the Trump presidency, every American presidential administration has advocated for free-trade policies, such as the **North American Free Trade Agreement** (*NAFTA*). The United States has even helped potential rivals join the international community to prevent large-scale conflict and achieve American goals. For example, the United States integrated China into the world economy by pushing through their membership in the **World Trade Organization**. American economists argue that free trade benefits everyone in the increasingly globalized world. The rise and proliferation of Internet-based technologies has also contributed to the rapid development of a global marketplace.

The Digital Revolution has increased globalization through technological innovations like the internet and social media. Now, people anywhere in the world, no matter how far away, can communicate instantaneously. This has resulted in greater appreciation of people's different backgrounds and lived experiences, an economic boom in the computer-based economic sector, and lower costs for many businesses. However, the Digital Revolution has also threatened privacy, increased automation's threat on employment, and allowed nation-states to spread dissent in foreign countries. For better or worse, the internet has been the most disruptive technological innovation since Gutenberg's printing press (1439).

Cold War Era

The United States and Soviet Union emerged from World War II as the world's preeminent superpowers. The Americans and Soviets considered each other to be an existential threat, but the development of nuclear weapon technology forced the two superpowers to avoid a direct conflict, which is why this period is known as the **Cold War**.

American Cold War era foreign policy followed the **domino theory**—the idea that if one country turned communist, so would neighboring countries. The United States also created the **North Atlantic Treaty Organization** (*NATO*) in 1949 to protect against Soviet aggression in Europe.

On June 25, 1950, North Korea invaded South Korea, and the United States entered the war to protect its ally and to prevent communism spreading across Asia. When General MacArthur disobeyed President Truman and crossed the 38th parallel, China entered the war on the side of North Korea, and the war turned into a bloody stalemate. All three combatants signed the **Armistice Agreement** on July 27, 1953, though the war never officially ended. The Korean War's brutality left a lasting impression on North Korea, heavily influencing the dictatorship's approach to the United States.

In August 1964, Congress passed the Gulf of Tonkin Resolution. Although Congress didn't declare war, the Resolution authorized President Lyndon B. Johnson's use of military force in Vietnam. As in Korea, the United States feared that if North Vietnamese communists prevailed, communism would spread across Asia. The Vietnam War was largely a military success, but it was a disaster politically. The draft proved to be enormously unpopular, and President Richard Nixon's decision to invade Cambodia and bomb Laos further buried public opinion. Domestic protests forced the United States to withdraw in 1973.

The United States also regularly conducted covert operations, trained rebels, and funded political opposition through the **Central Intelligence Agency** (*CIA*) to overthrow democratically elected socialist governments.

Non-exhaustive list of successful Cold War interventions:

- Iran (1953)
- Guatemala (1954)
- Dominican Republic (1961)
- Brazil (1964)
- Congo (1965)
- Indonesia (1965)
- Ghana (1966)
- Guyana (1968)
- Chile (1973)

Socialist governments were often replaced with strongman dictators, and the transition frequently resulted in civil wars. The United States also directly armed morally dubious proxy forces. President Ronald Reagan covertly sold weapons to Nicaraguan death squads, the "Contras," funding it through weapon sales to Iran in violation of an American arms embargo. Similarly, during the Soviet-Afghan War (1979-1989), the Carter and Reagan administrations gave $3 billion of taxpayer money to the mujahedeen (Islamic extremists) who fought alongside Osama bin Laden.

Besides military conflict and interventions, the United States competed with the Soviet Union in the Space Race. In addition to national pride, both superpowers hoped to advance their spying and rocket capabilities. As such, they both invested heavily in technology, some of which was enormously beneficial for society, like increased computing power. To the great surprise of Americans, on October 4, 1957, the Soviet Union launched the world's first satellite, **Sputnik 1**, into orbit. Less than four years later, Soviet cosmonaut Yuri Gagarin became the first human to enter space. On July 20, 1969, the United States struck back by landing the first humans on the Moon during the Apollo 11 mission. The Soviet Union repeatedly failed to land a crew on the Moon. Besides technological and scientific advancements, the Space Race also played a role in *détente*—the easing of tensions between the Cold War rivals—after the Soviets and Americans launched the *Apollo-Soyuz Project* (1971), their first joint space mission.

For domestic politics, the Cold War era was tumultuous. Throughout the 1950s, U.S. Senator Joseph McCarthy led a series of increasingly conspiratorial criminal investigations. McCarthyism exploited Americans' fear of the Soviet Union, commonly referred to as the "**Red Scare**." The public hearings mixed unsubstantiated rumors with total hysteria, and the accusations ruined thousands of lives.

The **Watergate scandal** was the most dramatic domestic political event during the Cold War era. Although Richard Nixon was a popular incumbent on track for a landslide victory over Democrat George McGovern in the 1972 presidential election, Richard Nixon ordered his cronies to burglarize and spy on the **Democratic National Committee** headquarters. Then Nixon attempted to use intelligence agencies to cover up his involvement. After a tape surfaced of Nixon ordering the coverup, the House of Representatives impeached Nixon, and he resigned before the Senate convicted him. On September 8, 1974, President Gerald Ford controversially pardoned Nixon. Besides Nixon, Andrew Johnson and Bill Clinton are the only presidents to be impeached, but no president has ever been convicted (removed from office) by the Senate.

Martin Luther King's civil disobedience is one of the most successful forms of nonviolent protest in world history, particularly his March on Washington in 1963, which concluded with King's famous "I Have a Dream" speech. King's leadership during the Civil Rights movement directly led to the **Civil Rights Act** of 1964 and Voting Rights Act of 1965. Despite these legislative victories, political assassinations, bombings, and riots were commonplace in the United States from the late 1960s to the 1970s.

Shocking the world, the **Berlin Wall** fell in 1989, and national revolutions swept out the Soviets' puppet government in Eastern Europe. By late 1991, the Soviet Union had collapsed. Considerable credit is attributed to President Ronald Reagan for escalating the arm's race, which the Soviets couldn't afford, while simultaneously holding talks with Soviet General Secretary Mikhail Gorbachev.

Post-Cold War Era
Following the collapse of the Soviet Union, the United States became an uncontested superpower. No country in the world could match its economic and military might until China's emergence as a world power in the 2010s.

September 11, 2001 was a traumatic day for the United States, as terrorists hijacked four commercial airliners and crashed three of them into the World Trade Center and Pentagon. Osama bin Laden, the Saudi-born leader of Al-Qaeda, plotted the attacks from his base in Afghanistan. President George W. Bush responded by launching the **War on Terror**, declaring the entire world as a battlefield. Less than one month after the attack, the United States invaded Afghanistan to dismantle Al-Qaeda's training camps, but Osama bin Laden evaded the American military for nearly a decade. American Special Forces killed bin Laden at his Pakistani compound during the Obama administration. The War in Afghanistan is the longest war in American history.

More controversially, Bush invaded Iraq in 2003 based on his administration's claims that Iraqi dictator Saddam Hussein collaborated with Al-Qaeda and was producing "weapons of mass destruction." Both were categorically false assertions, and the Iraq War was an unmitigated disaster for the nation's reputation and security. Shortly after toppling and executing Saddam Hussein, Iraq erupted in a bloody, sectarian civil war. One faction in the Iraqi Civil War would later form the Islamic State and carve a caliphate out of Syria and Iraq.

Opposition to the Iraq War became so fierce that it helped two American presidents, Barack Obama and Donald Trump—who share little else in common politically—win the presidency. President Obama withdrew from Iraq in December 2007, but an American-led coalition returned in 2014 to fight the Islamic State. President Obama expanded the War on Terror, largely relying on drone strikes and Special Operations forces to hunt terror cells in more than a half-dozen countries. In addition, Obama administration officials lied to Congress about the scope of a secret domestic surveillance program, which Edward Snowden exposed, angering civil liberty activists.

The financial crisis of 2007-2008 was a pivotal moment in recent American history. Large investment banks created an artificial bubble in the mortgage sector, and when it burst, the American government bailed them out at taxpayers' expense to prevent the economy from collapsing. Both sides of the political spectrum erupted in anger, directly leading to the anarchist and socialist *Occupy Wall Street Movement* and partially contributing to the growth of the libertarian Tea Party.

Civics/Government

Role of the Citizen in a Democratic Society

The rights enjoyed by Americans are rooted in the Bill of Rights, but no right is unconditional. In some specific circumstances, the government can restrict citizens' rights. The government is especially powerful during wartime. For example, President Abraham Lincoln suspended the right to due process during the Civil War. This is a major reason why the present War on Terror is controversial, as it creates a permanent state of war. Other common limits are defamation laws and limits on "fighting words."

Following the Civil War, the concept of civil rights was added to rights enjoyed by Americans. The Thirteenth, Fourteenth, and Fifteenth Amendments prohibited slavery, provided for due process and equal protection under the law, and protected the right to vote based on race, color, or previous condition of servitude, respectively. Inspired by the Civil Rights Movement, activists have fought for equal rights and protection under the law based on gender and sexuality, culminating in the landmark Supreme Court decision *United States v. Windsor* (2013), which established the right to same-sex marriage.

Americans also have responsibilities. First and foremost, Americans are expected to obey the civil and criminal justice system, or else they will face what can be severe punishment. The United States imprisons more of its citizens, in both the total prison population and per capita, than any other country in the world by a wide margin. Civil disobedience has been used in the United States to great effect, but there's typically strong pushback from the establishment and the public. Despite being fondly remembered, Martin Luther King, Jr. was opposed by the majority of the country for his entire life.

The other major responsibility for citizens is to participate in the political system. Although voting, donating to candidates, running for office, and other forms of participation are not legally mandated, the United States is a representative democracy, and a lack of public participation undermines the government's legitimacy. Yet, voter turnout is low in the United States. Approximately 55 percent of registered Americans vote in presidential elections, and local elections have even lower turnout.

The Structure and Functions of Different Levels of Government in the United States

The United States of America has three major branches in its federal government: the legislative branch, the executive branch, and the judicial branch. The legislative, executive, and judicial branches of government all must adhere to the laws and regulations set forth by the United States Constitution, which was first ratified in 1789 and last amended in 1992. The United States Constitution is a living document that has a section known as the Bill of Rights. **The Bill of Rights** ensured that the **Constitution** remained a living document by setting the precedent for making changes, known as amendments. The Bill of Rights established the first ten amendments. There have since been 27 amendments made to the United States Constitution. The Bill of Rights also helped further establish the roles of the three branches of the federal government. Additionally, it constitutionalizes the separation of powers between the branches.

These branches function as follows:

Legislative Branch
The **legislative branch** creates legislation; it makes the laws. It is comprised of the two houses of Congress: **The House of Representatives** and **The Senate**. This is known as a **bicameral legislature**. As illustrated in the following diagrams, Congress currently has 535 voting members. There are 435 representatives in the House of Representatives and 100 senators in the Senate. The number of senators is always restricted to two per state. The number of representatives per state is based on the population of the state. However, each state is guaranteed at least one representative. States are then split into house districts of about equal size. For example, California has a large population, and Delaware has a small population. Both states have two senators. However, because California is more populous, it has dozens of members in the House of Representatives, while Delaware only has one. The representative from Delaware is elected by and represents the entire state. However, the representatives from California are elected by and represent their district, which is a smaller area within California.

The Senate typically holds more prestige and has more media coverage. Senators serve six-year terms and representatives serve two-year terms. Senators, unlike representatives, can block legislative action through a process known as filibustering, which only requires forty-one votes.

	House of Representatives	**Senate**
Membership	435	100
How Representation is Determined	Population of the state	Two Senators for each state
Length of Term	2 years	6 years

Executive Branch
The **executive branch** helps the nation by carrying out the laws; it backs all legal aspects of the government with presidential power. It is composed of the president of the United States, the vice president of the United States, and the president's cabinet. The president of the United States has vetoing power, which means he or she can block the creation of laws with his or her constitutional right to reject a legislative proposal made by Congress.

Judicial Branch
The **judicial branch** analyzes or interprets the law, deeming whether a law created by Congress or order enacted by the president of the United States is constitutional. The federal judicial branch is comprised of the **Supreme Court** and all lower federal courts. Nine judges preside over the Supreme Court; they are

appointed by the president. See the following diagram for an illustrated explanation of the branches of government.

This diagram illustrates the constitutional system as it pertains to the three branches of government:

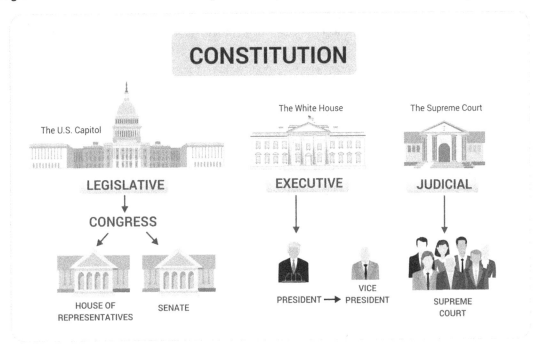

All three branches consequently keep an eye on one another through a process of **checks and balances**, ensuring that all power is separated and no branch gains too much authority over the citizens and government of the United States. The president has the power to appoint judges to the Supreme Court and veto laws made by Congress. Congress can overturn a presidential veto; it also controls the budget of the government. Additionally, Congress approves presidential court appointments and can remove judges if necessary. The Supreme Court can declare presidential acts and legislation as unconstitutional.

CHECKS AND BALANCES ON THE US GOVERNMENT

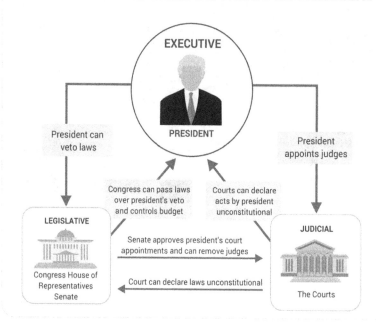

Outside of the federal government, there are also state and local (*municipal*) governments in the United States. Under this federalist system, local governments, state governments, and the federal government share the responsibilities of governing the people. The state and federal governments share the most responsibilities. Both collect taxes, possess their own courts, and have their own systems of punishment. Yet their powers are also separate in some ways. State governments cannot maintain a military force or declare war like the federal government, and federal governments do not maintain total control over state public education.

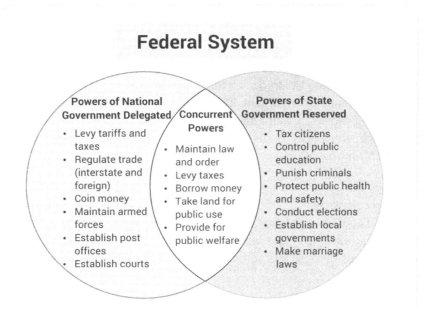

These shared responsibilities become even more complex when local government is included. Local governments focus mostly on local services (such as emergency response units, utilities, and public libraries). However, some responsibilities, such as education, are shared by state and local boards. Local municipalities are also responsible for carrying out the functions necessary to ensure fidelity in the federal voting process. Lastly, all three types of governments join forces to maintain roads and carry out law enforcement initiatives.

Purposes and Characteristics of Various Governance Systems

The history of humanity has been marked by various government systems that have different purposes, functions, and characteristics. Below is a list of governance systems that should be reviewed and understood. This list is not comprehensive. Moreover, most real governments do not fall neatly into just one of these categories. Typically, most governments—historical and contemporary—draw from more than one governance category. For instance, the Soviet Union under Joseph Stalin was both communist and authoritarian. The United States can be considered a **democratic republic**, but it also has characteristics of **federalism**.

Authoritarian
Authoritarian governments allow the state—or a symbol of the state (such as a dictator)—to control every aspect of the citizens' lives. The Soviet Union was known for its authoritarian practices. It was also communist in character. It created intense systems of control that disempowered its citizens in the name of equality.

Communist

Cuba is a classic example of a **communist** government that still exists today. While communism is mainly an economic system, it is very much a political economy that tries to create equality by eliminating private property. Communism in theory, however, is different than communism in practice. In Cuba, much like in the Soviet Union, the Communist government helped pave the way to dictatorial authoritarianism. In Cuba, Fidel Castro was dictator for life from the 1950s until his death in the early 2010s. Since the Cold War (and even before that, around the age of the Bolshevik Revolution in Russia), the United States has been staunchly anti-communist in character.

Confederacy

There are two classic historical examples of confederacies. The first is the Iroquois Confederacy, a coalition of Native American tribes during the precolonial and colonial eras. The second is the Confederate States of America, which seceded from the Union during the Civil War era in American history. **Confederacies** limit the federal power of a central government. It is usually a coalition of states, provinces, or tribes. These states, provinces, and tribes maintain their sovereignty, only turning to federal support for unique, previously agreed-upon circumstances. Under the Articles of Confederation, the United States of America borrowed heavily from this model of governance.

Democracy (Pure/Popular)

Pure or **popular democracy** places full voting and governance powers in the hands of the people. Many national governments in Latin America followed this model of governance following revolutions in the 18th, 19th, and 20th centuries. Many early leaders in the United States of America thought this form of democracy represented chaos because it granted the masses too much power.

Democratic Republic

The United States government is traditionally known as a democratic republic. **Democratic republics** allow citizens to vote for representatives who, in turn, vote on their behalf. The electoral college is a classic example of a democratic republic: during the presidential election, citizens vote, but the electoral representatives ultimately cast the final vote for each state.

Monarchy

This form of government was popularized during antiquity and the Middle Ages. It places full ruling power in the hands of a monarch—a king or queen—who is usually believed to be divinely appointed. A good example of a traditional **monarchy** was England prior to the creation of parliament. With the inclusion of parliament and a constitution, modern-day England now has a constitutional monarchy; the king and queen are customary figureheads, while parliament and the constitution wield the most power. Breaking from England during colonial times, the United States has historically avoided this model.

Dictatorship

When authoritarian governments place power into the hand of one leader, they are typically referred to as **dictatorships**. There are usually no constitutional checks and balances in this system of governance. Fidel Castro is a classic example of a dictator.

Socialist

Drawing from Marxist theory, many **socialist** governments still exist today. They believe in an equitable distribution of goods, which is not the same as the abolition of private property or 100% equality in distribution of goods. Many countries in Europe have socialist tendencies. The United States has historically leaned toward capitalism and away from socialism. However, with the creation of the New Deal, the US became more socialistic in character (in limited fashion).

Theocracy

A religious organization, not humanity, is the supreme power. In a **theocracy**, people such as caliphs or priests become earthly symbols of religious prowess. ISIS is a recent example of a theocracy, which emphasizes the creation of an Islamic Republic. A historical example of a theocracy is Roman Catholic Spain. The United States has tried to separate church and state, making it anti-theocratic.

There are also three larger frameworks that might dictate the legislative nuances of the types of governments listed above: unitary systems, federal systems, and confederations. **Unitary systems** have concentrated power controlled by a centralized authority. Unitary systems emphasize national government over state government. **Federal systems** divide the power of the state and federal governments; they also place more power in the hands of the people via certain voting rights. **Confederations** emphasize decentralized power and concentrated authority, which takes hold when states have more power than the federal government.

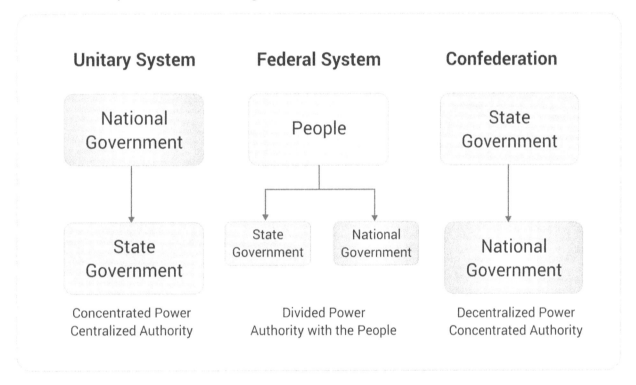

Economics

Fundamental Economic Concepts

Economists study the production, distribution, and consumption of wealth as expressed through commodities, goods, and services. **Commodities** are raw materials or agricultural products—gold, silver, gas, oil, corn, barley, etc. Goods are physical, tangible items, such as computers, bathing suits, refrigerators, televisions, etc. Services are activities based on labor and/or skills, like the work of lawyers, waiters, engineers, computer programmers, etc.

The government's role is to implement an economic system to distribute resources. In a centrally planned economic system, the government makes all decisions related to production and distribution. In addition, price controls are common, if not pervasive. Communist countries have a centrally planned economic system, and they attempt to abolish private property, transfer the means of production to the public, and distribute profits on an equitable basis.

In contrast, **capitalism** is a market economic system. The **free market** is what sets prices based on the existing supply and demand, and private firms make all decisions related to production and distribution. Market economic systems are based on merit, meaning that resources are distributed unequally. Under a pure laissez-faire free-market system, the government can't regulate, tax, subsidize, or otherwise interfere with the free market. The only rule in laissez-faire economics is *caveat emptor* (buyer beware).

Nearly all economies are mixed economic systems where the public and private sectors share the means of production and profits. As such, all governments with a mixed economic system collect taxes, set tariffs, and spend resources on public projects; however, they differ on the amount of control over the free market. For example, some governments implement price controls to regulate certain economic sectors for a designated period.

When a free market is present, the price and quantity are based on supply and demand. **Supply** is the amount of a commodity, good, or service that's available for consumption, and **demand** is the desire of buyers to acquire the commodity, good, or service. **Money** refers to the method of payment, whether it's a government currency, credit, digital currency, or precious metal.

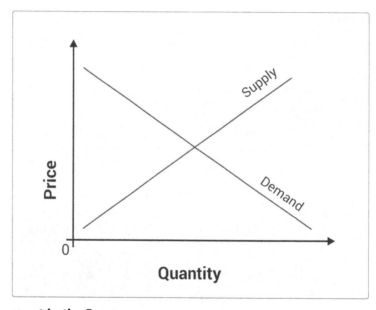

Government Involvement in the Economy

Governments can create their own currencies, and they can typically control the amount of currency in circulation, as well as the cost of short-term borrowing. Most countries designate a central bank or government agency to set and enforce monetary policies to control inflation and interest rates. **Inflation** is a period of increased prices and decreased currency values. **Interest rates** are what lenders charge borrowers to receive loans or credit. For example, governments sell bonds to borrow money for a designated period, and when the period ends, the buyer receives an interest payment plus the amount of the original debt.

In general, economics is a source of domestic and foreign conflict due to the scarcity (shortage) of resources like money, raw materials, and skills. Within a country, the needs and wants of the population are unlimited, while many resources are scarce. The same is also true in the context of international resources—the supply of resources is limited, while demand is unlimited.

Countries' domestic economic policy agendas involve raising revenue through taxes and spending money to enact and enforce policy decisions. Foreign economic policy agendas involve tariffs for imported goods. **Tariffs** are mostly used to protect industries with a special value to the homeland or to punish foreign countries.

Governments make production and trade policy decisions based on **comparative advantage**—the principle that countries produce the goods and services that have the least **opportunity cost**. Opportunity cost measures the loss of potential gain if the country had produced some other good or service. As such, in a free market, countries specialize at what they can do most efficiently relative to their competitors.

Most governments sign free-trade agreements with close allies and major trading partners. **Free-trade agreements** lower or eliminate tariffs. The agreements can be bilateral (2 countries) or multilateral (3+ countries). For example, the Gulf Cooperation Council is a multilateral free-trade agreement between Arab states in the Persian Gulf. In addition, many countries join intergovernmental organizations to borrow money, donate aid, or settle disputes. The **World Trade Organization** provides structural support for trade negotiations and offers dispute resolution services. The International Money Fund tries to increase global economic cooperation. The **World Bank** attempts to alleviate poverty by loaning money to developing countries for large projects.

Income inequality and perceptions of unfairness stir domestic unrest. In effect, worsening economic conditions cause political conflicts. The public might object to tax policies, stagnant or falling incomes, shortages of resources, underfunded social services, or any other specific policy. If the economic situation continues to deteriorate, the public might seek to overthrow the government. For example, during the French Revolution (1789-1799), King Louis XVI was executed after the monarchy amassed enormous war debts, enforced regressive taxes on their poorest citizens, and publicly flaunted their wealth in front of an increasingly poor public.

Technological innovation can both drive economic growth and send economies spiraling into a depression. Throughout history, technology has increased the efficiency of human labor. If the excess labor migrates to a different economic sector, the economy will grow. Many emerging technologies create new economic sectors, but some economic sectors are hurt by this transition. For example, the printing press expanded the markets for writers, records-keepers, and librarians. On the other hand, the printing press destroyed the market for scribes. Typically, free-market economic systems generate more technological advancements since the private firms are incentivized to pursue profits, which drives innovation. However, public spending can also lead to groundbreaking advances. For example, the Space Race revolutionized the computer-based economic sector, and the American military created **ARPANET**, the foundation for what would later become the *Internet*.

Due to comparative advantage in the free market, developing countries handle most of the world's labor-intensive work, like manufacturing. Developed countries benefit from the cheaper labor in lower costs, but there are fewer jobs for blue-collar workers. These workers often don't have the necessary skills to smoothly transition into a new field, especially as fields become increasingly specialized. Automation has a similar impact—decreased costs, increased unemployment, and a skills gap in the labor market. Consequently, governments must enact economic policies that maximize the advantages of cheaper labor and lower costs, while also protecting the economic system's legitimacy. The legitimacy of an economic system is context-specific, depending on historical and cultural forces. However, systems with higher standards of living and less income disparity enjoy more widespread legitimacy with the public.

Consumer Economics

Consumer economics focuses on the ways in which individuals (consumers) interact with the market economy at the local, regional, national, and global level. Consumer economics became more embedded within global systems following the market revolution of the early 1700s. This consumer turn that occurred only became catalyzed by the broader Industrial Revolution, which took hold across the globe from the mid 1700s to early 1900s. Both of these revolutions allowed for previously self-sustaining economies to be thrust into the consumer frameworks of the so-called market. People transitioned out of their traditionally self-sufficient agrarian lifestyles and into highly dependent consumer economies. Today, consumerism refers to the practice of buying and selling goods that exist within this larger international market, which was once industrialized but is swiftly becoming deindustrialized. Consumer habits affect our everyday lives, so economists, economics teachers, and students of economics spend a lot of their time analyzing consumer trends. These trends can focus on local consumer economies, regional consumer economies, national consumer economies, or international consumer economies.

Below are some key concepts for understanding consumer economics:

Free Enterprise Economy: A free enterprise economy—sometimes referred to simply as a market economy—helps producers determine what to produce, how to produce, how much to produce, and for whom to produce based on supply, demand, and competition. In a market economy, profit drives decisions; government does not control resources, goods, or any other major sector of a market economy. Businesses run by the people determine the economic structure, supply, and demand of a market economy. A free enterprise economy has five distinctive characteristics: 1) private property, 2) economic freedom, 3) economic incentives, 4) competitive markets, and 5) limited government interference.

Supply and Demand: In free enterprise systems, such as the US market economy, people are free to buy the goods and services they want/need. Since the consumer often dictates purchases/consumption in this free enterprise system, pricing and production often ebbs and flows with the economic principles of supply and demand. Supply refers to the amount of goods and services a business is willing to produce and distribute. Supply rises when businesses can sell products and services at a higher price. Supply decreases when prices are low. Demand refers to how many products and services consumers are willing to buy. When prices rise, demand tends to drop. When prices drop, demand tends to rise. Supply and demand, therefore, work together to dictate prices. The effects of supply and demand on prices can be seen, for instance, during times of national disaster. Drought can, for example, raise water rates for communities.

Competition: Competition refers to economic rivalries between companies. Competition can also affect supply and demand, and, in turn, production, pricing, distribution, and consumption. Competition can often force companies to lower their prices or strengthen the quality of their product or service.

Scarcity: The consumer economy is built upon the foundation and notion of scarcity, the limited availability of a particular commodity. Scarcity affects all economic agents within the consumer economy; it affects laborers, businesses, households, consumers, banks and other financial institutions, and creditors. Scarcity demands that all economic agents in the consumer economy must make difficult decisions about the ways in which they obtain resources, budget, save, and take on credit.

Budgeting: Buying one good or service means that these economic agents will have to sacrifice another good or service. Saving money entails sacrificing spending. Budgeting entails sacrificing excessive expenditures, savings accounts, or credits. Budgeting in a consumer economy mean that households, businesses, and governments must constantly make educated decisions about the necessity, quality, and satisfaction of the goods and services they choose to produce or purchase. There are only so many dollars available for each household, business, and government. All economic agents must, therefore, decide how to adequately spend their time and money.

Geography

Physical and Human Geography

A **physical system** refers to a region's landmass, environment, climate, and weather. **Human systems** are a group of people participating in a joint enterprise. Human systems adapt to and manipulate the physical system, and in doing so, they change the physical system. The extraction of resources is why physical systems change.

Human systems require the consumption of natural resources—air, fossil fuels, iron, land, minerals, sunlight, water, wind, wood, etc. Some resources are renewable, such as oxygen, fresh water, and solar energy, while others are finite. Nonrenewable resources, like fossil fuels, take millions of years to form; once they're depleted, they're gone. The variety and availability of resources plays an integral role in the human system's development, and the fight over resources also changes human systems when there's conflict.

The interaction between human and physical systems is most evident in economic development, particularly agriculture. Some regions are entirely barren, while others are rich in natural resources. Where a human system is located within that range will have an important, if not decisive, impact on its future. For example, an advanced human system has never developed in the Sahara Desert due to the lack of water and arable land. As such, the human system adapted to the harsh physical system by embracing a nomadic lifestyle. In contrast, nearly all ancient empires were located on or near a large river delta. The nutrient-rich soil enabled agriculture, and the resulting food production supported a large population. With a larger population, human systems based on agriculture could diversify their economies with the excess labor, form governments, and raise armies to conquer more land. All of this would be impossible in the Sahara without implementing some future technological innovation to moderate the physical system.

The interaction between the Native Americans and Europeans resulted in the rapid decline of the Native American system and drastic changes to the physical system. The European system's large-scale production of crops and livestock sustained permanent settlements, which facilitated the establishment of a centralized system of government. In contrast, the Native Americans were nomadic, making it easier for Europeans to drive native tribes off their land. Old World diseases also overwhelmed the Native American population, reaching some of the western tribes even before the arrival of European colonists. In addition, the Europeans were armed with superior weaponry—steel and firearms. As a result, the European system devastated the Native American system. The indigenous population in California dropped from 150,000 to 15,000 between 1848 and 1900.

The American takeover dramatically altered the physical system. Agriculture, industrialization, and urbanization have caused a sharp decline in wilderness, which now accounts for less than 5 percent of total American land. Similarly, the United States outside of Alaska has lost 60% of its natural vegetation. As wilderness and habitats decline, so does biodiversity. In the United States, 539 native species are extinct or missing, and more than one-third of all species are at risk. These trends are mirrored across the world as more countries industrialize. Only 23 percent of the world's wilderness remains, and the rate of habitat destruction has markedly increased over the last few decades. If current trends continue, two-thirds of all wild animal species will be extinct by 2020.

Furthermore, the human system is causing climate change by increasing the amount of carbon dioxide and methane in the atmosphere. Deforestation, land use changes, and the burning of fossil fuels release carbon dioxide into the atmosphere. Landfills, livestock, and rice cultivation produce methane. These gases blanket the atmosphere, preventing heat from escaping into space. The scientific data is undeniable. Sixteen of the hottest seventeen years on record have occurred since 2001, and global ice is melting at an unprecedented pace. Consequently, rising temperatures have increased the rate of extreme weather events and ocean acidification.

Using Geographic Concepts to Analyze Spatial Phenomena and Discuss Economic, Political, and Social Factors

The study of geography should not be confined to lifeless maps of counties, states, nations, and continents. Though these maps are, indeed, foundational to the study of geography, **geography** is also tied to the economic, political, and social trends of humanity. These geographic concepts play out over time and space, signaling changes in not only the spatial environments, but also the spatial phenomena that fill these environments. There is a reason why many scholars, educators, and students have turned to a field known as human geography: the field of geography centers much on the interaction of humanity with the spatial and physical environments.

Geographic concepts such as **nation-states** and **borders** have stirred economic, political, and social turmoil; they have spawned revolutions and genocides. The Mexican-American War was largely a war over imaginary lines we call borders. The American Civil War was largely over a geographic phenomenon we call **regionalism**. Even the war over words of the most recent presidential election in the United States had a lot to do with the concept of a rural-urban-suburban trichotomy in geography.

Other geographic concepts include migration, standard of living, and cultural diffusion. **Migration** occurs when people move from one place to another for economic, political, social, or environmental reasons. **Standard of living** describes the wealth, comfort, necessities, and luxuries that are available to people in a specific geographic area. **Cultural diffusion** happens when a certain culture and its corresponding traits, beliefs, and activities spread to other nationalities and ethnicities.

Put simply, geography is yet another interdisciplinary field that continues to be a highly influential contributor and outgrowth to human existence.

Thus, there are some basic geographical concepts that everyone should know (beyond the highly specified aforementioned terms such as border and nation state). Most geography teachers focus on what are called the seven major geographical concepts:

Space: The environmental and human characteristics placed on the physical arrangement, area, or pattern of a location. Space is also about the significance, organization, and distribution of this physical arrangement, area, or pattern of a location. Space, much like place, can have a personal or collective significance. It can be experienced in different ways. People studying geography analyze the connections and organizations of space, especially as it relates to human civilizations and global interactions.

Place: Places are absolute locations on the Earth's surface that are critical components of human identity, security, and culture. Humanity has the ability to actually alter the physical characteristics of place. They can also alter the existential characteristics that infuse meaning into these absolute locations.

Interconnection: Interconnection refers to the connectedness of human, natural, and environmental processes, which help create physical and cultural changes that are byproducts of human civilization. Interconnection is about the interplay between human and environmental forces, which are both considered part of geography. The agricultural revolution is a great example of the ways in which the interconnectedness of the environment and humanity created vast changes through symbiotic structures.

Change: Change is, at its core, about the environmental and human evolutions and evolutions that play out over time. Change is the ways in which cultures and physical places/spaces transform throughout history.

Environment: Environment is more than just space or place—it is about the biological, atmospheric, geological, and chemical processes that influence human decisions in a particular place. The environment is about the ways in which humans interact with waterways, food sources, chemical compounds, geological features, and climate changes.

Scale: Scale exists on a spectrum from extremely microcosmic to extremely macrocosmic; it is about minute details and vast superstructural generalizations. Geographers must decide the relative scale of a particular event, action, or phenomenon; they must think about how minor or major the impact.

Sustainability: Sustainability has become a crucial component of geographic study because it focuses on the ability of humanity (and life in general) to sustain or continue in the midst of wide-reaching geographical chain reactions that have severe consequences (i.e., pollution → global warming → loss of habitat → loss of food sources and/or animal species) in terms of biological persistence.

Uses of Geography

History is the story of interaction and conflict between interests, movements, cultures, societies, nations and empires. **Geography** is the key to understanding history. World leaders always include geographic considerations in their decision-making. As such, understanding geography helps explain the past, interpret the present, and plan for the future. This is especially true for any armed conflict due to the strategic importance of higher ground, supply routes, defensible borders, etc. The field of studying political power through the lens of geography is called **geopolitics**.

The three theories of geopolitical power with the most historic, contemporary, and future significance are the Heartland Theory, Rimland Theory, and Organic Theory.

Published in 1904, Halford John MacKinder's **Heartland Theory** emphasizes the strategic importance of Eastern Europe, Russia, and Western China. These three regions form the "Heartland," occupying a central position on the "World Island"—Africa, Asia, and Europe. MacKinder considered the rest of the world to be mere islands floating around the World Island. This reflects the Heartland Theory's Eurocentric approach. Due to their military power and geographical proximity to Eastern Europe, MacKinder correctly identified Germany and Russia as the greatest threats to global security in the early 20th century.

The Cold War also heavily featured the Heartland, as the Soviet Union controlled most of Eastern Europe and worked closely with China. As a result, American foreign policymakers' primary goal was to weaken the Soviets' dominance in the Heartland. American forces used covert operations to sow dissent and discord throughout the region. For example, the Central Intelligence Agency (CIA) covertly funded Radio Free Europe—an American government program that broadcast news across Eastern Europe. In addition, the United States courted China as an ally, which culminated in President Richard Nixon's state visit to China in February 1972. Following the Soviet Union's loss of the Heartland and subsequent dissolution, the United States encouraged Eastern Europe to join the North Atlantic Treaty Organization (NATO), an American-led military alliance.

Like the Heartland Theory, the **Rimland Theory** is Eurocentric, prioritizing the Eurasian continent above all other regions. However, the Rimland Theory differs in its emphasis on the "Rimland" (Eurasian coasts). Accordingly, the European coast, Middle East, and Asiatic coasts have the most geopolitical value. The Rimland Theory's author, Nicholas John Spykman, believed these regions served as a buffer between land and sea power. The theory predates the widespread use of airplanes, which detracts from some of the coastal regions' importance. However, the United States remains a military superpower through naval superiority.

Rather than name a strategically important region, the **Organic Theory** argues that nation-states need to consolidate political power and acquire new geographic territory in order to sustain themselves. In this concept, the nation-state operates like a living organism, and new territory provides the resources it needs to survive. First published by German Friedrich Ratzel in 1897, the Organic Theory was often combined with Social Darwinism—the idea that Darwin's natural selection applies to humans—to justify imperialism and racism. Nazi Germany was one of the Organic Theory's most vocal advocates.

Empires and nation-states have historically pursued their self-interest by consolidating political power and acquiring foreign territory. Mercantilism, colonialism, and imperialism are all related to conquering territory, and they were common European policies for centuries. The Roman Empire is another illustrative example. Rome started as a single city-state, conquered the Italian Peninsula, and then seized most of Europe, North Africa, the Balkans, and Asia Minor. Once a territory was captured, the Romans enslaved large populations of people, collected taxes, and extracted resources from the land. Each conquest increased Rome's political power, and Rome collapsed partially due to its inability to conquer and hold new territory.

The Organic Theory also helps explain why nation-states resist fracturing into smaller, independent states. If acquiring territory is the nation-state's lifeblood, then losing territory is a direct harm. In addition to undermining the government's legitimacy, losing territory could decrease the labor pool, natural resources, and military strength. It also means sharing a new border with what could be a hostile neighbor. This was the case in the American Civil War. President Abraham Lincoln didn't initially deploy military force to free the slaves. In fact, several of the border slave states remained in the Union for the entire conflict. Instead, President Lincoln wanted to preserve the Union and prevent the loss of territory, economic power, military power, and natural resources. Lincoln's reasoning was consistent with the Organic approach to geopolitics.

The present-day United States is not a traditional empire. Past empires have conquered and directly ruled over their territories. The United States hasn't added a major piece of territory since the Spanish-American War. Nevertheless, the United States' leverage over many nation-states is impossible to overstate. The United States has acquired political power by wielding military force to coerce countries and money to make them willing partners. The United States outspends its rivals on its military by a significant margin every year, though some of the difference can be attributed to higher labor costs.

American foreign policy revolves around geopolitics. The United States maintains approximately 800 military bases in 70 different countries, and the bases surround every potential challenger to American hegemony. The U.S. Navy operates nineteen aircraft carriers, while the rest of the world has nine altogether. Consequently, the United States can exercise control over five oceans, challenge any coastal region it desires, and send military planes anywhere in the world at a moment's notice.

Other than a genuine interest in combatting terrorism, the wars in Afghanistan and Iraq served American geopolitical interests in the Middle East, a region that holds most of the world's oil reserves. Some think the Iraq War was motivated by the goal of "stealing" Iraqi oil, but that theory goes too far. The United States was more interested in exercising American power vis-a-vis Iran, Iraq's geographical neighbor. The United States is allied with all of the Sunni Muslim oil-producing countries (Saudi Arabia, Qatar, United Arab Emirates, etc.), and they are Iran's geopolitical rivals. As a result, the United States is constantly trying to prevent Iranian influence from spreading outside its territorial borders, as has happened in Lebanon and Syria.

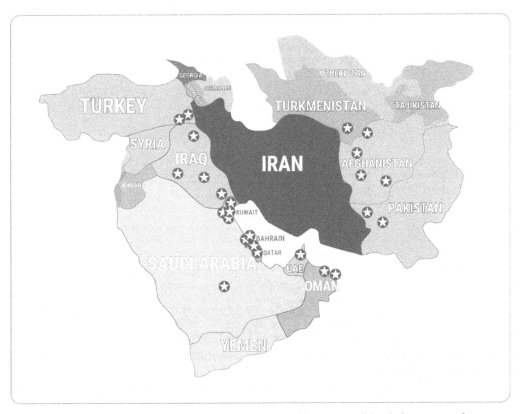

Like its impact on predicting and understanding armed conflict, geopolitics is important for understanding trade and foreign aid decisions. Human systems have emphasized trading with their geographic neighbors since the prehistoric era. In addition to generating wealth, trade improves diplomatic relations and avoids conflict. Depending on the measure, the United States is either the largest or second-largest (after China) economy in the world, and the U.S. leads the global financial industry.

Consequently, the United States can curry favor by promising greater economic cooperation and donates twice the amount of foreign aid as the next-largest donor.

Interpreting Maps

Contrary to popular perception, geography is not simply about the study and creation of maps (that is **cartography**). Nevertheless, geography is, without a doubt, a very visual field; it focuses on interpreting visual representations of environmental processes and human behaviors (and the interconnectedness between the two). These visual representations can range from simple maps to complex **geographic information systems** (*GIS*). In fact, the entire field is becoming more and more technological in the midst of the current digital revolution. Geographers no longer focus solely on two-dimensional, handcrafted maps. They now take advantage of both two-dimensional and three-dimensional mapmaking technologies. First launched in 1978, the **global positioning system** (*GPS*) has also revolutionized the field of geography. Think about how complicated and user-friendly maps have become as a result of GPS technologies. Systems like Google Maps, for instance, are a perfect example of the rapid changes brought about by new geographic technologies. In many ways, human beings are interacting with their environments more than ever with the advent of smartphone GPS capabilities. The whole landscape—thanks to virtual reality and augmented reality—has quite literally changed. We are studying geography on a daily basis without even knowing it!

Most maps—traditional or digital—use **legends** to help the audience navigate the meaning and content of the tool. Legends provide visual representations and explanations of the content on a map; they are key for conveying the unique features of every map. Each legend consists of a set of symbols and a short description of the information it conveys. Although the placement of the legend varies from map to map, they are normally found with the corner or margins of a map. Below is an example of a traditional map legend:

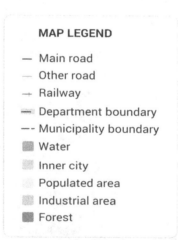

Specific types of geographic maps and graphics include contour maps, population pyramids, and climographs. **Contour maps** use contour lines to show elevations and slopes of land surfaces and features. A **population pyramid**, also known as an age-sex pyramid, is a graphical representation of a populations' distribution based on age and sex. **Climographs** describe a location's climate by displaying climatic elements on a graph or chart.

Regardless of technological changes, one thing remains the same about geography: it is still a field heavily devoted to case studies. Like many other scholars in the humanities and social sciences, geographers spend a lot of their time creating and analyzing geographic case studies. These studies try to consolidate and analyze data about particular geographical interconnections. The seven major geographical concepts are key for developing and examining case studies. Geographers analyze maps, geographic technologies, and data through the lenses of these seven major geographical concepts. These lenses assist geographers with better understanding the ways in which human beings bring changes to their spaces, places, and environments. It is up to the geographer to decide the scale of these changes and whether they are conducive to overall human or environmental sustainability.

Case studies usually focus on a particular process, space, place, or environment. In most instances, using visual and technological tools, geographers will cross-analyze multiple case studies. For example, a geographer may compare four major metropolitan cities that have similar population sizes or demographics. Or geographers may compare the educational outcomes of certain minority groups across different spaces, places, or environments. Geographers may even study the sustainability practices of one rural village over the course of a decade. The options are endless, but these case studies are crucial to expanding the academic canon of geographic understanding.

Below are two different case study maps created by geographers. The case study map analyzes and compares, at the state level, violent crime per 100,000 people. The second case study map simply identifies the states that have the death penalty. It is possible that a student of geography could cross-analyze these two maps (with relevant data) to create a third case study, one that might analyze whether or not the threat of capital punishment deters violent crimes in pro-death-penalty states. This is an example of the intellectual work one might do to create a geographic case study.

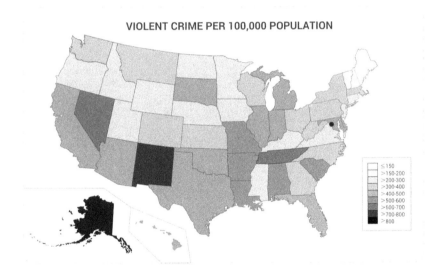

VIOLENT CRIME PER 100,000 POPULATION

317

**MAP SHOWING STATES
WITH AND WITHOUT THE DEATH PENALTY**

States with a valid death penalty statute
States without the death penalty

Practice Questions

The next three questions are based on the following passage from the biography Queen Victoria *by E. Gordon Browne, M.A.:*

The old castle soon proved to be too small for the family, and in September 1853 the foundation-stone of a new house was laid. After the ceremony the workmen were entertained at dinner, which was followed by Highland games and dancing in the ballroom.

Two years later they entered the new castle, which the Queen described as "charming; the rooms delightful; the furniture, papers, everything perfection."

The Prince was untiring in planning improvements, and in 1856 the Queen wrote: "Every year my heart becomes more fixed in this dear Paradise, and so much more so now, that *all* has become my dearest Albert's *own* creation, own work, own building, own laying out as at Osborne; and his great taste, and the impress of his dear hand, have been stamped everywhere. He was very busy today, settling and arranging many things for next year."

1. Which of the following is this excerpt considered?
    a. Primary source
    b. Secondary source
    c. Tertiary source
    d. None of these

2. It took _____ years for the new castle to be built.

3. What does the word *impress* mean in the third paragraph?
    a. to affect strongly in feeling
    b. to urge something to be done
    c. to impose a certain quality upon
    d. to press a thing onto something else

The next question is based on the following passage from The Federalist No. 78 *by Alexander Hamilton.*

According to the plan of the convention, all judges who may be appointed by the United States are to hold their offices *during good behavior,* which is conformable to the most approved of the State constitutions and among the rest, to that of this State. Its propriety having been drawn into question by the adversaries of that plan, is no light symptom of the rage for objection, which disorders their imaginations and judgments. The standard of good behavior for the continuance in office of the judicial magistracy, is certainly one of the most valuable of the modern improvements in the practice of government. In a monarchy it is an excellent barrier to the despotism of the prince; in a republic it is a no less excellent barrier to the encroachments and oppressions of the representative body. And it is the best expedient which can be devised in any government, to secure a steady, upright, and impartial administration of the laws.

4. What is Hamilton's point in this excerpt?
    a. To show the audience that despotism within a monarchy is no longer the standard practice in the states.
    b. To convince the audience that judges holding their positions based on good behavior is a practical way to avoid corruption.
    c. To persuade the audience that having good behavior should be the primary characteristic of a person in a government body and their voting habits should reflect this.
    d. To convey the position that judges who serve for a lifetime will not be perfect and therefore we must forgive them for their bad behavior when it arises.

5. One factor that contributed to the rise of consumer economics in the United States of America was:
    a. The market revolution
    b. The Mexican-American War
    c. The Enlightenment
    d. The stock market crash of 1929

6. Which of the following is NOT listed as one of the seven major geographical concepts?
    a. Space
    b. History
    c. Place
    d. Change

7. Why do geographers traditionally use case studies?
    a. Lobby for environmental sustainability changes
    b. Program GPS systems
    c. Collect and analyze important geographical data
    d. Communicate important theories to cartographers

8. What is the best definition for a unitary governmental framework?
    a. Power is centralized and balanced between parliament and the monarchy.
    b. National government is emphasized over state governments, with concentrated power and centralized authority.
    c. State governments are emphasized over national government, with decentralized power and concentrated authority.
    d. Power is placed in the hands of the people through voting rights, and power is balanced between state governments and national government.

9. What is the best definition of scarcity in a consumer economy?
    a. An increased supply of a product
    b. An increased demand for a product
    c. The limited availability of a particular commodity
    d. The excess production of a particular commodity

10. Which of the following types of government would most likely use Marxist philosophy?
    a. Socialism
    b. Monarchy
    c. Theocracy
    d. Federalism

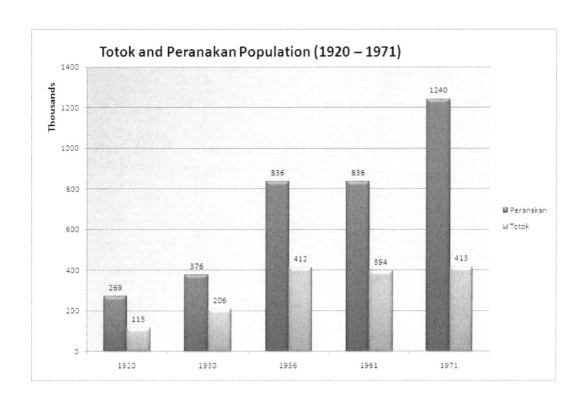

Totok and Peranakan Population (1920 – 1971)

11. According to the graph above, what is the population of Totok in 1961?
    a. 115,000
    b. 206,000
    c. 394,000
    d. 836,000

12. According to the graph, what is the difference between the population in Peranakan in 1971 and Peranakan in 1961?
    a. 394,000
    b. 403,000
    c. 404,000
    d. 412,000

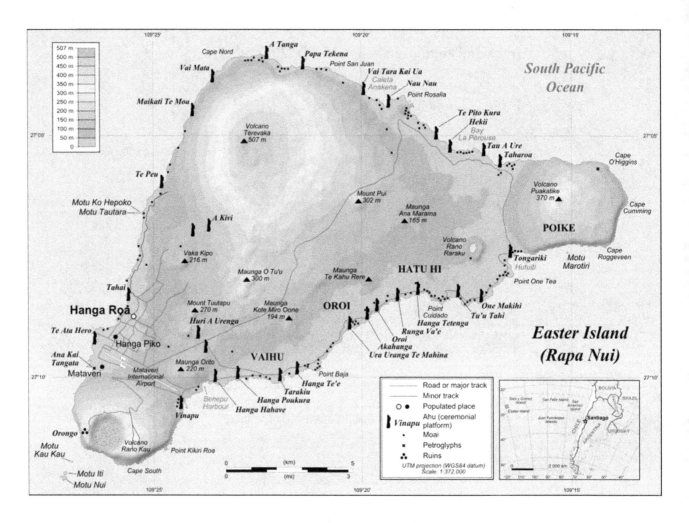

13. According to the map, which of the following is the highest points on the island?
    a. Volcano Terevaka
    b. Maunga Ana Marama
    c. Puakatike Volcano
    d. Vaka Kipo

14. Why are there more representatives than senators in Congress?
    a. Senators are more prestigious than representatives, so fewer are needed.
    b. Voters decide how many representatives they want each year.
    c. The number of representatives is dictated by a state's population size.
    d. The number of representatives and senators was established by the Bill of Rights, which states that there should be 100 senators and 435 representatives.

15. What countries formed the Triple Entente in World War I?
    a. The United States, Britain, and Canada
    b. The United States, Britain, and France
    c. Britain, France, and Russia
    d. Austria-Hungary, Germany, and Italy

16. What did President Franklin Delano Roosevelt call his legislative agenda?
    a. The War on Poverty
    b. The New Deal
    c. The Great Society
    d. Social Security

17. When it was founded, what was the Republican Party's primary issue?
    a. States' rights
    b. Tax cuts
    c. Manifest destiny
    d. Abolition of slavery

18. What is the Organic Theory's thesis?
    a. Naval and air power is more important than territorial control.
    b. Africa, Asia, and Europe form the World Island, which is the most important landmass for geopolitics.
    c. Countries need to acquire political power and territory to survive.
    d. Communism will grow organically unless there's early military intervention.

19. Which of the following maps is the most subjective?
    a. Climate maps
    b. Political maps
    c. Population density maps
    d. Topographic maps

20. How did Jim Crow laws impact the American South?
    a. African American slaves could vote for the first time in American history.
    b. The South diversified from a one-crop economy.
    c. The South industrialized, following the Northern example.
    d. The Southern states contravened the Reconstruction Amendments.

The following document is a section of the Constitution. This section of the Constitution focuses on the Senate. It includes directions about the methods and parameters for selecting a senator. Additionally, it provides an analysis of some of the leadership roles and duties of the US Senate.

Section. 3.

The Senate of the United States shall be composed of two Senators from each State, chosen by the Legislature thereof, for six Years; and each Senator shall have one Vote.

Immediately after they shall be assembled in Consequence of the first Election, they shall be divided as equally as may be into three Classes. The Seats of the Senators of the first Class shall be vacated at the Expiration of the second Year, of the second Class at the Expiration of the fourth Year, and of the third Class at the Expiration of the sixth Year, so that one third may be chosen every second Year; and if Vacancies happen by Resignation, or otherwise, during the Recess of the Legislature of any State, the Executive thereof may make temporary Appointments until the next Meeting of the Legislature, which shall then fill such Vacancies.

No Person shall be a Senator who shall not have attained to the Age of thirty Years, and been nine Years a Citizen of the United States, and who shall not, when elected, be an Inhabitant of that State for which he shall be chosen.

The Vice President of the United States shall be President of the Senate, but shall have no Vote, unless they be equally divided.

The Senate shall [choose] their other Officers, and also a President pro tempore, in the Absence of the Vice President, or when he shall exercise the Office of President of the United States.

The Senate shall have the sole Power to try all Impeachments. When sitting for that Purpose, they shall be on Oath or Affirmation. When the President of the United States is tried, the Chief Justice shall preside: And no Person shall be convicted without the Concurrence of two thirds of the Members present.

Judgment in Cases of Impeachment shall not extend further than to removal from Office, and disqualification to hold and enjoy any Office of honor, Trust or Profit under the United States: but the Party convicted shall nevertheless be liable and subject to Indictment, Trial, Judgment and Punishment, according to Law.

21. According to Section 3 of the US Constitution, how long is the term for US Senators?
    a. Two years
    b. Six years
    c. Lifelong
    d. Four years

The following document is an excerpt from Abraham Lincoln's "House Divided" speech, which was delivered to the Republican state convention on June 16, 1858.

A house divided against itself cannot stand. I believe this government cannot endure, permanently, half slave and half free. I do not expect the Union to be dissolved—I do not expect the house to fall—but I do expect it will cease to be divided. It will become all one thing or all the other. Either the opponents of slavery will arrest the further spread of it, and place it where the public mind shall rest in the belief that it is in the course of ultimate extinction; or its advocates will push it forward, till it shall become lawful in all the States, old as well as new—North as well as South.

22. When President Lincoln uses the phrase a "house divided," what is he most likely referring to?
    a. Political conflicts in the White House
    b. Tensions between North and South over slavery
    c. Dissolution of the Union
    d. Ideological differences in his family

The following graphic depicts a political cartoon poster used as propaganda during World War II.

23. What is the main message of the poster?
    a. American citizens should stop driving cars.
    b. People who drive alone are Nazi sympathizers.
    c. Joining a Car-Sharing Club is a sign of American patriotism.
    d. Carpooling helps save fuel for the war effort.

24. Which term best describes the relationship between the Japanese attack on Pearl Harbor and the United States' entry into World War II?
    a. Correlation
    b. Causation
    c. Opposition
    d. Indirect connection

The next question is based on the following bar graph, which depicts the violent crime rate in the United States between 1999 and 2005:

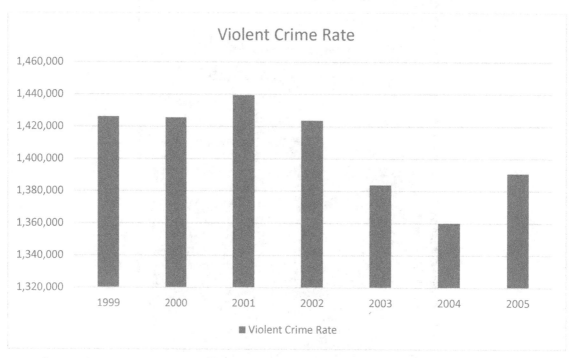

25. According to the graph, when was the United States violent crime rate at its highest?
    a. 1999
    b. 2001
    c. 2002
    d. 2005

The following graphic depicts a political cartoon that was possibly published in the Washington Evening Star, *and used as propaganda during World War I.*

26. Which historical event is most likely depicted by the political cartoon above?
    a. Mexican Cession
    b. Mexican-American War
    c. Zimmerman Telegram
    d. Immigration Act of 1965

Use this graph for Question 27.

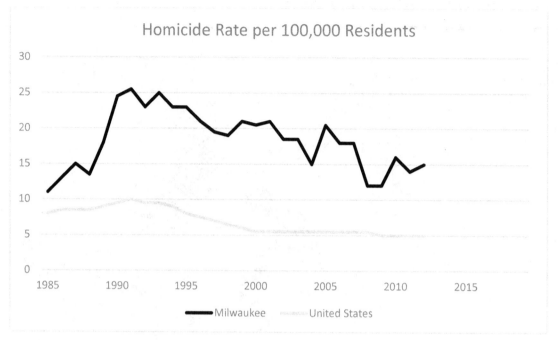

27. According to the graph above, at its peak, Milwaukee had how many homicides per 100,000 residents?
    a. 10.5
    b. 20
    c. 25.5
    d. 30

The following document is an excerpt from John F. Kennedy's "We Choose to Go to the Moon" speech in 1962.

> Those who came before us made certain that this country rode the first waves of the industrial revolutions, the first waves of modern invention, and the first wave of nuclear power, and this generation does not intend to founder in the backwash of the coming age of space. We mean to be a part of it—we mean to lead it. For the eyes of the world now look into space, to the moon and to the planets beyond, and we have vowed that we shall not see it governed by a hostile flag of conquest, but by a banner of freedom and peace. We have vowed that we shall not see space filled with weapons of mass destruction, but with instruments of knowledge and understanding.

28. Considering that this speech was delivered during the Cold War, which nation is John F. Kennedy alluding to with the phrase "hostile flag of conquest"?
    a. Great Britain
    b. Nazi Germany
    c. Soviet Union
    d. Iraq

29. Which of the following was the first European colony established in North America?
    a. St. Augustine
    b. Roanoke
    c. Jamestown
    d. Hudson's Bay

30. Which of the following best describes the 10th Amendment to the U.S. Constitution?
    a. The 10th Amendment reserves all non-specified powers to the states and people.
    b. The 10th Amendment establishes the existence of rights not named in the document.
    c. The 10th Amendment protects the right to bear arms.
    d. The 10th Amendment protects a series of rights for people accused of crimes.

31. What was the primary cause of widespread urbanization in the United States?
    a. The First Industrial Revolution
    b. The Second Industrial Revolution
    c. The Great Migration
    d. Successive waves of immigration

32. What is the government's role in a centrally planned economic system?
    a. The government distributes resources based solely on merit.
    b. The government allows the private market to set all prices based on supply and demand.
    c. The government makes all decisions related to production, distribution, and price.
    d. The government balances a private and public sector to stabilize the economy.

33. What was NOT included in the Missouri Compromise of 1820?
    a. Slavery was banned in Washington, D.C.
    b. Missouri was admitted as a slave state.
    c. Slavery was prohibited in future northern territories.
    d. Maine was admitted as a free state.

34. Which of the following best describes commodities?
    a. Physical, tangible items
    b. Activities based on labor or skill
    c. Raw materials or agricultural products
    d. High-demand products

35. When did immigration to the United States shift from Great Britain to southeastern Europe?
    a. Early 18th century
    b. Early 19th century
    c. Early 20th century
    d. Early 21st century

36. What was the Zimmerman Telegram?
    a. A famous telegram from Germany to the United States describing the experiences of a Jewish German family during the Holocaust
    b. An intercepted communication from Germany to Mexico stating Germany's willingness to help Mexico invade the United States
    c. A telegram that announced Germany's invasion of the Sudetenland
    d. A new type of telegraph invented during the Second Industrial Revolution

37. What caused the Korean War?
    a. Soviet and North Korean forces invaded South Korea.
    b. Chinese and North Korean forces invaded South Korea.
    c. North Korean forces invaded South Korea.
    d. South Korean forces invaded North Korea.

38. What treaty ended the American Revolution?
    a. Treaty of Paris
    b. Treaty of Ghent
    c. Treaty of Alliance
    d. Treaty of Versailles

39. What is the usual effect of inflation?
    a. Increased regulation and decreased growth
    b. Decreased regulation and increased growth
    c. Increased prices and decreased currency values
    d. Decreased prices and increased currency values

40. What was included in the Connecticut Compromise?
    a. Two legislative bodies with different methods of representation
    b. Two legislative bodies with the same method of representation
    c. One legislative body with representation based on population
    d. One legislative body with one vote per state

41. What happened to the global population during the 20th century?
    a. The population declined due to repeated global conflicts.
    b. The population declined due to worsening global poverty.
    c. The population increased rapidly in the first half of the century before declining in the second half of the century.
    d. The population increased the most in world history.

42. What is the Space Race's legacy?
    a. The United States dismantled its space exploration program after beating the Soviets to the Moon.
    b. The superpowers' relations improved, and the United States withdrew from the Vietnam War.
    c. The underlying technology led to the Digital Revolution.
    d. The Soviet Union collapsed due to its repeated failure to put a man on the Moon.

43. Which of the following powers is exclusive to the federal government in the United States?
    a. Regulate immigration
    b. Regulate local government
    c. Implement welfare and benefit programs
    d. Levy taxes

44. What was the domino theory?
    a. An economic theory that caused the Great Depression
    b. An American foreign policy approach to contain communism
    c. A framework for analyzing how military alliances caused World War I
    d. A British attempt to integrate French Canada into the British Empire

45. What was a major achievement during the Han dynasty?
    a. The Han dynasty installed the world's first representative government.
    b. The Han dynasty defeated Alexander the Great's legendary military.
    c. The Han dynasty invented a new, more powerful form of concrete.
    d. The Han dynasty connected the Far East with the West for the first time.

46. How did large-scale agriculture facilitate European colonization in the Americas?
    a. Large-scale agriculture increased American exports to Europe, and the colonies received military aid in return.
    b. Large-scale agriculture led to the development of new technologies that were applied to the military.
    c. Large-scale agriculture supported the development of major cities in the American South, and Native Americans couldn't pierce the city's defenses.
    d. Large-scale agriculture produced a food surplus that supported a larger population, permanent settlements, and a centralized government.

47. Which of the following laws was NOT a major cause of the American Revolution?
    a. Indian Removal Acts
    b. Proclamation of 1763
    c. Quebec Act
    d. Townshend Acts

48. In what way did Native American and European colonists' family structures differ?
    a. Men served as the primary provider of resources.
    b. European colonists lived more sustainably.
    c. Native Americans included broader kinship networks.
    d. European colonists were more likely to honor their elders.

49. What event exposed the Articles of Confederation as a deeply flawed system of government?
    a. Publication of The Federalist Papers
    b. John Brown's raid at Harper's Ferry
    c. Whiskey Rebellion
    d. Shays' Rebellion

50. The United States has NOT successfully overthrown the government in which of the following countries?
    a. Iran (1953)
    b. Cuba (1961)
    c. Dominican Republic (1961)
    d. Chile (1973)

51. Which of the following presidential administrations conquered and annexed the most territory from a foreign power?
    a. Polk administration
    b. Madison administration
    c. Monroe administration
    d. Jackson administration

52. What is the consequence of economic comparative advantage?
    a. Countries increase tax rates to subsidize more industries.
    b. Countries enforce high tariffs on products they can't produce efficiently.
    c. Countries specialize at what they can do most efficiently relative to their competitors.
    d. Countries increase spending to stimulate economic growth.

53. What country successfully launched the first satellite into orbit?
    a. Japan
    b. Soviet Union
    c. Germany
    d. United States

54. What two American presidential campaigns most featured the candidates' opposition to the Iraq War?
    a. Barack Obama and Hillary Clinton
    b. Barack Obama and Donald Trump
    c. Hillary Clinton and John McCain
    d. Hillary Clinton and Donald Trump

55. What contributed to the First Industrial Revolution in the United States?
    a. Cotton gins
    b. Steel
    c. Electricity
    d. Railroads

56. Which of the following was achieved during the Progressive Era?
    a. Emancipation of all slaves
    b. Direct election of United States Senators
    c. Annexation of Texas
    d. Victory in the Mexican-American War

57. Which of the following civilizations most influenced the drafting of the U.S. Constitution?
    a. China
    b. Egypt
    c. Greece
    d. Rome

58. President Andrew Jackson refused to enforce what Supreme Court decision?
    a. United States v. Windsor
    b. Worcester v. Georgia
    c. Dred Scott v. Sandford
    d. Plessy v. Ferguson

The following speech was given by Franklin D. Roosevelt on December 8, 1941. It is about the Japanese bombing of Pearl Harbor.

It will be recorded that the distance of Hawaii from Japan makes it obvious that the attack was deliberately planned many days or even weeks ago. During the intervening time the Japanese Government has deliberately sought to deceive the United States by false statements and expressions of hope for continued peace.

The attack yesterday on the Hawaiian Islands has caused severe damage to American naval and military forces. I regret to tell you that very many American lives have been lost. In addition, American ships have been reported torpedoed on the high seas between San Francisco and Honolulu.

Yesterday the Japanese Government also launched an attack against Malaya.

Last night Japanese forces attacked Hong Kong.

Last night Japanese forces attacked Guam.

Last night Japanese forces attacked the Philippine Islands.

Last night the Japanese attacked Wake Island.

And this morning the Japanese attacked Midway Island.

Japan has therefore undertaken a surprise offensive extending throughout the Pacific area. The facts of yesterday and today speak for themselves. The people of the United States have already formed their opinions and well understand the implications to the very life and safety of our nation.

As Commander-in-Chief of the Army and Navy I have directed that all measures be taken for our defense, that always will our whole nation remember the character of the onslaught against us.

No matter how long it may take us to overcome this premeditated invasion, the American people, in their righteous might, will win through to absolute victory.

I believe that I interpret the will of the Congress and of the people when I assert that we will not only defend ourselves to the uttermost but will make it very certain that this form of treachery shall never again endanger us.

Hostilities exist. There is no blinking at the fact that our people, our territory and our interests are in grave danger.

With confidence in our armed forces, with the unbounding determination of our people, we will gain the inevitable triumph. So help us God.

I ask that the Congress declare that since the unprovoked and dastardly attack by Japan on Sunday, December 7th, 1941, a state of war has existed between the United States and the Japanese Empire.

59. Which causal relationship in American history does this document capture?
    a. The Japanese attack on Pearl Harbor and the United States' entry into World War II
    b. The Japanese attack on Pearl Harbor and the United States' entry into the League of Nations
    c. The Japanese attack on Pearl Harbor and the end of World War I
    d. The Japanese attack on Pearl Harbor and the failure of Congress to declare war

60. Which of the following data visualization tools would best be used for illustrating the percentage of the African-American population in the US affected by unconstitutional searches during the civil rights era of the 1950s and 1960s compared to entire population of African-Americans in the US (part or percentage of a whole)?
    a. Pie chart
    b. Line graph
    c. Table
    d. Bar graph

Answer Explanations

1. B: Secondary source. This excerpt is considered a secondary source because it actively interprets primary sources. We see direct quotes from the queen, which would be considered a primary source. But since we see those quotes being interpreted and analyzed, the excerpt becomes a secondary source. Choice *C*, tertiary source, is an index of secondary and primary sources, like an encyclopedia or Wikipedia.

2. It took **two years** for the new castle to be built. It states this in the first sentence of the second paragraph. In the third year, we see the Prince planning improvements, and arranging things for the fourth year.

3. C: To impose a certain quality upon. The sentence states that "the impress of his dear hand [has] been stamped everywhere," regarding the quality of his tastes and creations on the house. Choice *A* is one definition of *impress*, but this definition is used more as a verb than a noun: "She impressed us as a songwriter." Choice *B* is incorrect because it is also used as a verb: "He impressed the need for something to be done." Choice *D* is incorrect because it is part of a physical act: "the businessman impressed his mark upon the envelope." The phrase in the passage is meant as figurative, since the workmen did most of the physical labor, not the Prince.

4. B: To convince the audience that judges holding their positions based on good behavior is a practical way to avoid corruption.

5. A: The answer is Choice *A*, the market revolution. The market revolution witnessed the United States transition from a mostly self-subsistence agrarian economy to an interdependent consumer economy. The Mexican-American War (Choice *B*) and the Enlightenment (Choice *A*) have no historical relation to the rise of consumerism. And the stock market crash of 1929 was actually the result of the excesses of consumer culture (Choice *D*).

6. B: The answer is Choice *B*, history. Space (Choice *A*), place (Choice *C*), and change (Choice *D*) are all elements of the seven major geographical concepts. History influences all these major concepts, and it is an important part of geography, but it is not included on this traditional list.

7. C: The answer is Choice *C*, collect and analyze important geographical data. While it can be argued that these case studies can influence sustainability lobbying efforts (Choice *A*), GPS systems (Choice *B*), and communications with cartographers (Choice *D*), they have most traditionally been used as a platform for data collection and analysis.

8. B: The answer is Choice *B*, national government is emphasized over state governments, with concentrated power and centralized authority. Choice *A* is not discussed as governmental framework, but describes a constitutional monarchy, like the one in England. Choice *C* is the definition of a confederacy. And Choice *D* is the definition of a federal system.

9. C: The answer is Choice *C*, the limited availability of a particular commodity. While too much demand (Choice *B*) might lead to scarcity, it is not the definition of scarcity. Choices *A* and *D* would be the opposite of scarcity.

10. A: The answer is Choice *A*, socialism. While any of the governments listed can technically incorporate Marxist philosophy, socialism is the political philosophy most closely associated with Marxism.

11. C: 394,000. Choice *A* is the population of Totok in 1920. Choice *B* is the population of Totok in 1930. Choice *D* is the population of Peranakan in 1956 and 1961.

12. C: The answer is 404,000. This question requires some simple math by subtracting the total population of Peranakan in 1971 (1,240,000) from the total population of Peranakan in 1961 (836,000) which comes to 404,000.

13. A: According to the map, Volcano Terevaka is the highest point on the island, reaching 507m. Volcano Puakatike is the next highest, reaching 307m. Vaka Kipo is the next highest reaching 216m. Finally, Maunga Ana Marama is the next highest, reaching 165m.

14. C: The answer is Choice *C*; the number of representatives is dictated by a state's population size. Although being a senator is typically seen as prestigious (Choice *C*), prestige has no effect on the number of representatives and senators in office. Voters choose representatives in elections, but they do no choose how many representatives there are in office (Choice *B*). The Bill of Rights does not discuss the necessary number of congressmen (Choice *D*).

15. C: Britain, France, and Russia formed the Triple Entente in World War I, so Choice *B* is the correct answer. Canada and the United States both joined the Triple Entente, but they were not original members. Austria-Hungary and Germany fought against the Triple Entente. Italy allied with Austria-Hungary and Germany before the war broke out, but the Italians held out and later joined the Triple Entente once it appeared they would be victorious.

16. B: Choice *B* is the correct answer. President Franklin Delano Roosevelt's legislative agenda was called the New Deal, and it created the modern-day welfare state—Social Security, unemployment insurance, disability, labor laws, and housing and food subsidies. The War on Poverty and Great Society were Lyndon B. Johnson's agenda, though they were influenced by the New Deal. Social Security was part of the New Deal, not Roosevelt's legislative agenda.

17. D: Northern Whigs and Free Soil Democrats created the Republican Party to advocate for the abolition of slavery, so Choice *D* is the correct answer. Abraham Lincoln was the first Republican to be elected president. The Republicans exposed the Southern theory of states' rights, and their pro-business agenda didn't develop until after the Civil War. Manifest destiny also wasn't the Republicans primary issue; it was slavery.

18. C: The Organic Theory argues that nation-states need to consolidate political power and acquire new geographic territory in order to sustain themselves. Under this conception, the nation-state operates like a living organism, and new territories provide the resources it needs to survive. Thus, Choice *C* is the correct answer. The Rimland Theory emphasizes the importance of naval power and coastal land, and the World Island is part of the Heartland Theory. In the Cold War era, American foreign policymakers created the domino theory to justify military intervention for limiting the spread of communism.

19. B: Choice *B* is the correct answer. Political maps are the most subjective, because territorial boundaries are frequently contested or otherwise in dispute. For example, 19th century political maps differed on the boundaries between the competing European colonies, as well as what constituted Native American territory. In the present day, there's dispute over who owns Crimea. In contrast, climate, population density, and topographic maps are all based on objective data.

20. D: Immediately after the federal government withdrew troops from the South, ending Reconstruction, the Southern states enacted Jim Crow laws. These laws contravened the Reconstruction Amendments by preventing African Americans from voting and enforcing segregationist policies. Therefore, Choice *D* is the correct answer. African Americans voted for the first time during Reconstruction, but following the passage of Jim Crow laws, they couldn't vote until the 1960s Civil Rights Movement. The South didn't diversify from a one-crop economy (cotton) or industrialize until the 20th century. In addition, it wasn't related to Jim Crow. If anything, Jim Crow contributed to the continued focus on cotton as sharecropping replaced slavery.

21. B: As discussed in this Section of the Constitution, Senators serve six-year terms. Choice *A* is an incorrect answer; members of the House of Representative serve two-year terms. Choice *C* is an incorrect answer; Supreme Court Justices hold lifelong terms, but Senators only serve six-year terms. Choice *D* is an incorrect answer; the President of the United States has a four-year term length.

22. B: The speech was made before the Civil War, which began in 1861; it captures the rising tension over slavery between the North and the South. Choices *A* and *D* are incorrect because Lincoln is not talking about an actual house. Instead, he is talking about a nation divided over slavery by using the metaphor of a house. Choice *C* is incorrect because the speech was given in 1858, before the secession of South Carolina and the dissolution of the Union in 1861.

23. D: The political propaganda poster is trying to get American citizens to carpool with one another in order to save fuel for the war effort. Choice *A* is incorrect because cars were an essential part of American culture and lifestyle at the time. If citizens stopped driving cars altogether, it would put a strain on the economy and the work flow. Choice *B* is incorrect because the message the poster is trying to convey is not that extreme. Even though the poster says, "When you ride ALONE you ride with Hitler!", it does not mean the government is accusing its citizens of treason. It was a hyperbole used to drastically encourage its citizens to understand the importance of conserving fuel to aid the war effort. Choice *C* is incorrect because, even though American patriotism is an important message behind all propaganda posters, it is not the main message of this particular poster.

24. B: The Japanese attack on Pearl Harbor and the United States' entry into World War II have a cause-and-effect relationship. Choice *A* is incorrect because the relationship is causation, not correlation. Choice *C* is incorrect because these two events are causally related rather than in opposition to one another. Choice *D* is incorrect because this particular cause-and-effect relationship is direct.

25. B: The line graph shows that the violent crime rate in the United States was at its peak in 2001. Choice *A* is incorrect because, despite its high numbers, the violent crime rate in 1999 was still lower than it was in 2001. Choice *C* is incorrect because it is almost equal to the violent crime rate in 1999. Choice *D* is incorrect because the graph shows a decrease in the violent crime rate in 2005.

26. C: This political cartoon symbolically depicts the Zimmerman Telegram, in which Germany tried to recruit Mexico to join the Central powers during World War I. The image shows a hand carving up a map of the southwestern United States, representing the portion of the country that Germany promised to give Mexico if they joined their cause. Choice *A* is incorrect because, although the territory depicted was included in the Mexican Cession, the incorporation of the hand wearing a gauntlet with the imperial German eagle on it clearly alludes to the Zimmerman Telegram. Choice *B* is incorrect because Germany had no involvement in the war between the U.S. and Mexico. Choice *D* is incorrect because, although the act aimed to be more permissive in terms of immigration by eliminating quotas, it had nothing to do with Germany.

27. C: According to the graph, Milwaukee had a peak of about 25.5 homicides per 100,000 in 1991. All other answers can be eliminated by finding the highest peak (designated by Milwaukee's darker line) on the graph.

28. C: The United States and the Soviet Union were the two superpowers of the Cold War, racing to go to the moon. Choice *A* is incorrect because the United States was an ally of Great Britain at this time. Choice *B* is incorrect because Nazi Germany had already fallen by 1962. Choice *D* is incorrect because tensions between Iraq and the United States did not begin until Desert Storm of the 1990s.

29. A: Spanish explorers established St. Augustine in 1565. All other colonies were founded at least a decade later. Sir Walter Raleigh founded Roanoke in 1585, making it the first British colony, but it was abandoned in 1590. Established in 1607, Jamestown was the first successful British colony. The British explorer Henry Hudson explored Hudson's Bay in the 1610s, and the French established trading outposts in the area during the 1670s. Thus, Choice *A* is the correct answer.

30. A: The Tenth Amendment reserves all non-specified powers to the states or people, establishing the principle of federalism. So, Choice *A* is the correct answer. The Ninth Amendment establishes the existence of unnamed rights. The Second Amendment protects the right to bear arms. The Fourth, Fifth, and Sixth Amendments include procedural protections for people accused of crimes.

31. B: Choice *B* is the correct answer. The United States first industrialized during the First Industrial Revolution to replace the decline in British imports caused by the War of 1812. However, this development was limited to towns in the Northeast. Urbanization didn't occur until after the Civil War, and it was caused by the Second Industrial Revolution's technological innovations, like steel production, railways, and more sophisticated factory machinery. The Great Migration and successive waves of immigration increased the pace of urbanization, supplying labor to the urban boom. Yet, urbanization wouldn't have occurred without the Second Industrial Revolution transforming America from an agricultural to industrial economy.

32. C: The government makes all decisions related to production, distribution, and price in a centrally planned economic system; therefore, Choice *C* is the correct answer. This allows communist governments to oversee price control, and, as a result, the production and distribution of resources is less efficient but more equitable. If a government distributes resources solely based on merit and allows private markets to set prices, then the economy is operating under a free-market system. A government that balances a private and public sector is characteristic of a mixed economic system.

33. A: The question is asking what was NOT included in the Missouri Compromise of 1820. The Missouri Compromise had three parts: Maine was admitted as a free state; Missouri was admitted as a slave state; and slavery was prohibited in new territories north of the 36°30' parallel. The only answer that doesn't name a part of the Missouri Compromise is Choice *D*, so it must be correct. The Compromise of 1850 banned slavery in Washington, D.C.

34. C: Choice *C* is the correct answer. Commodities are raw materials or agricultural products—gold, silver, gas, oil, corn, barley, etc. Physical, tangible items are goods. Activities based on labor or skills are services. High-demand products typically have a higher price in a free-market system, depending on the supply.

35. C: Choice *C* is the correct answer. American immigrants' homeland first changed dramatically in the early 20th century. The American colonists were mostly British, and immigration until the late 19th century was mostly confined to the British Isles. In the late 19th century and early 20th century, immigrants arrived from southeastern Europe en masse. This placed considerable strain on American society, as the current residents were Protestant, and the new immigrants were Catholic and Jewish.

36. B: Choice *B* is the correct answer. The Zimmerman Telegram was a message from Germany to Mexico that stated Germany's willingness to help Mexico invade America. Great Britain intercepted the Zimmerman Telegram and delivered it to the American government, hoping it would persuade the United States to join the Triple Entente. Germany didn't invade the Sudetenland until World War II, and the Zimmerman Telegram wasn't a type of telegram or related to the Holocaust.

37. C: The Korean War started with North Korea invading South Korea, so Choice *C* is correct. The United States entered the war to protect its ally and prevent communism spreading across Asia. China didn't enter the conflict until General MacArthur led the American military across the 38th parallel. The Soviets provided material support to communist North Korea, but they never formally entered the war.

38. A: The Treaty of Paris concluded the American Revolution, securing American independence. Thus, Choice *A* is the correct answer. The Treaty of Ghent ended the War of 1812. The United States and France signed the Treaty of Alliance during the American Revolution, and French support turned the tides of war for the Americans. The Treaty of Versailles ended World War I, and its failure to establish a lasting peace partially caused World War II.

39. C: Inflation is a period of increased prices and decreased currency values, so Choice *C* is the correct answer. Most countries designate a central bank or government agency to set and enforce monetary policies to control inflation and interest rates. Regulation typically correlates with growth, but that relationship is not inflation.

40. A: Delegates at the American Constitutional Convention agreed on the Connecticut Compromise, establishing two legislative bodies. Representation in one house was based on population (House of Representatives), and the other granted each state two votes (Senate). As such, Choice *A* is the correct answer. The New Jersey Plan proposed a single legislative body with one vote per state, but it was rejected in favor of the Connecticut Compromise.

41. D: The global population increased in both halves of the 20th century. From 1900 to 2000, the global population increased from 1.6 billion people to 6 billion people, marking the largest increase in world history. Therefore, Choice *D* is the correct answer.

42. C: During the Space Race, the United States and Soviet Union invested in computer-based technology to advance their rocket and satellite capabilities. These advancements in computing power led to the Digital Revolution, so Choice *C* is the correct answer. The United States never dismantled its space exploration program, and the Soviet Union collapsed more than a decade after they gave up trying to put a man on the Moon. Although the Space Race improved the superpowers' relationship, this didn't influence the American withdrawal from Vietnam. Instead, mounting domestic protests caused the United States to leave Vietnam.

43. A: The power to regulate immigration is exclusive to the federal government; so, Choice *A* is the correct answer. The powers to regulate local government and implement welfare programs are reserved to the states. Both the federal and state governments enjoy the power to levy taxes.

44. B: American Cold War foreign policy followed the domino theory, the idea that if one country turned communist, so would all of the neighboring countries. Consequently, the United States regularly intervened militarily to contain communism, which happened in Korea and Vietnam. Therefore, Choice *B* is the correct answer. None of the other answer choices are related to American foreign policy in the Cold War era.

45. D: The Han dynasty created the Silk Road, a series of trade routes that linked the Far East and West for the first time in world history. Thus, Choice *D* is the correct answer. The Greeks installed the first representative government; Indian rulers defeated Alexander the Great's legendary military; and the Romans invented a superior type of concrete.

46. D: The European system's large-scale production of crops and livestock produced a food surplus that sustained permanent settlements and centralized government. So, Choice *D* is the correct answer. The European colonies didn't trade agriculture for military aid; agriculture didn't lead to military technological advancements; and the American South didn't urbanize for several centuries.

47. A: The question is asking what was NOT a major cause of the American Revolution. The Proclamation of 1763 angered colonists by prohibiting colonial expansion west of the Appalachian Mountains. Likewise, the Quebec Act and Townshend Act increased resentment. The Quebec Act granted rights to French Canadians at the same time Great Britain passed a series of laws to increase taxes in the Thirteen Colonies, like the Townshend Acts. The Indian Removal Acts weren't passed until the Andrew Jackson administration; so, Choice *A* is the correct answer.

48. C: Native Americans traditionally live in extended kinship networks. As such, Native Americans define family as the nuclear family, extended family, tribal members, and/or entire nation. Fictive (non-blood related) kin are also commonly included as family members. In contrast, European colonists' families typically only included the nuclear and extended family. Therefore, Choice *C* is the correct answer. Men served as the primary provider of resources in both family structures; the European colonists didn't live more sustainably than Native Americans; and Native Americans also honored their elders.

49. D: From 1786 to 1787, Revolutionary War veteran Daniel Shays led an armed insurrection in western Massachusetts, and the government's inability to quash the rebellion exposed the Articles of Confederation's flaws. Thus, Choice *D* is the correct answer. The *Federalist Papers* were published during the ratification process, and, at that time, the Articles of Confederation was widely considered to be untenable. The question was what form of government would replace it. John Brown's raid at Harper's Ferry increased regional tension before the Civil War, and President George Washington's suppression of the Whiskey Rebellion demonstrated the strength of the U.S. Constitution.

50. B: The question is asking which government was NOT overthrown with American assistance. President John F. Kennedy attempted to overthrow the Cuban government in 1961, but his Bay of Pigs invasion was a disastrous failure. The United States provided military support, training, and/or funding in Iran, the Dominican Republic, and Chile. All three governments were successfully replaced with pro-American dictators. Therefore, Choice *B* is the correct answer.

51. A: James K. Polk won the 1844 presidential election based on his promise to complete America's manifest destiny. Polk annexed the present-day American southwest and California from Mexico, and his administration settled border disputes with Great Britain in the Oregon territory. None of the administrations in the other answer choices annexed a comparable amount of territory; therefore, Choice *A* is the correct answer.

52. C: Comparative advantage is the principle that countries produce the goods and services that have the least opportunity cost. When a country has a comparative advantage, they'll specialize in that industry because it's more efficient. Thus, Choice *C* is the correct answer. A country might implement the other three policies, but it's less likely when the country enjoys a comparative advantage.

53. B: On October 4, 1957, the Soviet Union launched the first satellite, Sputnik 1, into orbit; therefore, Choice *B* is the correct answer. The United States was shocked by the Soviets' achievement, and they redoubled their efforts, culminating in the Apollo 11 mission's Moon landing. The Japanese and German space exploration program weren't active until 1969, and they weren't major competitors to the American and Soviet programs.

54. B: Choice *B* is the correct answer. Barack Obama used his opposition to the Iraq War to differentiate himself from his opponent, John McCain, who wanted to increase the number of American troops in Iraq. Although it's unclear whether Donald Trump opposed the Iraq War before the invasion, he was one its earliest and most vocal public critics. Like Obama, Trump raised the issue to paint his opponent, Hillary Clinton, as the candidate more likely to start another war. Clinton had voted for the Iraq War as a Senator, but she had since publicly recognized it as a mistake.

55. A: Eli Whitney invented the cotton gin in 1793, and the cotton industry exploded. Following its invention, American cotton production doubled every decade until the latter half of the 19th century. As a result, New England textile mills enjoyed access to more cotton at cheaper prices, facilitating the First Industrial Revolution in the United States. Therefore, Choice *A* is the correct answer. The other three answer choices contributed to the Second Industrial Revolution.

56. B: The Progressive Era (1890-1920) strengthened American democracy in a variety of ways. One of those legislative achievements was the 17th Amendment, which established the direct election of U.S. Senators. Previously, state legislatures elected Senators. Thus, Choice *B* is the correct answer. The 13th amendment freed slaves after the end of the Civil War; Texas was annexed in 1845; and the United States defeated Mexico in 1848.

57. D: Choice *D* is the correct answer. James Madison used the Roman Republic's separation of powers as a model for the U.S. Constitution. Greece generally influenced the U.S. Constitution in its implementation of the world's first direct democracy, but the Roman influence is more specific. China and Egypt didn't influence the Constitution's drafting.

58. B: President Andrew Jackson refused to enforce the Supreme Court's decision in *Worcester v. Georgia* (1832), which established the concept of tribal sovereignty. Instead, Jackson encouraged the colonization of Native American land. *United States v. Windsor* (2013) established the right to same-sex marriage. *Dred Scott v. Sandford* (1857) overturned the Missouri Compromise's prohibition on slavery in northern territories. Plessy v. Ferguson (1896) upheld the South's "separate but equal" segregation.

59. A: The Japanese attack on Pearl Harbor and the United States' entry into World War II are causally related, as conveyed by this document. All other answer choices are factually incorrect and not cited in the text.

60. A: Pie charts are best for comparing parts or percentages of the whole. While all other data visualization tools could likely convey this comparison in some way, the pie chart is best because of its part-to-whole focus.

Greetings!

First, we would like to give a huge "thank you" for choosing us and this study guide for your HiSET exam. We hope that it will lead you to success on this exam and for years to come.

Our team has tried to make your preparations as thorough as possible by covering all of the topics you should be expected to know. In addition, our writers attempted to create practice questions identical to what you will see on the day of your actual test. We have also included many test-taking strategies to help you learn the material, maintain the knowledge, and take the test with confidence.

We strive for excellence in our products, and if you have any comments or concerns over the quality of something in this study guide, please send us an email so that we can improve.

As you continue forward in life, we would like to remain alongside you with other books and study guides in our library. We are continually producing and updating study guides in several different subjects. If you are looking for something in particular, all of our products are available on Amazon. You can also send us an email!

Sincerely,
APEX Test Prep
info@apexprep.com

FREE Test Taking Tips DVD Offer

To help us better serve you, we have developed a Test Taking Tips DVD that we would like to give you for FREE. **This DVD covers world-class test taking tips that you can use to be even more successful when you are taking your test.**

All that we ask is that you email us your feedback about your study guide. Please let us know what you thought about it – whether that is good, bad or indifferent.

To get your **FREE Test Taking Tips DVD**, email freedvd@studyguideteam.com with "FREE DVD" in the subject line and the following information in the body of the email:

    a. The title of your study guide.

    b. Your product rating on a scale of 1-5, with 5 being the highest rating.

    c. Your feedback about the study guide. What did you think of it?

    d. Your full name and shipping address to send your free DVD.

If you have any questions or concerns, please don't hesitate to contact us at freedvd@studyguideteam.com.

Thanks again!

DISCARD